Energy from Biomass

Solar Energy R&D in the European Community

Series E:
Energy from Biomass

Volume 5

Publication arrangements: D. NICOLAY
P. P. ROTONDÓ

Solar Energy R&D
in the European Community

Series E   Volume 5

Energy from Biomass

Proceedings of the Workshop and
EC Contractors' Meeting
held in Capri, 7-8 June 1983

edited by

W. PALZ and D. PIRRWITZ
Commission of the European Communities

D. Reidel Publishing Company

A MEMBER OF THE KLUWER ACADEMIC PUBLISHERS GROUP

Dordrecht / Boston / Lancaster

for the Commission of the European Communities

Library of Congress Cataloging in Publication Data
Main entry under title:

Energy from biomass.

    (Solar energy R&D in the European community. Series E, Energy from biomass ; v. 5)
    1. Biomass energy–Congresses. I. Palz, W. (Wolfgang), 1937– II. Pirrwitz, D. III. Commission of the European Communities. IV. Series.
TP360.E535   1983   662'.8   83–22926
ISBN 90-277-1700-1

Organization of the Workshop and Contractors' meeting by
Commission of the European Communities
Directorate-General Science, Research and Development, Brussels

Publication arrangements by
Commission of the European Communities
Directorate-General Information Market and Innovation, Luxembourg

EUR 8829
© 1984 ECSC, EEC, EAEC, Brussels and Luxembourg

LEGAL NOTICE
Neither the Commission of the European Communities nor any person acting on behalf of the Commission is responsible for the use which might be made of the following information.

Published by D. Reidel Publishing Company
P.O. Box 17, 3300 AA Dordrecht, Holland

Sold and distributed in the U.S.A. and Canada
by Kluwer Academic Publishers,
190 Old Derby Street, Hingham, MA 02043, U.S.A.

In all other countries, sold and distributed
by Kluwer Academic Publishers Group,
P.O. Box 322, 3300 AH Dordrecht, Holland

All Rights Reserved
No part of the material protected by this copyright notice may be reproduced or utilized in any form or by any means, electronic or mechanical, including photocopying, recording or by any information storage and retrieval system, without written permission from the copyright owner.

Printed in The Netherlands

PREFACE

The present proceedings of the European Community's R&D programme on biomass provide a good summary of the contractors' work performed over the past few years in European industry, universities and research institutions, particularly the agricultural research centres. The meeting concluded the Commission's four year biomass programme and gave an opportunity to pool results and achievements.

There was a reasonable balance in the programme between "energy cropping" and biomass conversion and utilisation. In a number of European countries interesting new results have been obtained on productivity, yields and biomass production costs for energy forestry and other energy crops such as arundo donax or algae.

On the utilisation side, R&D work on straw burners has come to an end because this area has become by and large commercial.

Gasification and in particular the production of synthesis gas from wood also shows excellent prospects for future implementation. For the first time CO and H_2 ready for conversion into methanol were produced from several pilot plants.

Remarkable results were also obtained on biogas and fermentation in general. As shown in a special investigation of this programme, biogas is becoming increasingly successful on the market.

After the general session, contractors met separately in various groups to assess the state of the art and to identify relevant research axes for the coming years. The reports of these groups are given at the beginning of this book as an overview.

The Commission wishes to thank the Italian hosts of the meeting, namely the Mayor of Capri, the University of Naples, regione Campania and Cassa del Mezzogiorno, for all their efforts to make the meeting a success. Special thanks go to the local organiser, Professor Alfani.

Dr. W. PALZ
Head of the Biomass Programme

CONTENTS

Preface   v

ASSESSMENT STUDIES

Impact of biomass energy use on employment, energy situation
and agrarian market problems in the EC
    Technical University of Munich, Bavarian National
    Department of Agricultural Engineering, Freising,
    Federal Republic of Germany   2

Technical stand and possibilities for conversion of biomass
into charcoal, pyrolytic oil and in comparison with other
conversion systems
    Bayerische Landesanstalt für Landtechnik Weihenstephan,
    Technische Universität München, Freising, Federal
    Republic of Germany   11

Assessment study on biogas plants in Europe : economical
aspects
    Catholic University of Louvain, Louvain-la-Neuve,
    Belgium   20

REPORT OF SECTORIAL SEMINARS

Wastes, energy crops and forestry
    C.P. MITCHELL and M.L. PEARCE   28

Review paper on algae
    J.G. MORLEY and D. PIRRWITZ   32

The status and future of thermochemical processing of
biomass
    A.A.C.M. BEENACKERS and A.V. BRIDGWATER   34

Biomass conversion : Biological routes
    E. COLLERAN   43

AGRICULTURAL WASTES - ENERGY CROPS

Energy production using straw and animal wastes as feedstocks - Economic analysis of the straw pelleting route
    Institut National de la Recherche Agronomique, Paris, France ... 48

An experimental assessment of native and naturalised species of plants as renewable sources of energy in Great Britain
    Natural Environment Research Council's Institute of Terrestrial Ecology, Swindon, United Kingdom ... 57

Growing catch crops for fuel
    University of Reading, United Kingdom ... 66

Joint research on Arundo donax as an energy crop
    Institut National de la Recherche Agronomique, Paris, France ... 74

Quality and quantity of latex which can be produced from natural vegetation in Greece
    University of Thessaloniki (Laboratory of Ecology), Thessaloniki, Greece ... 82

FORESTRY - WOOD WASTES

An experimental study of short rotation forestry for energy
    Aberdeen University, Forestry Department, United Kingdom ... 88

Coppiced trees as energy crops
    Forestry Commission, Research and Development Division, Westonbirt Arboretum, United Kingdom ... 96

Improvement of forest trees for short term biomass production
    Institut national de la Recherche agronomique, Station d'Amélioration des Arbres forestiers, Olivet, France ... 104

Utilization of coppice forests biomass for fuel and other industrial uses
    Ente Nazionale Cellulosa e Carta, Roma, Italy ... 112

The production of energy from short rotation forestry
    An Foras Taluntais (Agricultural Research Council), Dublin, Ireland ... 120

Short rotation forestry harvester chipper
    Bord na Mona, Dublin, Ireland ... 128

Design and building of a forestry wastes harvester
    Centre National du Machinisme Agricole, du Génie rural des Eaux et des Forêts, Antony, France ... 136

Harvesting before the fire for energy; Mediterranean-type ecosystems in Greece - Costs and benefits
    University of Thessaloniki (Laboratory of Ecology), Thessaloniki, Greece   142

ALGAE

Development of a production size system for the mass culture of marine microalgae
    Centro di Studio dei Microrganismi Autotrofi del CNR, Firenze, Italy   150

Exploitation of lagoon macroalgae for biogas production - Preparatory study for a comprehensive project to exploit the algae in the lagoon of Venice to produce energy and improve the environment
    CSARE, Rovigo, Italy;
    Landesanstalt für Landtechnik, Universität München, Federal Republic of Germany   159

Biomass from offshore sea areas
    University of Liverpool, Port Erin, Isle of Man, United Kingdom   168

Biomass from offshore areas
    University of Nottingham, United Kingdom   177

Solar biotechnology study and development of tubular solar receptors for controlled production of photosynthetic cellular biomass for methane production and specific exocellular biomass
    Association pour la Recherche en Bioénergie solaire, Saint Paul Lez Durance, France   184

Culture of a hydrocarbon producing alga, Botryococcus braunii, at pilot level
    INIEX, Liège; University of Liège, Belgium   194

Renewable hydrocarbon production by cultivation of the green alga Botryococcus braunii - Investigation of the factors affecting hydrocarbon production
    Ecole Nationale Supérieure de Chimie de Paris, France   201

Methane production by anaerobic digestion of algae, I. Pilot plant biomethanation of cultivated marine algae Tetraselmis for energy production in southern Italy
    Catholic University of Louvain, Louvain-la-Neuve, Belgium   210

Methane production by anaerobic digestion of algae, II. Production of algae
    Catholic University of Louvain, Louvain-la-Neuve, Belgium   218

BIOMASS CONVERSION (THERMOCHEMICAL ROUTES)

The development of furnace/heat exchanger systems in which chopped cereal straw is the fuel
    University of Nottingham, United Kingdom   228

The use of gas scrubbers for heat extraction from straw furnaces
    Jordbrugsteknisk Institut, Royal Veterinary and Agricultural University, Copenhagen, Denmark   235

Heat energy from animal waste by combined drying, combustion and heat recovery
    Jordbrugsteknisk Institut, Royal Veterinary and Agricultural University, Copenhagen, Denmark   240

High temperature straw granulation
    CEMAGREF, Antony, France   244

Mobile pyrolysis plant
    Fritz Werner Industrie-Ausrüstungen GmbH, Department "New Technology", Geisenheim, Federal Republic of Germany   250

Modeling of a fluidized bed wood gasifier
    Twente University of Technology, Laboratory of Chemical Reaction Engineering, Enschede, The Netherlands   255

Biomass gasification - programme status
    Foster Wheeler Power Products Limited, London, United Kingdom   265

Hydrogen absorption in $LaNi_5$ dispersed in liquid
    Twente University of Technology, Enschede, The Netherlands   270

Catalytic liquefaction of wood material
    Catholic University of Louvain, Groupe de Physico-Chimie Minérale et de Catalyse, Louvain-la-Neuve, Belgium   279

Study on the pyrolysis of agricultural wastes
    Centro Ricerca Industriale Tecnologia Avanzata (CRITA), Pisa, Italy   289

BIOMASS CONVERSION (BIOLOGICAL ROUTES)

The anaerobic digestion of farm wastes and energy crops
    University College, Cardiff, United Kingdom   298

Anaerobic filter digestion of agricultural wastes
    University College, Galway, Ireland   306

Two-phase process for the anaerobic digestion of organic wastes yielding methane and compost
    Institute for Storage and Processing of Agricultural Produce (IBVL), Wageningen, The Netherlands   315

The feasibility of thermophilic anaerobic digestion for the generation of methane from organic wastes
    State Agricultural University, Wageningen, The Netherlands   323

Alcoholic fermentation : improvement of the technology based on physiological phenomena
    Département de Génie Biochimique et Alimentaire, ERA CNRS 879, Institut National des Sciences Appliquées, Toulouse, France   331

Enzymatic saccharification of native cellulose : Effect of product inhibition and biomass pretreatment
    Università degli Studi di Napoli, Istituto di Principi di Ingegneria Chimica, Naples, Italy   336

Enzymic liquefaction and saccharification of agricultural biomass
    Agricultural University, Wageningen, The Netherlands   344

PILOT PROJECTS "METHANOL FROM WOOD"

Introduction :
The methanol from biomass Pilot plant programme of the Commission of the European Communities related to other possible routes
    W.P.M. VAN SWAAIJ, A.A.C.M. BEENACKERS, Twente University of Technology, Enschede, The Netherlands   354

Synthetic fuel from wood
    Lurgi Kohle und Mineralöltechnik GmbH, Frankfurt am Main, Federal Republic of Germany   358

Pressurized oxygen blown fluidized bed gasification of wood
    Creusot-Loire, Le Creusot, France   364

The oxygen donor gasifier - Continuous recycling of solids between two fluidised beds
    John Brown Engineers and Constructors Ltd., Wellman Mechanical Engineering Limited, Warley, United Kingdom   371

Gasification of biomass for the production of synthesis gas with the intention to produce synthetic fuel in a further process
    Agip Nucleare S.p.A., Milano; Italenergie Srl, Sulmona, Italy   379

Demonstration of the methanol synthesis
 Lurgi Kohle und Mineralöltechnik GmbH, Frankfurt am
 Main, Federal Republic of Germany   389

LIST OF PARTICIPANTS   393

ASSESSMENT STUDIES

Impact of biomass energy use on employment, energy situation and agrarian market problems in the EC

Technical stand and possibilities for conversion of biomass into charcoal, pyrolytic oil and in comparison with other conversion systems

Assessment study on biogas plants in Europe : economical aspects

IMPACT OF BIOMASS ENERGY USE ON EMPLOYMENT, ENERGY SITUATION AND AGRARIAN MARKET PROBLEMS IN THE EC

Authors	:	E. HEIDRICH, R. SCHÄFER
Contract number	:	ESE - R - 065 - D(B)
Duration	:	18 months 1 January 1982 - 30 June 1983
Total budget	:	DM 192 700 CEC contribution DM 192 700
Head of Project	:	Dr. A. Strehler, Bavarian National Department of Agricultural Engineering (Landtechnik Weihenstephan)
Contractor	:	Technical University of Munich, Bavarian National Department of Agricultural Engineering
Address	:	Bayer. Landesanstalt für Landtechnik Vöttinger Straße 36, D 8050 Freising

Summary

The rapidly increasing costs for the agrarian market, a number of 12 mio unemployed persons in the EC and the problems of providing energy led to a research project reviewing wether it is possible to reduce surpluses and to lower their costs replacing agricultural production by energy farming. Before answering this question the first part of this report concentrates on the possibilities of energy production from forestry wastes especially from rotation forestry.

In the second part is demonstrated that there is a surplus area-equivalent of at least 5 - 6 mio hectares, which could be used for energy farming. It can be expected that this surplus area-equivalent will increase to about 9 mio hectare in 1986 and about 12 mio hectares in 1990. But under todays conditions the subsidies necessary for competitive energy farming exceed the possible savings from a reduction of the land use for food production.

1 Problems of using dry and semi-dry biomass for energy production
1.1 Introduction

The use of dry and semi-dry biomass for energy production can be carried out in several ways. Figure 1 shows a survey on the hierarchy of the regarded biomass. Because there is little information about short rotation forestry it seemed to be necessary to concentrate on the microeconomics and labour requirement of forestry wastes and short rotation forestry in this paper. The whole problem is concerned with following questions:
- Which contribution to the energy demand of the EC, especially to the energy demand of the agriculture can be provided by energy from biomass?
- Which employment effects are to expect?
- Is it possible to unburden the agrarian market and to lower its costs replacing agricultural production by energy farming?
- What are the microeconomics of energy farming?
- Which macroeconomic effects are to be expected?

Fig. 1 gives a survey on the hierarchy of the biomass regarded.

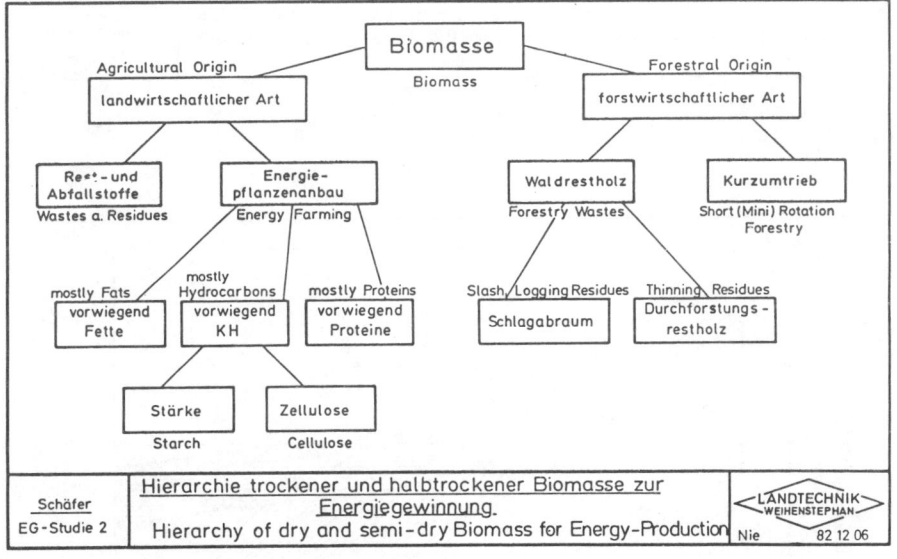

Fig. 1

1.1 Forestry Wastes

The labour requirement and the costs for wood chips production from forestry wastes (thinning residues) is reviewed by APFELBECK (1). Fig. 2 shows the total labour requirement to deliver spruce chips to the user depending on the breast height diameter (BHD) when chipping is performed on the forest road. The labour requirement ranges from 80 to 40 lab.min./m³ chips. Most of the time is required for manual dragging, while felling, cutting, whole tree skidding, chipping and transportation is of subsequent importance.

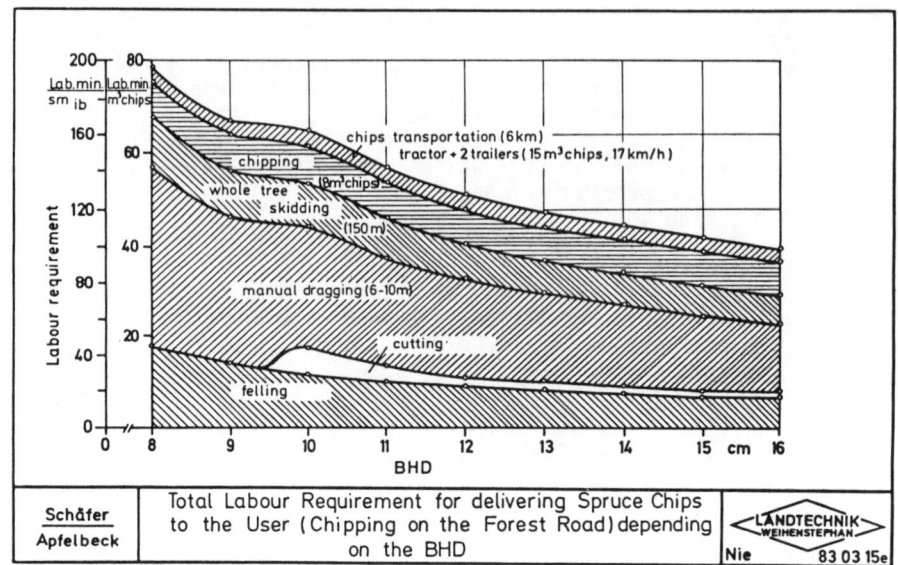

Fig.2

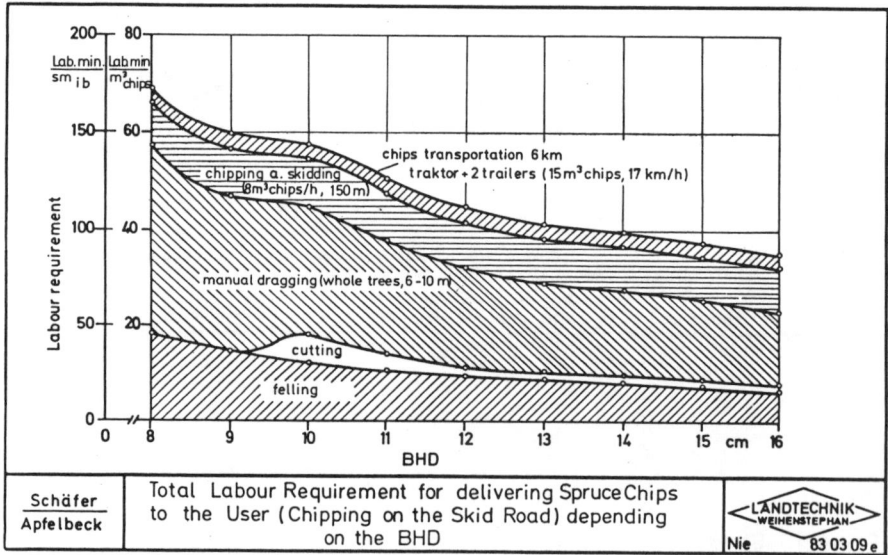

Fig.3

Regarding Fig. 3, which is reporting about the same production but chipping is performed on the skid road we can establish a slightly lowered labour requirement. Nevertheless, manual dragging is still the most labour intensive step. The lowering of the total labour requirement is due to the simplification of skidding the chips and not the whole trees. In some cases, if the ground surface is very uneven, chipping on the skid road is impossible and whole tree skidding to the forest road is absolutely necessary. Fig. 4 shows the total costs for chips production depending on BHD and salary level (chipping on the skid road).

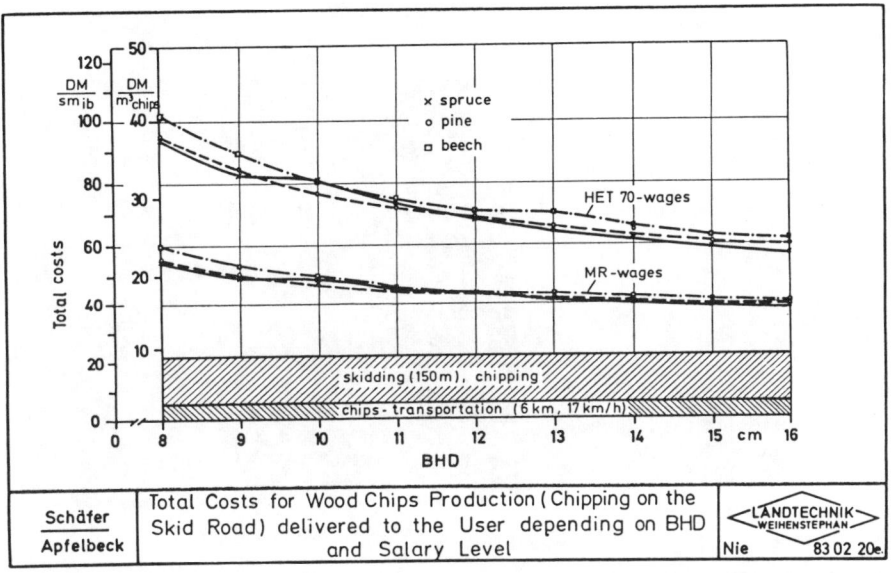

Fig.4

If the wages are on the level of state forests (HET 70), the production costs range from 23 to 40 DM/m³ chips. If machinery ring wages are applied, the production costs range from 16 to 24 DM m³/chips. Wood chips production from thinning residues in state forests seems not to be economically in Western Germany.

1.3 Short Rotation Forestry
As reported from BARNEOUD (2), some clones of poplar can achieve yields up to 35 t dm/(ha·a) on good agricultural soils. Normally a yield of 10 t dm/(ha·a) and somewhat more can be expected (3). The very fast growing of poplar is shown on Fig. 5. The microeconomic aspects of of wood chips production are reported on Fig. 6. Depending on the yield, the production costs range from 40 to 51 DM/m³ dry wood chips. Presently, a price of 39 DM/m³ dry wood chips could be accepted. Fig 6. shows, that the most important research work has to be carried out on harvesting the chips. In any case, continously harvesting systems (flow systems) have to be prefered against discontinuously (batch) systems this is a basic principle of agricultural engineering. Harvesting with a feller-bundler, transportation of the bundles to a store in order to lower the moisture content during storage and separate chipping with a still relatively high moisture content (about 30-35 % wet base) is not regarded as a succesful way.

1.4 Conclusions

The use of forestry wastes as an energy carrier is economically promising if the wages are not too high. In Western Germany, wages according to the machinery rings should be applied. Concerning short rotation forestry, the development of an effective harvesting and drying system is absolutely necessary.

Fig.5: Average BHD of a five year old poplar, grown at HANN. MÜNDEN (Hessen, Western Germany)

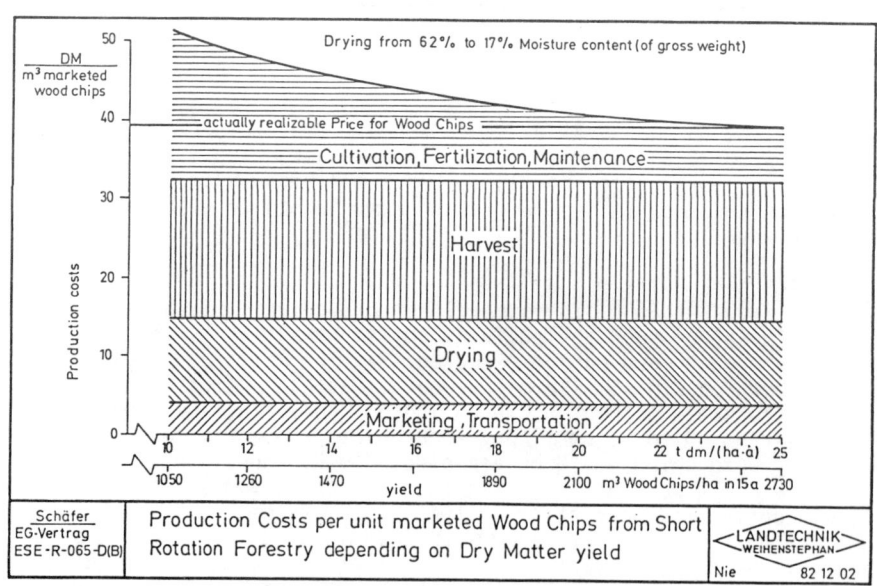

Fig.6

2 Energy farming and the agricultural market in the EC

2.1 Introduction

After the 1973 energy crises had demonstrated the high degree of European dependence on energy imports, the second hike of 1979 - 1981 again drew the attention of the political and scientific community to the possibilities of using plants as a renewable resource of energy. While a great number of research findings on the general technical conditions of energy farming and conversion is available, the first part of this report concentrates on the aspects of using forestry wastes and short rotation forestry (the latter is furtheron included in energy farming). Insofar as energy production does not take place in the form of using waste products or by-products, the question arises to what extent energy farming could be introduced at the costs of food production.

At the same time energy farming also attracts the attention of politicians and scientists in the field of agricultural policy. In the face of saturated food markets in the European Community, a market decline in real agricultural producer prices and increasing budgetary problems of the EC, the opening up of new markets could take away some of the pressure and contribute to increase the limited room for manoeuvre left to agricultural policy. However, success in this direction would presuppose that the savings from a reduction of the acreage used for food production exceed the cost incurred in the form of subsidizing the alternative use for energy farming. These subsidies must be large enough to be an incentive for the farmer to switch over to energy farming.

2.2 Potential area for energy farming

During the seventies the EC witnessed a more or less continual change from a position of a net food importer to one of an area producing increasing food surpluses. Looking at the overall accounts (expressed in terms of grain-equivalent) the degree of self-sufficiency - making allowance for imported animal feedstuff from third countries - could be raised from 94 % in 1972/73 to 107 % in 1980/81. Assuming a 100 % degree of self-sufficiency, this corresponds to "area-equivalents" of -6,3 million 1973 respectively +5,5 million hectares 1981. Since this kind of overall account is calculated by balancing surpluses and deficiencies while at the same time the surplus area consists of very diverse types of land, a more detailed study requires an analysis on the product and subproduct level. It should be noted in passing that the supply balance varies greatly among the EC member countries.

Considerable surpluses can be observed for: cereals, sugar, wine, milk and dairy products and beef. The corresponding area-equivalents and their development since 1973 are shown in figure 7. It should be mentioned that for the calculation of the area of green fodder (incl. grassland) required for beef production, allowance was made for the beef production of the calculated surplus dairy cows. All in all, assuming a 100 % degree of self-sufficiency, 3.5 million hectares of green fodder area could have been set free. The surplus area-equivalent of softwheat and barley amounted to 3.3 million hectares; taking into consideration the imports of other types of cereals, especially corn, that of cereals as a whole was only 1.2 million hectares. In the case of sugar-beets the equivalent would come to 0.4 million while in the case of wine it would amount to 0.5 million hectares (the destilling area-equivalent is regarded as surplus).

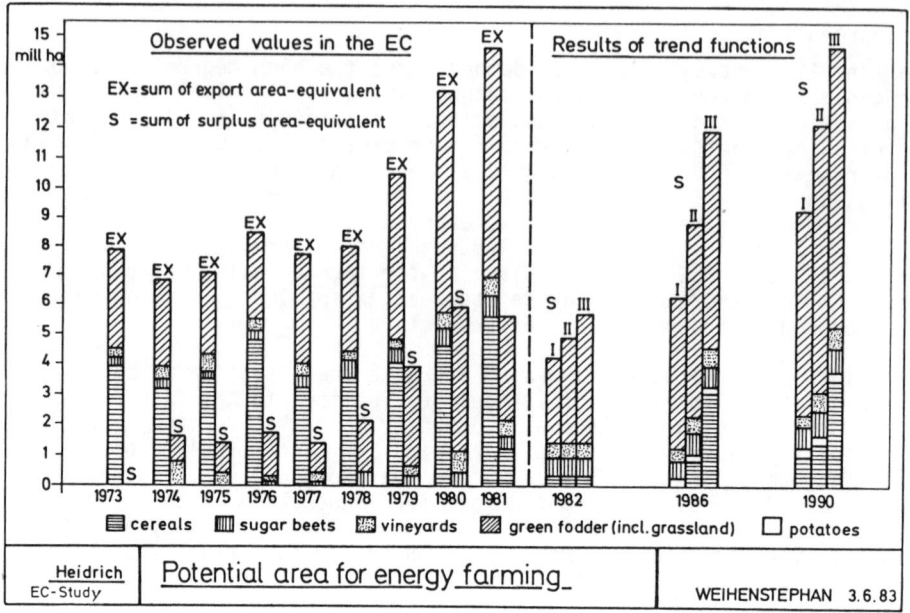

Fig. 7

A comparison of the surplus area-equivalents on the one hand, and of the export area-equivalents on the other hand, shows that the volume of production which cannot be sold inside the EC, is even larger than the calculated balance between usable production and domestic use.

The projection of surplus areas by 1986 and 1990 was carried out on the basis of a linear trend function for each individual product and subproduct with varying assumptions for each case. The results were added for higher, medium and lower estimate as given in figure 7. The difference in the projection of the cereals surplus-area depends on the assumptions one makes about the imported feedstuff (substitutes of cereals). In the case of the green fodder area (incl. grassland) the amount of land to be set free for alternative uses is influenced by the assumed fodder yields of the abandoned land.

2.3 Comparison of savings and subsidies in case of energy farming

The possible savings from a reduction of the land use for products with high expenses for export compensations or market interventions inside the EC is shown in figure 8. The given amounts can be saved in the agricultural budget of the EC by reducing the production area up to the export area-equivalent. As the figure shows the possible savings depend wether one takes into account only the export compensations or one includes the expenses for market interventions with or without storage. In the case of rape and olive growing the average expenses per hectare were calculated while for the other products the expenses per "export-hectare" were taken. If production is reduced only to the surplus area-equivalent, the savings which can be made consist mainly of export compensations. If more land is set free more and more of the expenses for intervention can be saved. However, it seems to be reasonable to maintain the storage of foodstuff to a certain extent, especially when the degree of self-sufficiency falls below 100 %.

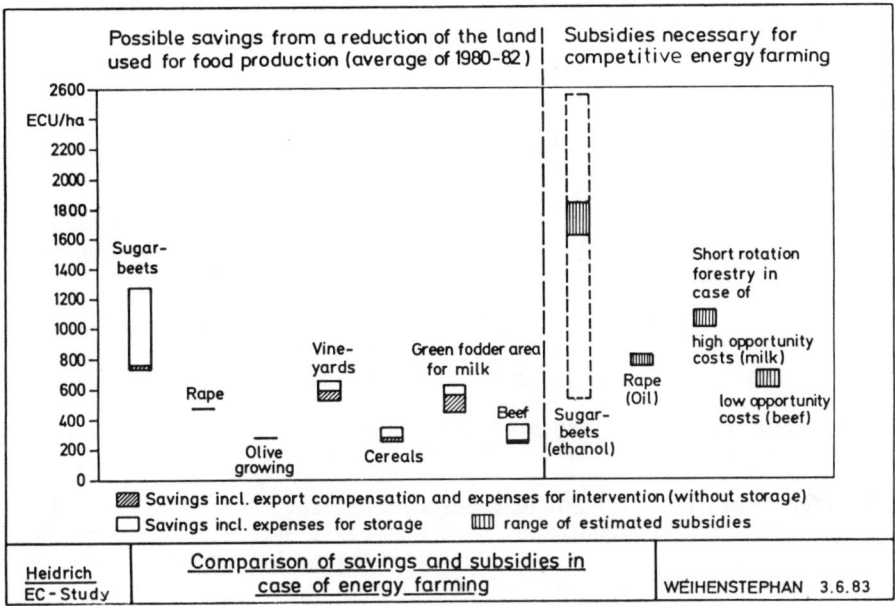

Fig. 8

Opposite to these savings the amounts of subsidies, which are necessary to make different kinds of energy farming competitive, are shown in figure 8. To make the production of ethanol instead of sugar from sugarbeets interesting for the farmer a subsidy of 1600 to 1800 ECU/ha would be necessary (assumed opportunity costs of 1400 ECU/ha). On the other hand the savings in case of not producing sugar are about 800 ECU/ha. This amount would be enough only under very optimistic conditions (very high price for ethanol and very low production costs). In literature these prices and costs still vary in a wide range (5). The necessary subsidy to use rape for energetic purposes (combustion of the oil in Diesel-engines) is about 300 ECU/ha larger than the subsidy for rape when used in the conventional way. In the case of the short rotation forestry the amount to which this sort of land use has to be subsidized (assumed opportunity costs 400 ECU/ha and 800 ECU/ha) does also not bring about savings for the agricultural budget of the EC.

2.4 Conclusion

The calculations with the supply balance sheets for food products show that there could be set free a large area for alternative uses up to about 12 million hectares inside the EC at the end of this decade. But the savings from a reduction of the acreage for food production are not large enough to allow subsidies for energy farming which would make energy production from biomass attractive to European farmers under todays conditions. Rising energy prices or/and declining prices for food products in the future together with technical improvements in the field of energy farming and of conversion might well falsify this conclusion. For that purpose it is necessary to carry out more research work in order to arise yields of energycrops by breeding better plant-varieties and improving farm-methods and to decrease costs for energy farming by construction of new powerful and better adapted machines for harvesting, processing and energy conversion.

References

1 Apfelbeck, R.: Arbeitszeitbedarf und Kosten der Bergung von Waldrestholz. Diploma thesis, unpublished, Bavarian Institute of Agricultural Engineering, 1983

2 Barneoud, C.: Les Taillis a courte Revolution. La Foret privee 1980, No. 133, S. 42-46

3 Dimitri, L., C. von Bismarck, P. Boettcher, J.C. Schulze: Produktion und Verwertung von Pappelschwachholz für die Spanplattenherstellung. Die Holzzucht, Juni 1981, S. 1-7

4 Kommission der Europäischen Gemeinschaften: Die Lage der Landwirtschaft in der Gemeinschaft, Berichte 1975 - 1982

5 Philipp, W.: Energiegewinnung aus nachwachsenden Rohstoffen - Möglichkeit und Konsequenzen, dargestellt am Beispiel der Kurzumtriebsplantagen. Unpublished diploma thesis, Lehrstuhl für Agrarpolitic, München-Weihenstephan 1983

6 Statistisches Amt der Europäischen Gemeinschaften: Agrarstatistisches Jahrbuch, Several years.

Abbreviations

a	year
BHD	Breast Height Diameter
DM	Deutsche Mark
dm	dry matter
h	hour
HET	Holzerntetarif
km	kilometer
lab. min.	labour minute
m	meter
m³	m³, loose volume
MR	machinery ring
smib	solid meter (included bark)
t	metric ton
ha	hectare
ECU	European Currency Unit
mio	million

TECHNICAL STAND AND POSSIBILITIES FOR CONVERSION OF BIOMASS
INTO CHARCOAL, PYROLYTIC OIL AND IN COMPARISON WITH OTHER
CONVERSION SYSTEMS

Author : Dr. E. M. Hofstetter

Contract number : ESE-R-065-D

Duration : 18 months 1. January 1982 - 30. June 83

Head of project : AOR Dr. A. Strehler

Contractor : Bayerische Landesanstalt für Landtechnik
Weihenstephan, Techn. Universität München

Address : Bayerische Landesanstalt für Landtechnik
Vöttingerstr. 36
D-8050 Freising

Summary

The paper gives an overview about the existing possibilities and systems for the conversion of biomass into charcoal, pyrolytic oil and gas. As pyrolysis is only one way to convert biomass into useable energy holders it has to be compared with other systems like direct combustion, ethanol production or methanol synthesis. All these conversion pathways have the disadvantage that under economic consideration they are not only in concurrence with fossil fuel or gas but also with coal. In order to improve the biomass systems it is therefore necessary to reduce the costs for the raw materials that means new planting and harvesting methods should be developed. In future it should be also taken into account that biomass derived fuels don't emit additional CO_2 and SO_2 to the atmosphere in opposite to fossil fuels. Direct combustion of biomass in power plants seems to be a promising possibility as the development of new technologies is not necessary. The production of ethanol is connected with ecological problems especially when it is carried out in a large scale. For the production of methanol it is necessary to reduce size of plants for the conversion of Syngas into methanol. Concerning the pyrolysis the possibilities to use pyrolytic oil should be investigated more detailed.

1. Introduction

The expenditures of the agricultural section which have to guarantee a sufficient income for the farmers even under surplus production are the greatest part by far of the total expenditures of the EC. Therefore there are deliberations to use some of the agricultural areas to produce biomass as energy holders instead of food and feedstuffs. As the produced biomass - no matter of which kind it is - generally is not useable directly as energy holder it has to be transferred into another form. Pyrolysis is one possibility for that as it is transforming biomass into charcoal, pyrolytic oil and gas.

The aim of this study is according to the literature on the subject to represent the possibilities of pyrolysis of biomass for the production of fuels with higher value than the original material. During the research it has become evident that it will give a better overview when pyrolysis is compared with other conversion systems like direct combustion, ethanol and methanol production. Therefore beside the pyrolysis these conversion systems have been also considered and their possibilities have been demonstrated.

2. Possibilities of thermochemical conversion of biomass

2.1 Combustion

In the meantime the combustion of biomass like wood and straw in small scale plant for heat production is well known and working although partial problems like combustion quality and automatic feeding systems have still to be improved. These plants are commonly used in farm houses and wood manufacturing factories where the fuels are either by-products or they can be collected cheaply from own agricultural and forestal areas.

But in this connection only the large scale production of biomass for energetic purposes is of interest. In this case the energy from biomass has to be converted into electric power via combustion and steam production as it is the only way to transport the energy for larger distances without considerable losses.

If you make the calculation for a small scaled power plant of 20 MW_{el} you will need a biomass production area of 23600 ha which are necessary to supply the plant all over the year with fuel when the efficiency of conversion is supposed with 25 %, the working time with 6000 hours/a and biomass yield of 5 t/ha and year. If this area would be arranged in an ideal manner around the power plants the radius of the area would not be more than 18,6 km. As these approximate figures show projects like these could be realized by all means, if it will succeed to produce and process the fuel in an low cost way. If we suppose that 5 Mio ha are available in all the EC-countries than power plants with a total capacity of 4 200 MW could be installed and runned.

Another major point of view is that it would not be necessary to develop a complete new technology as the existing technology of electric power generation by coal combustion should have just to be adapted.

2.2 Ethanol production

Ethanol with a calorific value of 26,8 MJ/kg (20,4 MJ/l) represents an high valuable energy holder which can be used as gasoline substitute or additive because of its physical and chemical properties. In this case it would not be necessary to modify the existing storage and distribution systems for fuels.

For the production of ethanol from biomass there has to be made a difference between lignocellulosic materials and sugar respectively starch containing plants. In the case of lignocellulosics the cellulosis has to be transformed into sugar before by application of hydrolytic processes with several types of acids. The lignins don´t participate on these reactions and have to be supplied to other chemical processes like production of synthetic resins. As the products from hydrolysis represents high valuable materials, e.g. dextrose this process should be more considered as provision of chemicals than for the production of energy holders.

In opposite to the lignocellulosics the starch and sugar containing plants don´t need the process of hydrolysis and they can be converted directly into ethanol. Whereas in tropical areas sugarcane is commonly used as raw material under european conditions sugar beets, potatoes, maize and also wheat are of interest. According to the average yields of these plants in the EX there may be produced from

sugar beet	2,86 t/ha
potatoes	2,24 t/ha
maize	1,44 t/ha and
wheat	0,89 t/ha of ethanol.

In these figures the energy input for planting and harvesting is already taken into account while the energy consumption for the conversion is covered by using byproducts like distiller´s wash. Whereas the technology of ethanol production from biomass is well known for a long time ago, nowadays there are efforts undertaken to improve the conversion especially to receive a better efficiency. The use of distillery columns, thicker mashes and the biogas production from distiller´s wash has to be mentioned.

Nevertheless the system is connected with some specific disadvantage which might become serious under large scale application. One of them is that the necessary plants can be produced just in an monocultural system with all the ecological problems of it.
Furthermore they have to be harvested during a few weeks and afterwards stored up to the moment of consumption. Another point of view is the distiller´s wash which has to be removed,

no matter if it is processed by biogasification or not. All
these facts may be the reason why in Brazil for the long
term energy planning the ethanol production will be reduced
in favour to the methanol synthesis.

2.3 Methanol production

In the same way like ethanol methanol is a high valuable fuel
too which may be used as a gasoline substitute. Producing
methanol from biomass two separate processes have to be considered. The first step is the production of so-called synthesis gas which consists of CO and H_2 via thermal gasification with addition of H_2O. In opposite to the conventional
gasification processes special technologies are necessary
because on the one hand no partial combustion with addition
of air is allowed and on the other hand high gasification
temperatures are necessary to result an almost complete conversion into CO and H_2. In order to solve these problems several projects are financed by the EC and the results of them
are also reported during this meeting.

During the second step in Syngas will be transformed into
methanol and CO_2 under pressure up to 300 bar and temperatures
up to 500 °C. The high temperatures and pressures require
containers which are relatively expensive. Large scale plants
with a better ratio of surface to volume are therefore necessary to reduce the costs but they require the availability
of a large amount of biomass. At the moment plants with a
capacity of 75000 t/year of methanol which means 150 000 t
of biomass are discussed to be realistic. Consequently the
methanol synthesis from biomass has the same problems in
provision of biomass like the electric power generation.
Another point of view is that methanol can be produced from
coal. The price for methanol from coal was in 1981 0,32 DM/l,
while at the same time according to approximate calculations
for methanol from biomass production costs where about 0,43
DM/l. One might remark that the european coal production is
state subsidized and the real costs for coal would be much
more. But the intention that the rate of selfsufficiency in
energy should be as high as possible and the serious number
of jobless people in coal producing areas make it unprobable
that the subsidizing of coal production will be reduced
in favour of biomass production. This may be only happen when
the positive secondary effects of a large scale energy production from biomass like reduction of total SO_2-emission
from fossil origin will be more considered in future.

2.4 Pyrolysis

Strictly speaking pyrolysis is understood as the thermal
decomposition of organic matter into char, pyrolytic oil and
gas. This technology well known a long time ago is in processing fossil coal was noticed again during the seventies
as a process to handle communal residues and wastes. But
the heterogeneous constitution of these above residues showed
up several processing problems. As biomasses have at least
rather homogeneous chemical properties pyrolysis is discussed

at the moment to be suitable to convert biomass into useful energy holders. The most common technologies are described in the annex. As average outputs from the input 30 % of charcoal, 30 % of pyrolytic oil and 30 % of gas can be noted even if there are differences in systems and raw materials. The produced charcoal has an calorific value of about 28 MJ/kg and represents a high valuable energy holder which is commonly sold as barbecue coal or basic material for chemical industries for rather high prices of 800 - 900 DM/t. If the raw material is available at low prices - e.g. as wastes from agroindustry or forestry - even under european conditions the production of charcoal may be economic as long as the high prices for charcoal will be paid. The value of pyrolytic oil is depending mainly from the degree of gas condensation. If water is still remaining as vapor, pyrolytic oil has an calorific value of about 20 - 25 MJ/kg.

According to the belonging literature this oil can be added to fuel oil with a percentage of 20 % for the use of conventional oil burners.

If the condensable components of the gas will be condensed totally pyrolytic oil consist more or less of contaminated water and is therefore useless. As there exists some uncertainity about the value of pyrolytic oil further research should be usefull to get detailed information. Generally pyrolysis can be considered as a technology which may be applied in rather smaller units than other conversion systems like e.g. methanol synthesis.

3. Starting points for further investigations

To all the above conversion technologies is common that an economic production under european conditions is not possible at the moment. The reasons for it are on the one hand that the costs for raw materials are too high and on the other hand that they are in competition not only with petroleum but also with coal. Therefore even if the prices for petroleum will increase considerably coal and their derivates may be cheaper than biomass derived fuels.
But the situation will change remarkable when the ecological advantages of biomass will be put into consideration too. As already mentioned the use of biomass will not emit further CO_2 or SO_2 originated from fossil origines. At the moment these advantages are not put into economic consideration and in order to do this in future corresponding political decisions should be made. Considering the forest diseases in central Europe there is time for it.

To give biomass a better chance the costs for production and harvesting have to be reduced. Therefore the development of new efficient harvesting technologies is necessary. These technologies have also to be able to supply large scale conversion units with raw material.

Regarding the methanol synthesis the running syngas pilot

projects will show if this will be a useable pathway for the future. Furthermore the development of the technology to convert syngas into methanol is necessary to achieve economic systems for smaller units.

Concerning the pyrolysis there exists uncertainity because of the oil. Further research in the production and use of the oil could give more detailed information.

Last not least there should be mentioned again that the ecological advantages of biomass in comparison with fossile fuels should be taken into account in future more than it happened up to now.

Annex: Schemes of different systems of pypolysis

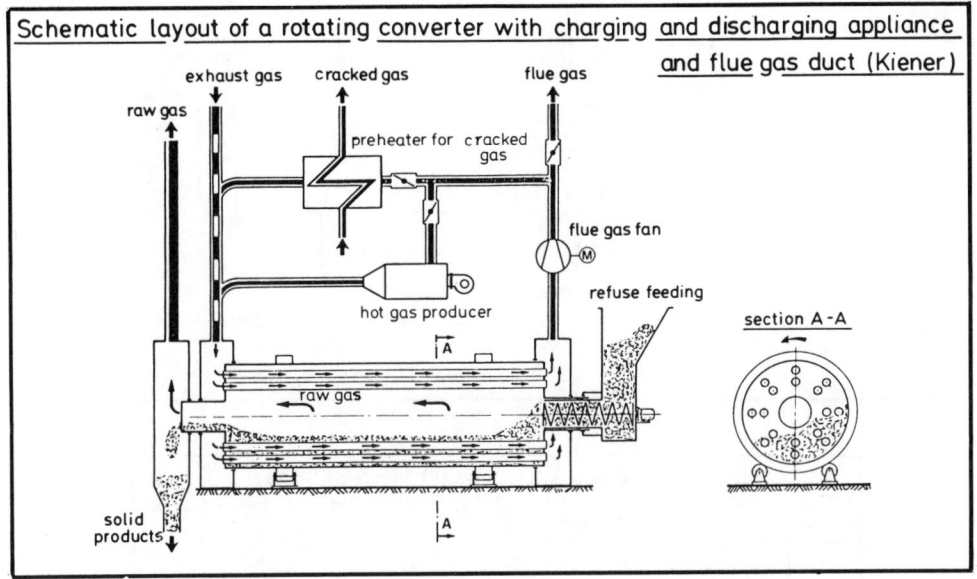

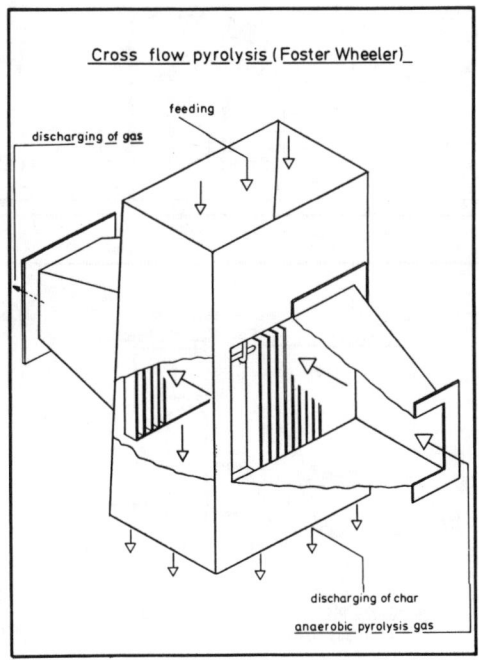

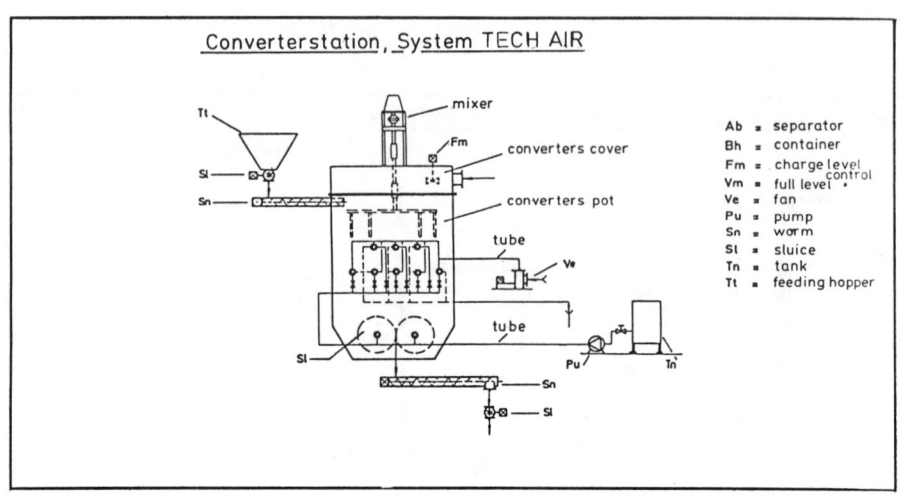

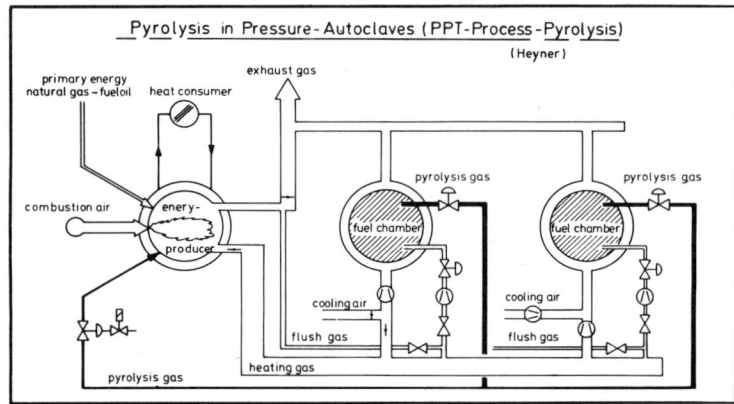

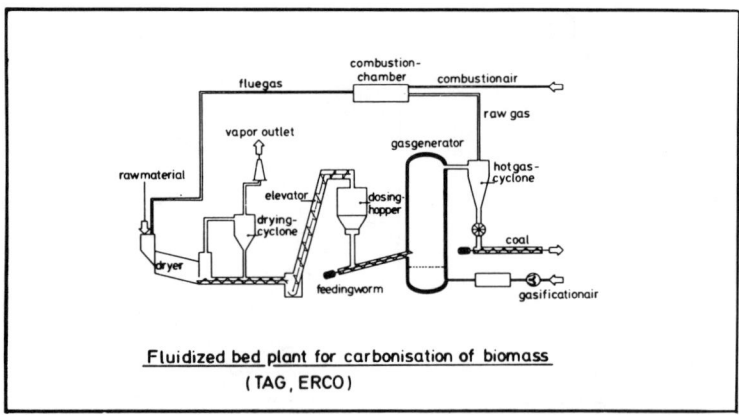

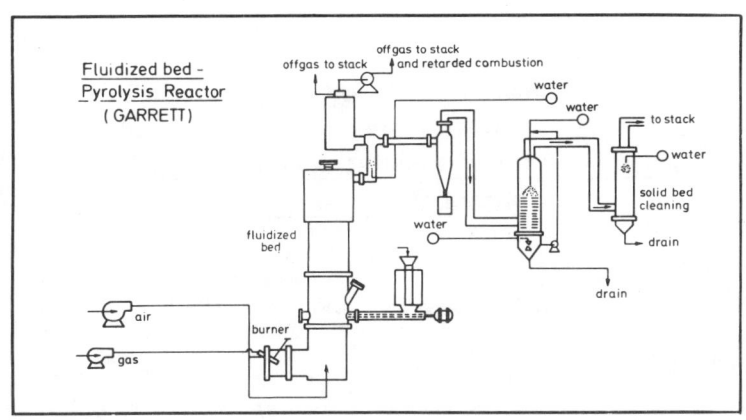

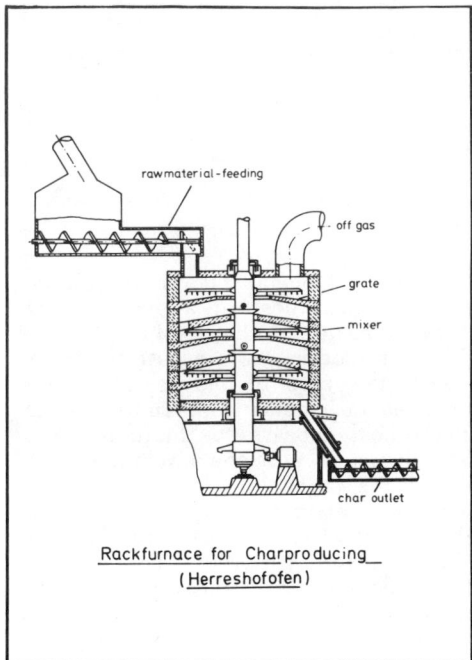

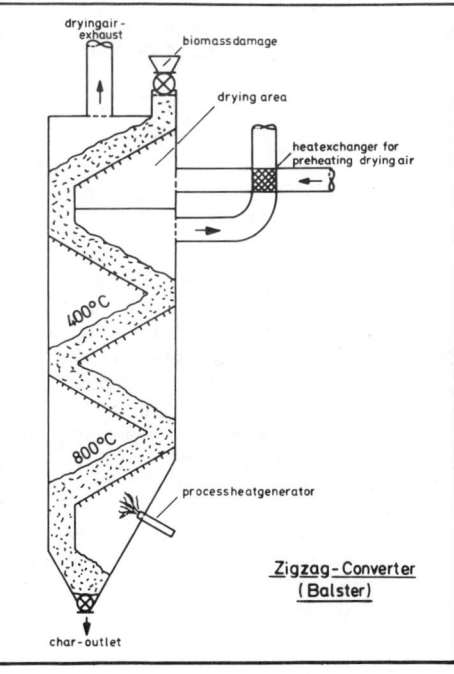

ASSESSMENT STUDY ON BIOGAS PLANTS IN EUROPE : ECONOMICAL ASPECTS

Authors	: M.G. DEMUYNCK, H.P. NAVEAU, E.-J. NYNS
Contract number	: ESE/R/050/B (RS)
Duration	: 24 months 1st July 1981 - 30th June 1983
Total budget	: F.B. 3.075.000 C.E.C. contribution : 100 %
Head of project	: Prof. E.-J. Nyns, Unit of Bioengineering
Contractor	: Catholic University of Louvain
Address	: Place Croix du Sud, 1 bte 9 B-1348 Louvain-la-Neuve, Belgium

Summary

Within the frame of the assessment study on biogas plants in Europe, economical data were obtained. The mean investment cost of a biogas plant on farm varies from 350 to 550 ECU per m^3 digester working volume. Fluctuations are essentially due to the type of construction. For industrial biogas plants, the mean investment cost varies with the type of biomethanation system. However, a wide range of performances i.e. methane productivities is encountered. As a result, whereas, investment cost appears to fluctuate from 250 to 2150 ECU per m^3 digester working volume, in fact the investment per unit of productivity does not vary. Concerning the profitability of 32 agricultural biogas plants, 5 of them are economically attractive with Pay-Back periods for the investment between 3 and 6 years. The two main reasons for the non-profitability of the 27 remaining biogas plants are first a poor operation of the methane digester and secondly a too high investment cost of the biogas plant.

1. Introduction

The survey performed from mid 1981 to the end of 1982 (1,2) allowed to identify 546 biogas plants in the European Community and Switzerland. Biomethanation involves mostly agricultural wastes since 77 % of the biogas plants are located on farm, whereas only 16 % of the biogas plants treat industrial wastes. From the evolution of the number of methane digesters as a function of time, it appears that the learning curve for biogas plants on farm started in 1979-1980, followed by the learning curve for biogas plants in industry in 1980-1981. Municipal wastes are often disposed off in landfills which can easily be transformed in "natural" methane digesters. Such transformed landfills represent already 7 % of the plants notwithstanding that they have been exploited only recently in Europe. Indeed, whereas there were only 2 sites in operation in 1980, a total of 36 sites is identified in 1983 in 5 of the states covered by the survey.

Since the biogas plants on farm are the "oldest existing" plants, valuable economical data exist for some of them and were obtained through the survey. On the other hand, nor for industrial biogas plants nor for landfills sufficient economical data could already be gathered. Only isolated investment cost was obtained for some industrial methane digesters. This cost is examined together with the one of the agricultural biogas plants in the first part of this paper. The second part is devoted to the Pay-Back periods of some agricultural methane digesters.

2. Economics of biogas plants

2.1. Investment cost of biogas plants

Investment cost expressed in actualized ECU (European Currency Unit) per m^3 of working volume was obtained for about 150 biogas plants. As far as the investment cost of agricultural biogas plants is concerned, the prices could be classified in 4 categories depending whether the plant was or was not constructed by the farmers themselves and depending whether there is or there is not an engine included in the methane digestion system. The mean investment cost of the agricultural biogas plants is presented in Table 1.

Table 1 :   Mean investment cost of agricultural biogas plants : actualized prices expressed in ECU/m^3 digester working volume

Categories	Prices (ECU/m^3)
1° Construction by the owner without engine	352
2° Construction by the owner with engine	416
3° Construction by a firm without engine	479
4° Construction by a firm with engine	555

Prices of methane digesters constructed by the owner are on an average 26 % lower than the prices of methane digesters constructed by a firm. This can be explained by the fact that manpower is not included in the former price and that often used material is recovered for the construction of the digester. The integration of an engine for generating electricity increases the investment price by 12 % but it allows of course the transformation of biogas into electricity which is a more noble energy.

From the analysis of the investment prices it appears that the prices per m^3 digester working volume are clearly decreasing with the increasing scale of the methane digester, at least until 100 m^3. Beyond this scale, the investment cost per m^3 becomes more or less stable.

As far as the investment cost of industrial biogas plants is concerned, it is classified per category of biomethanation systems. Three categories are considered : the anaerobic contact system (completely-mixed system with recirculation of active biomass after settling), the french I.R.I.S. (Institut de la Recherche de l'Industrie Sucrière) system (anaerobic contact in one tank) and the up-flow without carrier process (up-flow non-mixed digester with accumulation of settled active biomass by floculation). Depending on the biomethanation system, the gas production rate varies from about 1 to about 5 m^3 gas per m^3 of digester and per day. As an example, for the biomethanation of sugary wastewaters a mean gas production rate of 5.4 $m^3 \times m^{-3} \cdot d^{-1}$ is obtained with the up-flow without carrier system whereas 0.64 and 0.88 $m^3 \cdot m^{-3} \cdot d^{-1}$ are respectively obtained for the anaerobic contact and the I.R.I.S. systems. The investment cost per m^3 of working volume digester varies also in the same direction. A mean investment cost of 2159 ECU . m^{-3} is observed for the up-flow without carrier system whereas mean investment costs of 436 and 248 ECU . m^{-3} are respectively observed for the anaerobic contact and I.R.I.S. systems. Fig. 1. depicts this observation.

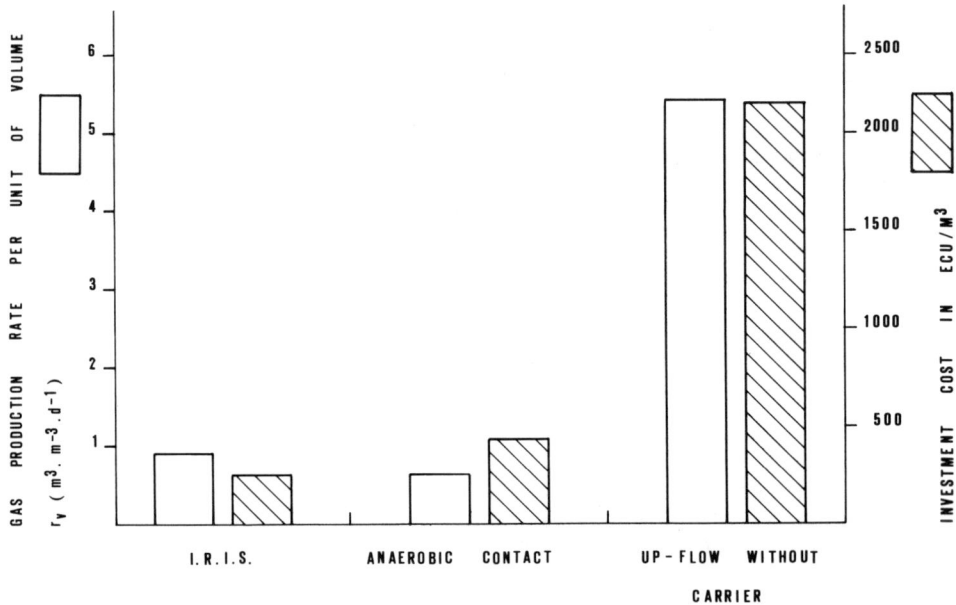

Fig. 1. :   Investment cost/m^3 related to gas production rate for 3 biomethanation systems

It can so be concluded that the investment cost relative to the performance is relatively the same for the three biomethanation systems. The I.R.I.S. system appears to be a little less expensive. The principal reason for this is that

I.R.I.S. methane digesters are hitherto entirely or partly constructed in polymeric material instead of steel or concrete which leads to lower investment cost. The tendancy is now to construct the I.R.I.S. methane digester in more durable material so that finally its investment cost is identical to the one of the anaerobic contact methane digester.

2.2. Pay-Back period of biogas plants

Limited information is as yet available concerning the economics of the biogas plants existing in the countries covered by the survey by the end of 1982. Not many methane digesters have already been in routine operation for more than one year and, if so, their economics have not always been pro perly evaluated. Nevertheless, valuable data were obtained for 32 methane digesters on farm.
For this economical analysis, the Financial Model Processor, perfected as processing tool by Schepens (Faculté Notre-Dame de la Paix, Namur, Belgium) has been used. This model is essentially based on the concept of cash flows and calculates the Net Present Value (N.P.V.) and the Internal Rate of Return (I.R.R.) for each scenario. The Internal Rate of Return (I.R.R.) is defined as the discount rate that equates the Present Value of the expected future cashflows, or receipts to the initial cash outlay. If the I.R.R. is greater than the cost of capital (k), the project will be able to use the cash-flows generated by the investment to repay the funds obtained, including the costs of the funds. The project can then be accepted. For this economical analysis, a reference capital cost of 15 % has been assumed. This rather high value was choosen so as to allow a certain factor of security in cases of temporary poor operation of the methane digester.
This economical analysis allows also the calculation of the simple Pay-Back Period which is the number of years it takes a project to recover its original investment from net cash-flows.

In this economical analysis, the yearly cash-flows were directly calculated by summing the three following elements :
- investments cost of the whole biogas plant including the digester itself, the storage tanks if newly build, the installation for biogas utilisation, including the manpower except for "do it yourself" biogas plants,
- yearly operation cost including the maintenance and repair costs. For "do it yourself" methane digesters, the cost of manpower brought by the farmer is again not taken into account,
- yearly returns which are provided first by the replacement of energy demands of the farm by biogas, or electricity, and eventually the sale of gas or electricity.

It has to be noticed that the values of odor reduction and/or depollution were not taken into account in these calculations, although in some circumstances, the latter were priority motivations.

In this economical analysis, 5 of the 32 biogas plants are "do it yourself" plants. Three of them are clearly profitable. Their simple Pay-Back period is 3 or 4 years and their Internal Rate of Return is higher than 30 % which is again higher than the capital cost of 15 %. The essential reason for their profitability is their low investment cost, from 100 to 160 ECUS/m^3. Another biogas plant more costly at the investment is nearly profitable (7 years of simple Pay-Back period). Finally the fifth biogas plant is uneconomic with a simple Pay-Back period of 10 years and an I.R.R. of 3.8 %. However, the latter biogas plant is not costly at the investment, but it operates only 6 months per year so that the gas production obtained does not cover sufficiently the investment cost.

The 27 other biogas plants analysed economically are plants constructed by a firm. Out of these 27 plants, only 3 are economic with simple Pay-Back periods of 5-6 years and I.R.R. higher than 15 %. Six of these 27 digesters have 8 to 9 years of simple Pay-Back period but I.R.R. lower than 15 %. The 18 remaining plants are clearly uneconomic with simple Pay-Pack periods equal or higher than 10 years and I.R.R. nearly equal to zero or even negative.

Table 2 :   Main reasons for non-profitability of biogas plants

Reason	% (share of total)
- Poor operation of the methane digester	42
- Too high investment cost	38
- Poor gas utilisation	8
- Too high maintenance costs	4
- Too important heat requirements	4
- Too high investment cost & poor gas utilisation	4
	100

The main reason for the non-profitability of the 24 plants varies from one plant to another. Nevertheless, it appears from Table 2, that the two main reasons are first a poor operation of the methane digester and secondly too high an investment cost. A poor gas utilisation influences also the non-profitability of the biogas plant but to a lesser extent. The three remaining reasons (too high maintenance costs, too important requirements for the heating of the methane digester, and too high investment joined to bad gas utilisation) are more particular.

The first reason i.e. a poor operation of the methane digester is sometimes temporary. It is during the two first years of operation, that the gas productivity remains low whereas afterwords it improves. These two years of poor operation are sufficient for lowering the profitability of the plants. Supposing that they had their cruising gas productivity from the beginning of the operation, their simple Pay-Back period would so be reduced from 8 to 6-7 years with an I.R.R. of 15 %.

As far as the second reason is concerned, unfortunately nothing can be changed. At least, from the survey, it appears possible to construct biogas plants for a price which does not exceed 300-400 ECUS per m^3 of working volume of methane digester.

3. Conclusion

That biogas plants can be economic is proven through the survey. Indeed 5 of 32 biogas plants for which data were available have been located with Pay-Back periods for the investment between 3 and 6 years.
Two major factors have to be respected for an economic methane digester and can be respected so as demonstrated by this economical analysis. The first one concerns the reliable performance of the methane digestion. An average daily productivity of about 1 m^3 of biogas per m^3 of methane digester appears to be the lower limit for economy. Of course, a proper use of the biogas produced is on par with the good productivity of biogas. On the other hand, the investment cost for the overal biogas plant must not exceed 300 to

400 ECU per m^3 digester working volume. When the owner is taking part in the construction of his biogas plant, such prices can easily be attained.

The complete survey will be published as a book by the end of 1983 on behalf of the Commission of the European Communities. The executive summary of the study will be presented on the third Anaerobic Digestion Symposium held in Boston on 14-20th August 1983.

References

1. Demuynck, M.G., Naveau, H.P., and Nyns E.-J., (1982)
    Assessment study on biogas plants in Europe. In P. Chartier and W. Palz (Eds), Energy from Biomass, Series E, Vol. 3, Reidel D Publishing Company, Dordrecht, Holland. pp. 17-22.

2. Demuynck, M.G., Naveau, H.P. and Nyns E.-J. (1983).
    Biogas plants in Europe. In A. Strub, P. Chartier and G. Schleser (Eds), Energy from biomass, 2nd E.C. Conference, Applied Science, London. pp. 620-625.

REPORT OF SECTORIAL SEMINARS

Wastes, energy crops and forestry

Review paper on algae

The status and future of thermochemical processing of biomass

Biomass conversion : Biological routes

REPORT OF SECTORIAL SEMINAR ON WASTES, ENERGY CROPS AND FORESTRY

C P Mitchell and M L Pearce

The following is a summary account of discussions held on 8 June in Capri by representatives of 11 of the 13 projects in the group Wastes, Energy Crops and Forestry. The French projects on straw (ESE-R-014-F) and forestry wastes harvester (ESE-R-049-F) were unfortunately not represented.

The aim of the seminar was to produce the main conclusions of the present research programme and outline areas for research and development in the near future.

This is a fairly heterogeneous group covering a wide range of activities. The discussion was divided into three main areas; natural systems, agricultural wastes and energy crops, and forestry (production and harvesting). These divisions are not exclusive but were used to aid discussion.

Natural Systems

The projects in this group considered biofuel production from natural vegetation in Great Britain (ESE-R-015-UK), a new method of cultivation of Arundo donax in France (ESE-R-042-F), harvesting Turkey oak coppice in Italy (ESE-R-035-I), harvesting Mediterranean-type ecosystems in Greece (ESE-R-063-GR) and the quality and quantity of latex which can be produced from the Euphorbiaceae family in Greece (ESE-R-064-GR).

The important point to note about all these projects is that there are relatively large resources technically available and that the use of natural vegetation for energy is an important management alternative for such ecosystems. Maintenance of production following continued harvests may be difficult and hence limit the suitability of these systems.

In Britain Callaghan has identified and studied 3 species with a high potential of use for energy. These are bracken, cordgrass and Japanese knotweed. Of these, the latter should more properly be considered as an energy crop. Bracken and cordgrass occur naturally and can be harvested annually with yields of 7 and 11 dt/ha/yr respectively. Continued harvesting of bracken may lead to its eradication and consequently its use for energy could be coupled with an eradication programme.

The 2 projects in Greece started later than the other projects and so full results are not available. However, it seems likely that yields of 50 dt/ha on a 10 year rotation are possible for the maquis ecosystem. Management of this system for energy will have an additional advantage in preventing the ravages of fire.

The work on Arundo donax is particularly interesting with productivities of up to 20 dt/ha/yr over 10 years. The ease with which it can be integrated into agricultural systems must make it an attractive proposition in southern France. Problems still remain concerning the best management system and the optimum frequency of harvests to maintain productivity.

Harvesting of natural Turkey oak coppice using mechanization is the first time that it has been done in Italy. There is a large potential resource and the indications are that it can be done without unduly damaging the cut stools and so allowing coppice re-growth. Here productivities of 2 dt/ha/yr might be expected on rotations of 20 years. We still await cost figures before it can be said that such a practice is economically viable. Distance from likely market appears to be a major determinant of viability.

A possible problem with the use of natural vegetation for energy is the low density of supply and often the remoteness of the resource from the point of end use making transport costs the limiting factor. However, if small-scale localized use is envisaged this may not necessarily be a large problem. It is felt that there is a strong need to identify possible uses and conversion processes for which natural vegetation is suited as this may effect management and harvesting regimes. That natural vegetation can be harvested (technically and/or economically) is as yet unproven.

Future Work

It was felt that there was a need for more fundamental work on plant productivity and its limiting factors along with a continued search for new wild species suitable for energy cropping.

Large scale trials are required of systems already identified to establish reliability of yield, harvesting and management. Such trials should also be used to produce cost estimates as these so far have been generally lacking from the programme.

Multiple cropping systems such as agroforestry as well as crop fractionation deserve more attention.

Agricultural Wastes and Energy Crops

This group contains projects on the use of straw and animal wastes (ESE-R-014-F) and growing catch crops for fuel in Great Britain (ESE-R-034-UK). Straw appears to be an economically viable fuel for combustion on farms for use in space heating and grain drying but its use off-farm will be limited due to unfavourable competition with coal.

The study on catch crops indicates that there is a technical potential production of about 1 to 7 dt/ha/yr. It has stressed the technical aspects (particularly in relation to crop production) of catch crops but has not examined in detail the economic potential. Catch crops are considered suitable for anaerobic digestion.

As with natural vegetation these sources have a low density of supply but if used with other energy crops or animal wastes should be suitable for on-farm use but are unlikely to be used off-farm.

Future Work

There is scope for more work on the locational aspects of catch crops particularly in relation to species and yield.

The economics of the use of catch crops and straw need to be further refined. This might best be accomplished by using a farm case study approach. This should also encompass a survey of farmer attitude to catch crops, availability of land and possible end uses for the energy crops.

Forestry

There are 6 projects in this group. Four are primarily concerned with production and 2 with harvesting.

Forestry Production

For production, coppice receives the most attention with work in Great Britain (ESE-R-017-UK), France (ESE-R-018-F), Ireland (ESE-R-036-EIR).

Single stemmed short rotation forestry is studied in Great Britain (ESE-R-016-UK). The French project also involves work on the production of inter-specific crosses within the genus Alnus.

Productivity data is beginning to emerge from these projects. For coppice which is limited to the more fertile and sheltered sites productivities of 12 to 15 dt/ha/yr appear possible. Whether this level of production can be obtained consistently in practice has been called in question from the results of the Irish programme although others felt that with good species selection to fit the site and environment coupled with good management these production levels should be possible.

Single stem plantations, which are likely to be on poorer ground than coppice and of which more will be available, have productivities of between 6 and 12 dt/ha/yr. The results in Ireland of 8.5 dt/ha/yr after 5 years growth with Lodgepole pine on virgin peat appear particularly promising for such a difficult site.

The management of forest biomass plantations is becoming better understood and the importance of good weed control in the two years following planting has been recognized. Information on costs and logistics of establishing plantations is beginning to emerge although this is, in the main, limited to small areas.

The potential of short rotation forestry to supply wood for fuel appears promising particularly if good quality land will be released from agriculture following revision of the Community's agricultural policy.

Future Work

The trials laid down in the current programme should continue to be monitored. Further small scale trials are needed to gain more information for species and clone selection. New trials will be necessary to further examine the silvicultural aspects of energy plantations such as spacing, rotation length, stool longevity, cultural techniques, effects of mechanization and mineral balance in relation to productivity. There is a strong need to establish large scale trials so that management and maintenance of productivity can be studied at a scale that might be met in practice. In conjunction with such studies there is a need for detailed systems studies to examine the planning and management of forest energy plantations.

In order that a long term strategy for forest energy can be developed within the Community detailed studies of land availability such as that carried out in Great Britain are necessary (1).

Forest Harvesting

In this programme 3 harvesting machines are being developed; one for forestry wastes and small trees (ESE-R-049-F) and 2 for coppice (ESE-R-016-UK and ESE-R-019-EIR). In addition a desk study has been undertaken to examine ways of harvesting forest biomass for energy (ESE-R-016-UK).

The French forestry wastes harvester appears to be a promising design but it is too early to be confident of its performance.

Of the two coppice harvesters that have been built the Bord na Mona machine was designed for harvesting the produce of large plantations on cut-over peat bogs and the Loughry machine for harvesting coppice on small areas of disadvantaged farm land. The Bord na Mona machine produces billets and the Loughry machine bundles. Both machines have suffered through not having sufficient material available for extensive harvesting trials.

In performance trials the Bord na Mona machine was found to be wanting in its ability to produce billets consistently. The main problem appears to be the inability to adequately billet the thin lateral branches of the coppice. This coupled with the end users' dissatisfaction with the billeted material as a fuel and as a boiler feed has led to the conclusion that a chipped end product would be more suitable.

The Loughry machine has been through several stages of development and now that the problems experienced with the bundling mechanism appear, hopefully, to have been resolved is ready for extensive field trials.

The desk study has identified the potential importance of Integrated Harvesting Systems to recover forest residues in a 'one pass' operation. This is particularly important considering the vast resource of small roundwood available in the short to medium term in north-west Europe.

Future Work

Extensive trials are needed to obtain experience in using the harvesting machines so that production and reliability data can be gathered particularly for the Wastes and Loughry coppice harvesters. Plantations need to be established so that adequate test material will be available for future harvesting trials.

The Bord na Mona machine should be re-assessed and perhaps re-designed to handle the lower than anticipated growth rates of the willow coppice plantations on cut-over peat and to produce chips or some intermediate form as an end product.

The potential importance of Integrated Harvesting Systems to produce both an energy and a conventional forestry product in a 'one pass' operation should be studied further. Desk studies can be carried out to examine the design criteria leading to system development and trial. This need not necessarily involve the development of new equipment but could use existing machines with modifications to their use. Field trials are needed to assess the production rates and economic viability of the systems across a range of crop and terrain types.

General Considerations

Results to date have led to a more realistic attitude towards levels of productivity to be expected from forest energy plantations and problems of management. Previously the former have been very high and the latter dismissed as not important. The opinion, however, is still that natural vegetation, agricultural wastes, energy crops and forestry can make a significant contribution to future energy needs. The recognition that biomass is more difficult to exploit and manage than some protagonists of biomass energy have led us believe and that even so it can be done is a great step forward in establishing the credibility of energy crops.

The uncertainty over likely end-users for some of the energy crops, their often low density of supply and problems of implementation leads to the conclusion that it is essential that resource location/allocation (or matching studies) should be undertaken as a matter of high priority.

Reference

1   Mitchell, C P; Brandon, O H; Bunce, R G H; Barr, C J; Tranter, R B; Downing, P; Pearce, M L; Whittaker, H A (1983) Land Availability for Production of Wood for Energy in Great Britain. In Proc 2nd EC Conf 'Energy from Biomass'. Ed A Strub et al. Appl Sci pp 159-163.

REVIEW PAPER ON ALGAE.

J.G. Morley, D. Pirrwitz.

The various projects concerning the use of algae as biomass can be devided into two groups utilising either micro algae or macro algae.

Micro Algae have been studied in the context of :

(a) closed systems or photoreactors (Paris and Cadarache)
(b) open systems consisting of fresh water ponds (Liege)
(c) open systems consisting of sea water ponds (Calabria)

Macro Algae have been evaluated in the context of

(d) growth on artificial substrates (North East Atlantic)
(e) direct harvesting under natural conditions (Venice Lagoon).

The objective of micro-algae closed systems is to study the feasibility of industrial production of hydrocarbons using algae.
To this end factors affecting hydrocarbon production of the green unicellular algae Botryococcus braunii such as the morphology and metabolism and the effects of the biological and physico chemical parameters of its culture have been investigated.

It has been shown that at least two species of Botryococcus exist, one producing linear hydrocarbons, the other producing branched hydrocarbons. The proportion of hydrocarbons is high in both species. It is considered that productivity may be increased further by strains selection, by culture parameter optimisation and through the use of a bacteria associated culture.

A first step towards the realisation of industrial production of hydrocarbons by Botryococcus cultivation has been taken by the setting up of a tubular photoreactor from which initial data have been obtained.

Micro Algae in open systems have been grown successfully in large artificial lagoons (ponds of 2200 m2), supplied by water from a river, heated up from cooling water of a power plant. The culture produces about 10 tonnes dry biomass hectar^{-1} year^{-1}.
The production of biogas with a methan content of 65 % has been demonstrated.
The changes which occur during the course of the year have been studied also.

In Southern Italy seawater ponds have been used for the culture of algae, the growth of which is facilitated by the high ambient temperature. Various algae species have been studied and Tetraselmis tetrathele has been observed to give the highest yields. It has been shown, that micro algae in seawater fertilised with recycled sludge from a digester, or by using polluted water as a source of carbon, nitrogen and phosphorous, gives an efficient system for biomass production.

Yields of 60 tonnes of dry biomass hectar^{-1} year^{-1} have been obtained and yields as high as 90 tonnes hectar^{-1} year^{-1} are expected in appropriate locations. The algae (Tetraselmis tetrathele) can be converted efficiently into methan (0,35 m^3 of CH_4 per Kg of volatile solids).

In open sea mariculture studies have been made of the growth of various types of macro algae on horizontal and vertical polypropylen ropes moored offshore.
Growth rates approaching 15 Kg per meter (fresh weight) have been measured during natural colonisation in sheltered conditions. Seeding techniques have been devised to grow particular species of plants on ropes and the effect of various factors, e.g. time of seeding, depth of rope, etc. have been shown to be important. In order to estimate the likely effects of tidal currents and wave action on the rope structures, measurements have been made of the hydro-dynamic drag forces developed for groups of plants of various species.

Macro algae growing naturally in the lagoon in Venice have been harvested using boats fitted with suitable equipment. Harvesting is facilitated by tidal currents which remove macroalgae from the bed of the lagoon and produce concentrations in particular areas. The annual crop is estimated to be approximately 10^6 tonnes fresh weight of macroalgae.
Mesophilic and thermophilic anaerobic fermentation to produce methane has been demonstrated using either one or two stage processes.

In the future further work of Botryococcus cultivation in a tubular photoreactor should include the selection of high productivity strains and genetic engineering and a search for suitable biological and chemical agents to protect Botryococcus cultures against contamination.
Studies of the feasibility of producing immobilized cell reactors for hydrogen production should also be done.

Concerning the open systems it should be envisaged to extend the present lagoons up to a size of about one hectar and to submit the harvested algae to a plant of biomethanisation.
The micro-algae cultivation in seawater should also be tested on a larger scale in order to demonstrate the industrial potential of the technology. For the macro-algae in open sea systems the maximum growth rates achievable and the factors which affect it should be established.

THE STATUS AND FUTURE OF THERMOCHEMICAL PROCESSING OF BIOMASS

A.A.C.M. BEENACKERS and A.V. BRIDGWATER

INTRODUCTION

Participants in the Fourth Contractors Meeting of the EC Biomass Programme E who are concerned with Thermochemical Conversion, met on June 8th., 1983 to discuss the achievements of the Programme to date; remaining problems and constraints; and R and D requirements for the future. While it is hoped that all views are accurately represented, this report is a personal presentation of the discussion and conclusions of the meeting, and individual views may occasionally differ.

The objectives of the workshop were to:

. firstly identify the achievements of the last four years in the various thermochemical conversion areas sponsored by the EEC;

. secondly identify the constraints inhibiting implementation of the technologies developed; and

. thirdly suggest promising or necessary research and development areas that deserve support in a future programme.

Consideration was given to combustion in small scale systems; pyrolysis to gas, liquid and solid products; small scale gasification giving low heating value fuel gas; large scale gasification giving synthesis gas for subsequent upgrading to liquid fuel; and finally high pressure hydrogenation to liquid hydrocarbons.

Liquefaction

There are several ways of producing liquid products by direct and indirect methods from biomass and it is helpful to differentiate between them.

Pyrolysis at, or near to, ambient pressure directly produces a pyroligneous liquor consisting of water, highly oxygenated water miscible organic compounds such as carboxylic acids, and oxygenated water immiscible materials such as tars. It is often referred to as pyroligneous acid or liquor. The material is corrosive and unstable, and at best is a low grade fuel according to the water and oxygen content. Little work has been performed in upgrading this liquid.

High pressure liquefaction at high hydrogen partial pressures but relatively low temperatures (usually below 300°C) gives moderate yields of slightly oxygenated and more homogeneous organic compounds that are often referred to as an "oil". This process is usually referred to as just liquefaction or direct liquefaction. Although currently investigated mostly on laboratory scale, potential applications for fuel products are seen as large scale processes.

The third main method of liquefaction, but by indirect means, is production of synthesis gas. After refining this can be converted or synthesised into a liquid hydrocarbon such as methanol. Production of gasoline is also possible, either directly from synthesis gas (Fischer-Tropsch) or indirectly from methanol (Mobil).

CONVERSION TECHNOLOGIES

The various conversion technologies mentioned earlier are now discussed, with four summary tables of achievements, constraints, and opportunities for further work at the end of this paper.

Combustion

Extensive work has been carried out on straw combustion in Denmark, Germany and the U.K. with agreement that continuous feeding is preferred to batch operation; combustion in a brick furnace is preferred to a water jacket; and pellets or chopped straw preferred to bales. All aid higher temperatures and more controlled residence time, giving better combustion efficiency and lower pollution levels. Most technical problems have now been solved, with the major disincentive to wider adoption being the high initial cost of such units. The primary requirement now is for low cost, high efficiency furnaces with automatic charging systems and adequate safety controls. A compromise between quality and cost may be necessary. The differing environmental regulations of member countries is an inhibitory factor in some locations.

A current problem is that many furnaces now in use are of poor design with a low efficiency and with consequent environmental problems. These factors tend to discourage wider adoption. However, over 500,000 combustors were claimed to have been sold in Germany and 1 in 6 Danish farmers have a unit. Systems of up to 10 MW supplying district heating schemes were reported in four locations in Denmark, although generally 1000 kW was believed to be the limiting capacity. A further limiting factor might be feedstock in some locations, for example Holland, where straw production has virtually disappeared in 5 years. Denmark, however, still has a surplus of 2 million t/y straw.

Alternative feedstocks: wood and bark can be satisfactorily burned in high qualtity, low cost, systems; and a comparative study of gasification and combustion of urban waste has recently been commissioned by the EEC.

It was generally felt that biomass combustion systems were sufficiently developed and available commercially to remove the need for further substantial R & D support. There are, however, improvements in efficiency and pollution levels that are desirable; and interesting possibilities exist for developing fluidised bed combustor systems.

Pyrolysis

Although some systems based on pyrolytic conversion of biomass to gas, liquid and solid fuels are available commercially, there was little enthusiasm expressed for this mode of operation. Most concern centred on problems of utilising and handling the liquid phase product. There are few uses for this product due to its corrosive and unstable nature, other than combustion or internal recycle. It can also give rise to an intractable waste disposal problem.

Some work has been carried out on maximising the liquid fraction, for example by flash pyrolysis, and an overview on the status and progress of work in this area would be valuable. There is a definite incentive for R and D on upgrading the product to improve stability, reduce corrosive constituents and produce a more uniform and compatible fuel. A low cost and efficient process might be favourably compared to direct liquefaction and/or indirect formation of liquid products via synthesis gas.

Gasification

The EEC programme has, over the last eight years, seen projects develop from fundamental and laboratory investigations to substantial demonstration units, and it is therefore opportune to examine the current status of activity and possible developments. It is clear that the European market for such systems will not develop rapidly, but the world market has a much greater potential to assimilate the technology being developed. What, then, has been learned; what are the problems; and what should subsequent work programmes consider?

A general point concerns feedstocks. In particular the question of omnivorous gasifiers versus specific fuel dedicated gasifiers aroused much discussion with the conclusion that, especially for moving beds, shorter term development should be orientated to plants dedicated to a specific feedstock. This requirement would place greater emphasis on a closer interface between biomass production and conversion. Feedstock preparation is a fertile area for further investigation in terms of size control, drying, handling and storage. The nature of biomass being essentially fibrous is likely to cause more problems in feeding than, for example, coal; and there is thus scope for more work in this area particularly for pressurised reactor systems.

Further discussion centred on either low heating value gas from small scale systems, or medium heating value gas from large scale systems with subsequent conversion to high grade fuel; and these are discussed separately.

Small scale gasifiers

These produce a low heating value gas and generally range in size up to 1MW. Attention focussed on the relative merits of combustion to heat and thence power, against gasification to fuel gas, and thence heat, power or electricity. Also between these two options there is two stage combustion or starved air incineration. Although gasifiers are now available commercially, there is little information on their suitability for different applications by energy form or industrial sector, and a market survey was believed to be valuable to identify and differentiate opportunities for different systems.

On technical developments, a need was expressed for study and evaluation of low cost and effective high temperature gas cleaning systems. The possibility of close coupling gasifiers with turbines, particularly if pressurised systems can be effectively developed, is an interesting possibility as turbines of 30 kW upwards are increasingly becoming available for which a range of applications exist.

The European market is now seen to be at a critical stage, and, although there are believed to be at least ten companies offering commercial gasifiers, most market growth potential is seen to be in developing countries. Probably the most useful activity that could be undertaken is monitoring the performance of units that have been sold and are being used in practical situations. This is perhaps the most effective way of identifying real problems requiring solution.

Large scale gasifiers

The two four-year programmes of the EEC have culminated in four contracts to build substantial demonstration units to produce synthesis gas for methanol synthesis. It is however, too early to comment on these activities.

Discussion centred on the next steps to be taken. Technical developments should include a variety of exotic and/or novel systems that could produce a suitable synthesis gas at low cost and high efficiency. In this connection, catalysis within the gasifier or as a second stage offers many possibilities and is seen as a fertile area for fundamental and applied research. An early start in this substantial area is considered advantageous. Other possibilities should not only consider the gasifier but other interactions within the system such as development of new gas separation technology, new synthesis technology to cope with the economics of scale in methanol production, and new and/or alternative fuel products. The existing pilot plants that have been supported by the EEC could be used for further investigation or optimisation studies.

The size and complexity of likely integrated biomass to fuel plants in a European or Third World context makes feasibility studies a necessary subject of investigation in order to establish capital costs, production costs, and economic sensitivities to feesdtock, gasifier technology, synthesis technology and location. Case studies concerning a specific biomass resource coupled to a specific technology would complement the feasibility studies and lead to optimised process systems.

Direct liquefaction

This is achieved by low temperature high pressure hydrogenation of biomass. Although most activity in the USA has now terminated, and an investigatory programme is about to start in Scandinavia, the technology is far from being commercial. Current work is at a very early stage of development and it is recommended that a monitoring exercise be maintained, complemented by fundamental experimental work. Some reservations were expressed on the economic viability of the process in principle, but it was felt that some contact with this subject was necessary.

Substitute Natural Gas Production

Although this was not discussed at the workshop, some progress is being on direct catalytic gasification to SNG (Battelle Columbus and Pacific North-West Laboratories). This is currently seen as an attractive option in the UK.

If an indirect route to SNG through synthesis gas is followed, the

achievements, constraints, and opportunities for further R and D are
identical to those for producing synthesis gas for methanol, as production
technology for SNG from synthesis gas is at a similar status. If, however,
the direct route is considered, considerable work is necessary on
fundamental and pilot plant levels to develop stable and selective
catalysts.

CONCLUSIONS

The following tables summarise the achievements, constraints and opportunities for each of the areas discussed above. Although much valuable progress has been made and many of the objectives have been met, it is clear that more problems have been identified than have been solved. The necessarily long term nature of research into conversion of biomass into useful fuel products means that with continued R & D support there is adequate opportunity to develop thermochemical systems that will enable the potential of biomass to be realised when it will serve the best purpose.

Table 1   COMBUSTION : Summary of Achievements, Constraints and R & D Opportunities for Thermochemical Processing of Biomass

Achievements to date	Constraints against implementation	Opportunities for further R & D
Commercialisation realised depending on location	High capital cost below 1 MW	Further design improvements Retrofitting applications
Some design improvements	Current models poorly designed	Efficiency improvements
Combustion of wet material with heat recovery	Temperature needs to be high	Environmental improvements with gas and liquid effluents
Low grade heat recovery for domestic applications	Residence time needs to be high	Application of fluid bed technology
Material handling improvements e.g. pelletising	Excess air needs to be low	Comparison of combustion and gasification; technically and economically
500 000 units sold in 5 years in Germany	Environmental requirements (site specific) Feedstock supply problems in some locations (e.g. Holland)	

— 38 —

Table 2   SMALL SCALE SYSTEMS FOR PYROLYSIS AND GASIFICATION : Summary of Achievements, Constraints and R & D Opportunities for Thermochemical Processing of Biomass

Achievements to date	Constraints against implementation	Opportunities for further R & D
Pyrolysis		
Available commercially Some design problems solved	Condensibles inhibit any application except direct burning Multiple products (gas liquid and char) Evironmental requirements Novelty precludes wider acceptance	Fundamental studies Maintain overview of activity Minimise liquid yields Modelling Effective utilisation of all products + All other opportunities as for Low Heating Value gasification (q.v.)
Low Heating Value Gasification		
Available commercially up to 1 MW. More than 10 companies offering systems in Europe	Feedstock flexibility and difficult feedstocks particularly with moving beds	Monitoring performance of units already sold Feedstock preparation and requirements (including drying)
Retrofitting possibility Some design problems solved Developments in design procedures and modelling	Capital costs on small scale (typically up to 1 kWe for moving beds and up to 1 MWe for fluidised beds) Environmental requirements from wet cleaning systems Gas cleaning necessary for internal combustion engine applications Novelty precludes wider implementation	Interface between biomass production and conversion Hot gas cleaning: (market survey, techno-economic evaluation, technical developments) Performance improvement Pressurisation alone, and with turbines for power generation Secondary gasification/ reforming for control of tars etc. Adaptation of moving beds to small particles and high ash solids Gas quality specification Ash behaviour including fusion for wastes above 5% ash Interfacing between manufacturer and consumer Optimisation (Continued)

Table 2 (Continued) SMALL SCALE SYSTEMS FOR PYROLYSIS AND GASIFICATION : Summary of Achievements, Constraints and R & D Opportunities for Thermochemical Processing of Biomass

Achievements to date	Constraints against implementation	Opportunities for further R & D
		Modelling
Improved design procedures
Export potential (adaptation to feedstock and environment)
Flexibility - feedstock, operation, turndown
System evaluation and application assessment
Comparison of gasification and combustion

Table 3 LARGE SCALE SYSTEMS FOR LIQUID PRODUCTS : Summary of Achievements, Constraints and R & D Opportunities for Thermochemical Processing of Biomass

Achievements to date	Constraints against implementation	Opportunities for further R & D
Indirect Liquefaction		
Four pilot plants in progress sponsored by EEC plus others. Too early for assessment	Imbalance of airings and economic fuel synthesis
Capital cost	Further reactor developments including exotic and/or novel ideas
Small scale economical fuel synthesis operation up to 300 t/d
Pressurisation
Control systems
Catalysis (with fundamental work) synthesis gas upgrading
Secondary processing operations for control of tars etc
(Continued)

Table 3 (Continued) LARGE SCALE SYSTEMS FOR LIQUID PRODUCTS : Summary of Achievements, Constraints and R & D Opportunities for Thermochemical Processing of Biomass

Achievements to date	Constraints against implementation	Opportunities for further R & D
		Techno-economic feasibility studies and optimisation
		Evaluation of hybrid systems
		Case studies of complete process from biomas production to fuel end-product
		Reactor modelling
		Control systems for complex reactors such as double fluidised beds

Direct Liquefaction (Pressure hydrogenation)

Possible but too early to judge	Yield Selectivity Feasibility	Fundamental investigations
		Monitoring ongoing activity
		(Later)- Feasibility studies
		Characterisation of products

Pyrolysis

Some recent progress on increasing liquid yield	Stability, corrosivity, and carcinogenicity of liquids Product yield Feasibility and viability	Upgrading of condensibles to stable non-corrosive products
		Maximising liquid yield
		(Later) Scaling up, modelling, and feasibility studies

Table 4 SUBSTITUTE NATURAL GAS (SNG) : Summary of Achievements, Constraints and R & D Opportunities for Thermochemical Processing of Biomass

Achievements to date	Constraints against implementation	Opportunities for further R & D
Indirect Route via Synthesis Gas		
Identical to indirect liquefaction	Identical to indirect liquefaction	Identical to indirect liquefaction
Direct Route via Catalytic Gasification		
Some recent progress in USA on selective catalysis at laboratory scale only	Catalyst stability and selectivity + All other constraints for indirect liquefaction	Developing selective and stable catalysts (Later) + All other opportunities as for indirect liquefaction

BIOMASS CONVERSION: BIOLOGICAL ROUTES

Rapporteur's Report of Contractors' Seminar: Capri, 8/6/'83

Participants: Chairman - H.P. Naveau, (Louvain-la-Neuve, Belgium);
Rapporteur - E. Colleran (Galway, Ireland); F. Alfani and M. Cantarella
(Naples, Italy); M.G. Demuynck and A. Legros (Louvain-la-Neuve, Belgium);
S.P. Etheridge, (Cardiff, U.K.); B.A. Rijkens and J.W. Woetberg (I.B.V.L.,
Wageningen, Holland); G. Beldman, G. Lettinga, A.G.J. Voragen and
W.M. Wiegant (Agricultural University, Wageningen, Holland) and G. Goma
(Toulouse, France).

The seminar was concerned with the research reports of nine contracts involved with energy production from biomass by biological routes. Four main research areas were included in this section of the Energy from Biomass Programme: (i) biogas production by anaerobic digestion of purpose grown terrestrial biomass and effluents and residues from the food industry and from agriculture (5 contracts); (ii) biogas from marine and fresh-water algae (1 contract); (iii) enzymic liquefaction and saccharification of cellulosic wastes (2 contracts) and (iv) fermentation alcohol from agricultural biomass. The purpose of the seminar was to summarise the progress achieved during the four years of the programme and to identify continuing and new areas of research in the context of any future programme on energy from biomass.

1. Anaerobic Digestion
   Two-phase anaerobic digestion of solid substrates was shown to be technically feasible by the independent research of Rijkens and Voetberg in Wageningen, Colleran and co-workers in Galway and Nyns, Naveau and co-workers in Louvain-la-Neuve. The Wageningen group demonstrated the feasibility of the process at laboratory and pilot scale for a variety of solid residues currently causing considerable disposal problems in the Netherlands. In Galway, the process was regarded as a viable means of generating additional energy at farm-level by the digestion of crop wastes in season through linking a hydrolysis reactor to an anaerobic filter operating on a year-round basis for animal slurries and other liquid wastes. Two-phase digestion of algal substrates was also shown to be feasible and to operate with a high methane conversion efficiency at Louvain-la-Neuve. Considerable differences with respect to the design and mode of operation of the first phase, liquefaction reactor were reported.

In Wageningen and Galway, the liquefaction reactor is operated in batch mode with either continuous or intermittent percolation of effluent from the methanation reactor through the packed bed of hydrolysing solids. Operation as a closed system was shown by the Wageningen group to sometimes lead to high ammonia and alkalinity levels in the circulating liquid phase. In such cases, the introduction of additional water may be required. Both the rate and the extent of volatile solids liquefaction achieved were shown to depend on the degree of comminution and crushing carried out and, more importantly, on the lignin content of the solid substrate. Both the UASB and anaerobic filter designs were shown to operate satisfactorily as the Stage II methanation reactor although the UASB sludge was found to have poor settling characteristics and to be easily washed out of the reactor under

thermophilic conditions. A two-stage system consisting of a CSTR reactor
linked to an upflow methanation reactor was investigated at Louvain-la-
Neuve and shown to give high methane yields for filamentous algal substrates.

Research on the performance of high rate digesters treating wastewater feeds
was also continued. The results of Lettinga and co-workers showed that the
UASB reactor, when operated under thermophilic conditions, was very easily
started up with simple sugar substrates, was capable of handling very high
loading rates and showed the same type of adaptation to toxicants as
demonstrated previously under mesophilic conditions. The results obtained
suggest that thermophilic UASB digestion is a viable option for wastewaters
which are discharged at temperatures in excess of 50^oC. Studies on the
effect of varying the support matrix in upflow anaerobic filters treating
pig slurry supernatant showed that fired clay fragments in random packed
filters allowed more rapid start-up and more efficient performance at
different loading rates than a variety of plastic and other support
materials. Initial studies on flow direction in random-packed clay filters
showed that the start-up and the COD conversion efficiencies achieved were
very similar in upflow and downflow feed mode. The downflow system appeared
to be less prone to clogging and to be technically easier to operate.

The results of comparative pilot scale studies on the digestion of farm
wastes and energy crops using a high rate contact digester, a plug flow
digester, a hydraulic digester and an anaerobic filter were reported by
Stafford and co-workers. High methane yields at a retention time of 3.3
days were obtained with the hydraulic digester treating pig waste. Naveau
and co-workers presented data on a continuous single stage system for algal
digestion. High CH_4 yields at a loading rate of 0.4 kg $VS_o.m^{-3}.d^{-1}$ were
reported and efficient integration of the various steps from cultivation
to digestion was achieved. The economic aspects of full scale digester
operation, as elucidated by the EEC and Switzerland Biogas Survey commissioned under the Solar Energy Programme, was presented by M. Deymunck.
Five full-scale farm digesters were shown to be economically viable. The
poor economics of other existing plants was attributed to too high initial
investment costs, inefficient operation and poor integration of the digester
on farm.

1.1 Recommendations for future work.

The need for further study of two-phase systems for solid substrate
digestion was clearly identified. In particular, detailed information is
required on the operation of the hydrolysis reactor with respect to the
microbiology of the hydrolytic and fermentation processes involved; the
factors governing the rate and extent of polymer hydrolysis; the control
of H_2 and CH_4 production in the hydrolysis reactor; operation in continuous
as well as batch mode; identification of the optimum liquid percolation and
recycling rates for different substrates; development of effective solids
dewatering systems; evaluation of the benefits of seeding the reactor with
specifically developed seed cultures and development of appropriate systems
for pumping high solids slurries. Comparison between the operational
efficiency and methane productivities of single stage and two-stage reactors
is also necessary for a variety of solid substrates, particularly for easily
digestible wastes which may be more economically digested by single than
by two-phase systems.

Further in-depth study of high rate digesters such as the UASB and anaerobic filter is also required. More basic research on matrix effects and flow rate patterns in anaerobic filters is clearly needed as are comparative studies on UASB and filter digestion of a variety of wastewaters with a view towards establishing the optimum design for a given waste. Further studies on thermophilic and low-temperature mesophilic operation of high rate digesters and on the adaption and acclimatization of these digesters to the toxic compounds present in certain wastes is also essential.

A lack of basic information on the microbiology of the overall anaerobic digestion process was also highlighted. A need for further information on the interaction of different trophic groups; on the role played by sulphate reducing bacteria; on the effect of hydrogen and other inhibitory metabolites on the acetogenic and methanogenic species and on the bacterial distribution in two-phase digestion systems was clearly identified. Alkalinity and redox potential effects on the microbiology of anaerobic digestion also require investigation.

The development of stable and reliable ancillary equipment for anaerobic digesters was seen to be an urgent prerequisite for full scale adoption of anaerobic digestion technology. In particular, suitable pumps, flow meters, heat exchangers and sensors to measure COD and VFA levels in feed, effluent and within the reactor are required. Further study on modelling and automation of anaerobic digestion is needed. The utilisation of biogas, liquid effluents and solid residues from operational digesters requires continued investigation and studies on further treatment of digester effluents by denitrification and other processes should be integrated with ongoing digestion studies where necessary. It was also considered essential to continue monitoring existing digester installations in Europe with a view to ascertaining common operational problems and with the aim of achieving better integration of digesters with existing farm management and crop production systems. Legislative and safety aspects of anaerobic digestion also require investigation.

2. Hydrolysis and Liquefaction of cellulosic wastes

The utilisation of commercially available cellulolytic and pectolytic enzyme preparations for the liquefaction and saccharification of beet pulp, potato fibre, tomato plant and wood wastes and a turnip catch crop was investigated by Beldman, Voragen and co-workers in Wageningen. A very high degree of conversion was obtained for beet pulp, potato fibre and stubble turnips. Tomato plant and wood wastes proved to be more resistant to hydrolysis due to their lignin content and pretreatment was found to be necessary. Extrusion increased the enzymic digestibility of both substrates but may be non-viable economically. The importance of the role played by polygalacturonase in the initial decrease in substrate viscosity was highlighted by this study as was the need for maximising the synergism between endo- and exoglucanase components of the cellulase complex. A high degree of enzyme stability was reported when the liquefaction of beet pulp was carried out in a column reactor connected to a hollow fibre ultrafiltration unit for enzyme recovery. The solid state fermentation of sugar beet to ethanol was also shown to be improved by the addition of cellulolytic and pectolytic enzymes.

The studies of Alfani, Cantarella and co-workers in Naples also highlighted the importance of synergism between endo- and exoglucanases in cellulose liquefaction and saccharification. The observed strong inhibition of cellulose hydrolysis by the glucose end-product could be relieved by performing the hydrolysis in an ultrafiltration membrane reactor which continuously removes glucose, thereby preventing β-glucosidase deactivation. Pretreatment was shown to be necessary for effective hydrolysis of cellulose in biomass substrates. A combined acid pretreatment at 80°C with H_2SO_4 followed by alkaline pretreatment at 90°C was shown to reduce cellulose polymerization and to break down the lignin barrier. Optimum reagent to biomass ratios were identified for the pretreatment and enzymic steps.

2.1 Recommendations for future work.

A clear need was identified for additional fundamental research on the polymer degrading enzymes involved in biomass saccharification. In particular, more information is required on the synergistic activity between different enzyme complexes and between the components of individual complexes. Control of product inhibition and techniques for the immobilisation of cellulase components and other saccharifying enzymes require detailed research and development. Research on reactor design was seen to be of importance with respect to removal of fermentation byproducts and optimisation of saccharification rates. Efficient and low-cost pretreatment methods must be developed in order to achieve saccharification of lignocellulosic biomass. Priority was also given to the need to develop efficient solid state fermentation systems which would allow liquefaction and saccharification to be combined with subsequent fermentation processes in a single reactor. A need was also identified for further research on the fermentability of pentose sugars and uronic acids formed during biomass saccharification.

3. Fermentation Ethanol

Goma and co-workers reported on a method for solid state fermentation of Jerusalem artichoke which yields 10% ethanol after 12 hours fermentation. A synergistic effect between ethanol and sugar inhibition was demonstrated. Detailed studies on ethanol inhibition revealed that a dual effect was involved - (i) inhibition as a result of ethanol accumulation within the cell and (ii) inhibition by an unknown excreted co-metabolite which is more inhibitory than ethanol. Inhibition was shown to be correlated with a decrease in cell-bound water. Technological studies on the use of immobilised cells; the inclusion of various additives to enhance cellulose hydrolysis and the development of an extractive fermentation system which couples solvent extraction of ethanol with fermentation in the same reactor were also carried out. The results obtained are also applicable to acetonobutylic and acidogenic fermentations for liquid fuel production.

3.1 Recommendations for future work.

It was agreed that a critical reappraisal of the ethanol by fermentation route for energy production from biomass should be carried out. Further laboratory studies are necessary on inhibition effects in ethanol fermentation, particularly with a view to ascertaining the identity and mode of action of the co-metabolite inhibitor. Technologically innovative concepts need to be explored with regard to continuous fermentation at high cell density. In this regard, cell recycling systems and the coupling of ultrafiltration and fermentation and flocculating devices require further investigation. Computer control of fermentation processes also merits additional study.

AGRICULTURAL WASTES - ENERGY CROPS

Energy production using straw and animal wastes as feedstocks - Economic analysis of the straw pelleting route

An experimental assessment of native and naturalised species of plants as renewable sources of energy in Great Britain

Growing catch crops for fuel

Joint research on Arundo donax as an energy crop

Quality and quantity of latex which can be produced from natural vegetation in Greece

ENERGY PRODUCTION USING STRAW AND ANIMAL WASTES AS FEEDSTOCKS

ECONOMIC ANALYSIS OF THE STRAW PELLETING ROUTE

Author   : V. REQUILLART

Contract number   : ESE-R-014-F

Duration   : 36 months   1 July 1980 - 30 June 1983

Total budget   : F 1 750 000   CEC contribution : F 700 000

Head of project   : P. CHASSIN   C.N.R.A., route de St-Cyr   78000-Versailles

Contractor   : Institut National de la Recherche Agronomique

Address   : 149, rue de Grenelle   75341-PARIS Cédex 07

Résumé

Dans ce papier, l'auteur présente un rapide bilan de l'utilisation énergétique des pailles en France. Après ce constat il analyse la filière "paille granulée" :

- Depuis la paille en andains jusqu'au granulé arrivé chez l'utilisateur le coût des différentes opérations techniques est analysé.

- Deux utilisations énergétiques du granulé sont étudiées : le chauffage d'habitation individuelle et le chauffage de bâtiments collectifs.

Summary

In this paper the writer sets out a short assessment of energetic uses of straw in France. Then he analyses "pelleted straw" route :

- From straw in swath to pellets delivered to the consumer cost of different processes is determined.

- Two energetics utilisations of pellets are studied : heating of individual houses and heating of buildings.

STRAW PELLETS

Before analysing the straw pelleting route I want to set out a short assessment of energetic uses of straw in France :

. grass and lucerne drying : 6-7 plants use straw as fuel. Their consumption is less than 50,000 tonnes.

. grain drying : there is about 30 corn driers using straw as fuel. Their consumption is about 5,000 tonnes.

. glasshouse heating : a few installations

. individual houses heating :

- 300 to 400 boilers burn straw pellets. Their consumption is about 3,000 to 5,000 tonnes.

- a more important number of boilers use straw in bales. In the "Centre" there are about 300 to 400 boilers (a boiler uses about 15 tonnes per year).

. buildings heating pellets. : one installation burns straw

So, in 1982, less than 100,000 tonnes of straw are required for energetic uses.

1. Presentation of the straw pelleting route

Straw pelleting which facilitates the transportation and storage of straw as well as the automated operation of heating units opens significant perspectives regarding the use of straw for energy. To evaluate the advantages and development potential of this product, the whole "pelleting" route starting from raw material (straw in swath) to the end product (pellets) shall be analyzed.

The "pelleted straw" route can be illustrated as follows :

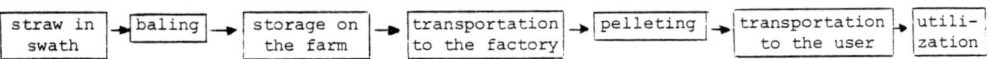

2. A short note on some economic concepts

2.1. Threshhold cost

The cost price of a product or a service is not an absolute magnitude but depends on the structure of the firm which produces it. The computation of this cost price is arbitrary in particular as to the estimation of fixed costs which cannot be ascribed to one produced item (labour, common implements, overheads ...).

Generally speaking, the decision to produce will only be taken if the new activity results in an increased income. The minimum price which yields a profit is called threshhold cost.

2.2. Opportunity cost

The opportunity cost or the maximum valorisation of a product is the maximum price that this product can attain while remaining competitive compared with substitute products.

2.3. Economic surplus

The economic surplus of a route is the total gain of all economic agents in this subsetor. It is obtained by comparing the opportunity cost with the threshhold cost.

The existence of a positive surplus is a condition for the development of a route but not sufficient in itself.

The share of the surplus between the different agents is determined by the internal transaction price of the products.

3. Pellet production from the swath to the user

3.1. Straw in swath

The removal of 1 tonne of straw corresponds to both mineral and humic removal.

In 1982, the mineral value of wheat and barley straw amounted to 30 FF and 55 FF (1) per tonne respectively (see table 1). Straw was found to have no nitrogen value.

The humic value is more difficult to assess. Due to the contradictory effects of straw manuring on the soil, it is difficult to evaluate straw humic value. In the present state of the art, straw removable without agricultural hazard is allocated a zero humic value.

The estimated threshhold cost of straw in swath is therefore 30 FF per tonne of wheat straw and 55 FF per tonne of barley straw.

---

(1) FF = French Francs

3.2. From the swath to the factory

The factory's purpose is to minimize its total procurement costs. Thus, the cost analysis should encompass the entire baling-storage-transportation chain (see diagram 1).

As soon as the distance between field and factory exceeds 5-10 km or 20-30 km depending on whether the straw is stored on the field or on the farm, the total procurement cost is smallest when straw is compressed into parallelipipedal bales of 500 kg each.

Besides, the use of Hesston bales in the factory minimizes the storage and handling costs.

3.3. Pelleting

There are two pelleting processes :

a) Straw pelleting in factories that dehydrate lucerne and pulps. In this case, straw pelleting enables the factories to operate over a greater number of months, thereby depreciating the installations over a longer period of the year. This process requires capital investment for straw pelleting. Production in such factories will remain limited.

b) Pelleting in factories that specifically process straw. The attempt will be made here to analyze the threshhold cost of pelleting in such a plant. The data are summarized in table 2.

Nevertheless, the cost estimate remains inaccurate due to the evaluation of the productivity of the plant.

3.4. Transportation to the user

Straw that is a product of low density cannot be transported over long distances. As a result, the factory should be located in production areas that may be relatively remote from the consumption areas. Therefore, pellets will occasionally be carried over considerable distances.

3.5. Threshhold cost of the pellets upon delivery to the user

Considering the above results and the following hypotheses :

- Hesston baling
- 25 % of the bales stored on the field
- 75 % of the bales stored on the farm
- farm and factory are 10 km apart
- factory and user are 40 km apart.

The cost of pellets at the user's is roughly 400 FF per tonne, tax not included, i-e 11 centimes per thermie   (1 tonne of pellets = 3,600 thermies) (see table 3).

Direct energy expenditures (oil, electricity) related to the manufacturing and distribution of the pellets represent 15 to 20 % of the lower heating value of the pellets.

4. Pellet utilization

Two utilizations are considered :

- heating of individual dwellings
- heating of buildings.

4.1. Heating of individual dwellings

In a new house that required a heating unit, the opportunity cost of the pellets relative fuel oil varied from 620 to 650 FF per tonne, tax not included, in 1982.

If the heating unit still in working condition was disposed of due to excessive cost, the opportunity cost amounted to 370-430 FF per tonne, tax not included.

These opportunity costs are calculated in relation to fuel oil, the most commonly used but rather costly fuel. State aid is not taken into account.

4.2. Heating of buildings

Coal is the major competitor in this case. Usually, the opportunity cost of the pellets is roughly 450 FF per tonne, tax not included (taking into account various forms of financial aid).

The competition is expected to continue in the future. Coal prices will slowly increase annually by 1 or 2 % (in deflated Francs).

5. Economic surplus

The use of straw pellets for collective heating yields a rather low economic surplus of about 50 FF per tonne of pellets (see table 4), whereas their use for individual heating provides a significant surplus (200-250 FF per tonne in a new house).

However, only collective heating provides a sufficiently large market enabling a straw pelleting factory to operate. Sales for individual heating remain an auxiliary resource.

The economic surplus to be split between producers and distributors (farmers, companies responsible for baling and transportation, pelleting factory) would roughly amount to 50 FF per tonne. It is not high enough to enable such units to operate. Indeed, multiannual contracts for straw deliverery by farmers are negociated at prices as high as 80 FF per tonne of straw in swath, so that the entire surplus is annihilated.

6. Conclusion

There again, coal appears to be a serious competitor to straw for energy, since its production is backed by state aid whereas straw production relies on private companies that need to make a profit.

On farm energetic uses of straw are expected to increase. Indeed straw is a cheap fuel (100 FF per tonne i-e 0,033 FF per thermie) and can be used to heat an individual dwelling or to dry grain. In agricultural cooperatives (corn drying, lucerne dehydratation) straw is also a competitive fuel. On the other hand, notably due to competition of coal, off-farm energetic uses of straw will probably increase rather slowly.

Bibliography

Lévy-Lambert L., Dupuy J.P. - Les choix économiques dans l'entreprise et l'administration - Tome 1 - Dunod - Paris

Requillart V. - La filière paille granulée - INRA Economie Rurale Grignon - 21 p. + annexes - Sept. 1982.

Requillart V., Sourie J.C. - Production de paille de céréales comme source de combustibles et produits associés - Analyse technico-économique de quelques opérations - INRA Economie Rurale Grignon - C.E.E. 15 p. - 1982.

Requillart V., Sourie J.C. - Valorisation énergétique des pailles - Aspects économiques - Biomasses Actualités n° Spécial Paille - Juillet 1932.

Sourie J.C. - Production de paille de céréales comme source de combustibles et produits associés - Méthodologie micro-économique - C.E.E. - INRA Economie Rurale Grignon - 38 p. + annexes - 1981.

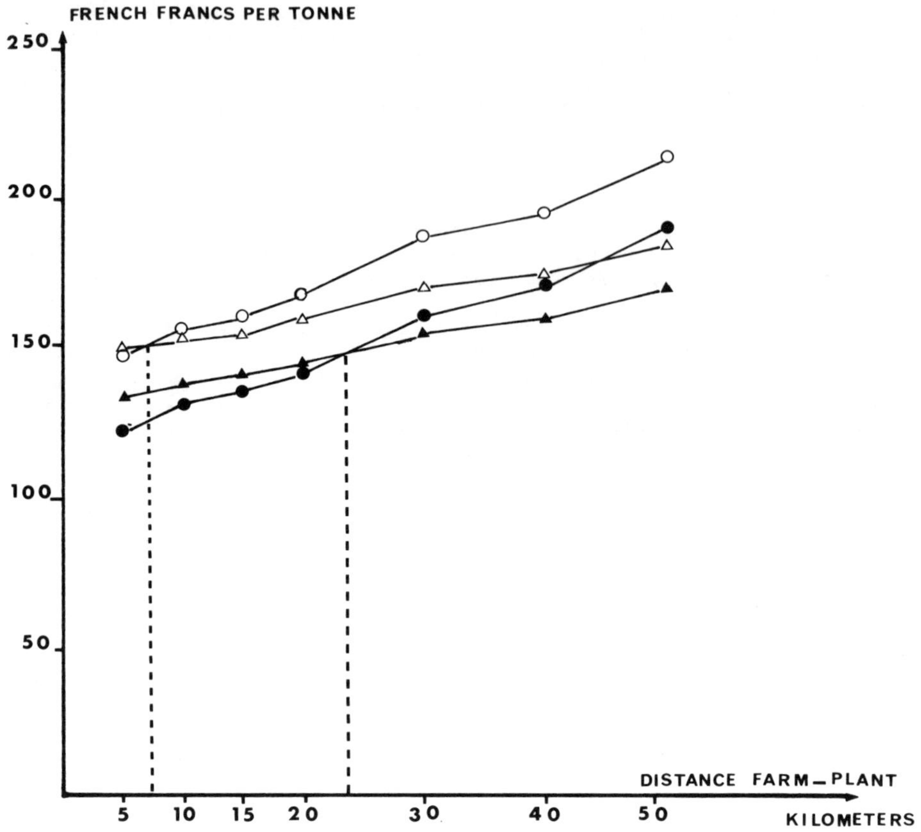

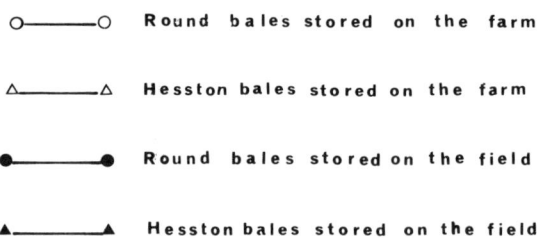

THRESHOLD COST OF BALED STRAW UPON DELIVERY TO THE PLANT - 1982

○────○ Round bales stored on the farm

△────△ Hesston bales stored on the farm

●────● Round bales stored on the field

▲────▲ Hesston bales stored on the field

Table 1 : Value of nutrients contained
in one tonne of straw - 85 % of dry matter
(FF 1982) (1)

	wheat straw		barley straw	
nutrients	quantity kg	price FF	quantity kg	price FF
P_2O_5 : 3,50 FF per unit	1.5	5.3	2.	7.0
K_2O : 2,00 FF per unit	10.	20.0	20.	40.0
CaO : 0,70 FF per unit	4.2	2.9	4.2	2.9
Trace elements	-	3.2		4.2
Total		31.4		54.1

Table 2 : Some characteristics and cost of
a straw pelleting plant (1982)

Total investment	5.10^6 FF
Hourly output	4 t/h
Installed power	850 KW
Electric consumption per tonne	210 KWh
Duration of use	3,500 h
Pelleting cost	200 - 220 FF/t
- fixed cost	100 - 110 FF/t
- variable cost	100 - 110 FF/t

(1) FF = French Francs

Table 3 : Cost of pelleted wheat straw
upon delivery to the user   (1982)

process	cost FF/tonne
Mineral value	30
Baling	80
Storage	30
Transportation	20
Pelleting	220
Transportation	20
Total	400

Table 4 : Economic surplus   (1982)

utilisation	economic surplus FF/t
Required heating unit (4,000 l fuel oil)	220
Required heating unit (6,000 l fuel oil)	250
Substitution of heating unit (4,000 l fuel oil)	-30
Substitution of heating unit (6,000 l fuel oil)	30
Collective heating	50

AN EXPERIMENTAL ASSESSMENT OF NATIVE AND NATURALISED SPECIES OF PLANTS AS
RENEWABLE SOURCES OF ENERGY IN GREAT BRITAIN

Authors: T.V. Callaghan, R. Scott, G.J. Lawson, A.M. Proctor
Contract number: ESE-R-015-UK
Duration: 36 months   1 July 1980 - 30 June 1983
Total budget: £ 254 826   CEC contribution: £ 48 000
Head of project: Dr. T.V. Callaghan, Institute of Terrestrial Ecology
Contractor: Natural Environment Research Council's Institute of
                   Terrestrial Ecology.
Address: Natural Environment Research Council
                   Polaris House
                   GB - SWINDON, Wilts SN2 1EU

Summary

Natural vegetation was known from previous studies to have significant potential as a source of energy because some species were widespread and represented an existing resource whereas others were highly productive but more restricted in distribution. However, little was known about the effects of harvesting and managing natural vegetation for biofuel production.

A three year experimental assessment of three species which showed particular potential as energy crops suggests that bracken (*Pteridium aquilinum*) and cordgrass (*Spartina anglica*) could be harvested annually in autumn for at least three years to give mean yields over this period of between 19.4-21.4 and 33.8-38.4t ha^{-1}. These species occur on underutilized land in Great Britain totalling over 3000km^2 and the use of bracken as a biofuel would obviate the costly practice of chemical control.

Yields of bracken and cordgrass varied by 37 and 81% according to weather conditions in the years 1980-82, and fertilizer applications were found to be unnecessary during the first three years.

Japanese knotweed planted in 1980 showed very low yields initially but a threefold increase in yield during the third year after planting suggests that high yields may soon be obtained, permitting annual harvesting to proceed.

The efficiency of the anaerobic digestion of natural vegetation varied between 35 and 57% and was generally costly. However, protein fractionation and the use of residues may increase profitability. The technological feasibility of producing and using pellets of bracken has been demonstrated and tentative costings suggest economic feasibility. Further research is required to optimise the mechanisation of biofuel production and to close the gap between research and commercial exploitation.

1. INTRODUCTION

Natural vegetation is an important source of biofuels. Some species are widely distributed and could be harvested immediately with no energy requirement for land cultivation and planting. Other species, such as invasive weed species may be highly productive when compared with traditional crops from agriculture and forestry. Little is known about the productivity of these species and their use as an energy crop would involve cultivation, planting and the management of new crops on good quality land. The integration of these crops with existing patterns of agriculture and forestry, and crop fractionation to give protein from leaves and biofuels from stems, could reduce the competition for land between fuel, food, feed and fibre production.

Although data were available on the extent, distribution, yield and chemical composition of natural vegetation in the United Kingdom (Callaghan *et al.* 1978; Lawson *et al.* 1981; Callaghan *et al.* 1981), little was known about the sustainability of yields after harvesting natural systems or the management of new energy crops. Consequently, a three year project was initiated to evaluate the effects of harvesting on three herbaceous perennial species ie. bracken *(Pteridium aquilinum),* cordgrass *(Spartina anglica)* and Japanese knotweed *(Reynoutria japonica).* Sites were established in existing natural areas of bracken and cordgrass whereas Japanese knotweed was planted on agricultural land. The main objectives of the study were:
a) to assess the effects of fertilizer applications on yield;
b) to assess the time of harvesting on yield and regrowth;
c) to assess the frequency of harvesting on yield and regrowth;
d) to assess the effects of weather on yield;
e) to assess the rate of establishment of a planted monoculture;
f) to compare methods of converting natural vegetation to useable fuels;
g) to develop strategies for managing natural vegetation for biofuel production;
h) to estimate levels of natural biofuel production at local, regional and national levels.

Three 0.17 ha sites were established in 1980 and plant performance was assessed over a three year period in relation to the various replicated treatments. A subsidiary site was also established to compare planting density of Japanese knotweed with the closely related, but more productive species, giant knotweed *(Reynoutria sachalinensis).* Plant material collected in summer was subjected to anaerobic digestion by Dr. D. Stafford of Cardiff University, whereas dry senescent biomass of bracken, obtained in autumn was pelletised and combusted.

This paper presents a summary of the experimental results relating to objectives (a) to (f) and interprets these to suggest management practices, further research and development requirements, and likely impacts of natural biofuels.

2. RESULTS

2.1 <u>Effect of weather on yield</u>

The main effects of weather on yield are due to temperature, spring drought and late frosts, which have particularly severe effects on non-woody

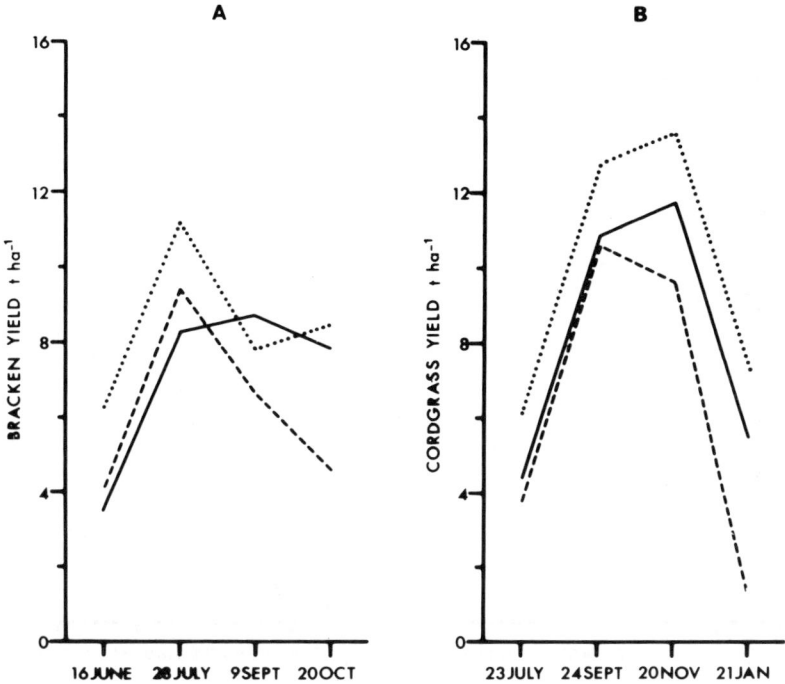

Fig. 1. Variations in the yields of bracken (A) and cordgrass (B) in 1980 (———), 1981 (....) and 1982 (----) due to different weather conditions.

perennial energy crops. Over the three years of the experiment, the weather in 1981 resulted in the highest yields of bracken (Figure 1A) and cordgrass (Figure 1B), whereas lower yields, particularly in autumn, were obtained in 1982. At peak yield, variations in yield due to weather were 37% for bracken and 29% for cordgrass, and in autumn these variations were 46% and 81% respectively.

It is important, therefore, to consider variations in weather when assessing the yields of such short rotation energy crops. Longer rotation crops i.e. trees, integrate the effects of weather on growth over a long period and yields would be expected to be significantly reduced only by exceptional weather conditions.

2.2 Effect of establishment on yield

It has not yet been possible to assess the effects of weather on a planted energy crop because variations in yield between the years 1980, 81 and 82 resulted predominantly from the establishment of the crop. Four rootless rhizome fragments of Japanese knotweed (mean weight 224g) were planted in Spring 1980 and maximum yields of only 0.7t ha^{-1} were attained in that year. In 1981, however, yields increased to a maximum of 0.9t ha^{-1} and in 1982 a mean yield of 2.7t $ha\ yr^{-1}$ was achieved (Figure 2A). It is

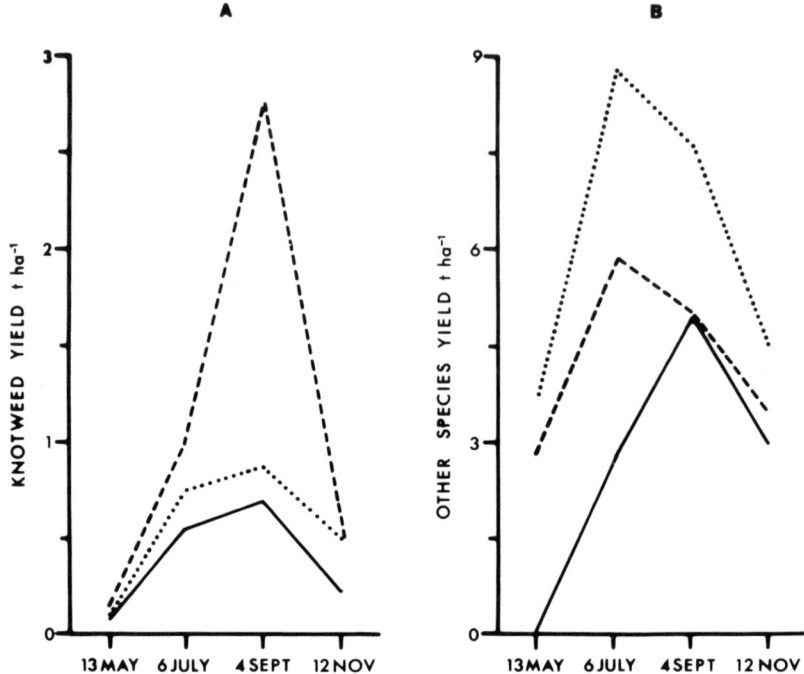

Fig. 2. Effect of establishment on the yields of Japanese knotweed planted in Spring 1980 (A) and weed species which invaded the site (B). Yields shown were obtained by harvesting for the first time in 1980 (———), 1981 (....) and 1982 (----).

unfortunate that the present experimental period has not been long enough to allow the complete establishment of a new perennial energy crop, particularly as the present annual rate of increase in yield suggests that high levels of production similar to those measured in the field might soon be obtained. Species which naturally invaded the knotweed site showed a high yield of 8.75t ha^{-1} during the second year: thus total site production can reach 9.5t $ha^{-1} yr^{-1}$ from bare ground by almost completely natural means during a 2 year period.

One reason why initial yields of Japanese knotweed were low was the low density of planting as little previous experience of planting such a crop existed. However, results from a separate experiment comparing planting densities of two species of knotweed - Japanese knotweed and giant knotweed - suggest that higher yields could be obtained more quickly by increasing the density to 16-25 plants m^{-2} (Table I).

2.3 Effect of fertilizer applications on yield

There are two reasons for applying fertilizers to natural energy crops: i) these crops may have been limited by nutrient supply and increases in yield might be expected; ii) the harvesting of these crops for the first time will result in the removal of biomass containing nutrients which must be replaced.

Table I. *Effects of planting density on the establishment of monocultures of two species of knotweed from March 1981 to November 1982.*

	planted density (plants m^{-2})	final density (stems m^{-2})	plant height (cm)	estimated yield (t ha^{-1})
Japanese knotweed	4	4.0	75.3	0.23
	9	10.7	73.0	0.29
	16	25.3	103.7	2.59
	25	27.7	197.3	1.09
Giant knotweed	4	4.0	150.7	2.46
	9	11.0	159.7	2.43
	16	11.3	157.7	1.28
	25	16.3	201.0	5.41

Surprisingly, applications of fertilizer did not significantly affect the yield of bracken, cordgrass or Japanese knotweed on plots which had not been harvested previously, even though the highest level of fertilizer application was twice that currently applied to agricultural root crops. Perhaps it is even more surprising that the yields of bracken plots which have been harvested annually for three years showed a significant decrease in yield as the level of fertilizer application increased (Figure 3A). Cordgrass (Figure 3B) showed a significant response to fertilizers only after three years of harvesting. Japanese knotweed, however, showed a significant response to fertilizer applications with a sixfold increase in yield with the two highest levels of application (Figure 3C). It would appear, therefore, that natural areas of vegetation are quite resistant to the removal of nutrients in harvested biomass, initially at least. New monocultures established from weed species may, however, have significant requirements for additions of fertilizers equivalent to those currently used in agriculture.

2.4 Effect of harvesting a yield

Callaghan *et al.* (1981) predicted that perennial energy crops harvested in autumn, when nutrients had been translocated from shoots to belowground storage organs, would show better regrowth than those harvested in summer. These predictions were validated in the present three year experiment. Yields of bracken harvested for 2 years in summer were reduced by 35% and 15% in autumn; yields of cordgrass were reduced by 60% with annual summer harvesting and 19% with annual autumn harvesting, while yields of Japanese knotweed were reduced by 50% and increased by 10% respectively (Callaghan *et al.* 1983 a; b; c). On the third annual harvesting occasion, the same seasonal patterns were observed but the magnitude of the decrease in yield was slightly greater. However, even on the third harvesting occasion, significant yields were obtained from the two species occurring on under-utilized land. Thus, a total yield of between 19.4 and 21.4t ha^{-1} of senescent bracken could be harvested over a three year period (Figure 4A). At the cordgrass site, a total of between 33.8t ha^{-1} and 38.4t ha^{-1} could be obtained over this three year period which gives high mean yields of 11.25 to 12.9t ha^{-1} yr^{-1} (Figure 4B) for a totally unused area of vegetation. However, the decreasing yields in these natural sites suggest that occasional rest years would be required to sustain the vegetation as a permanent energy crop. In the case of bracken, this rest would be

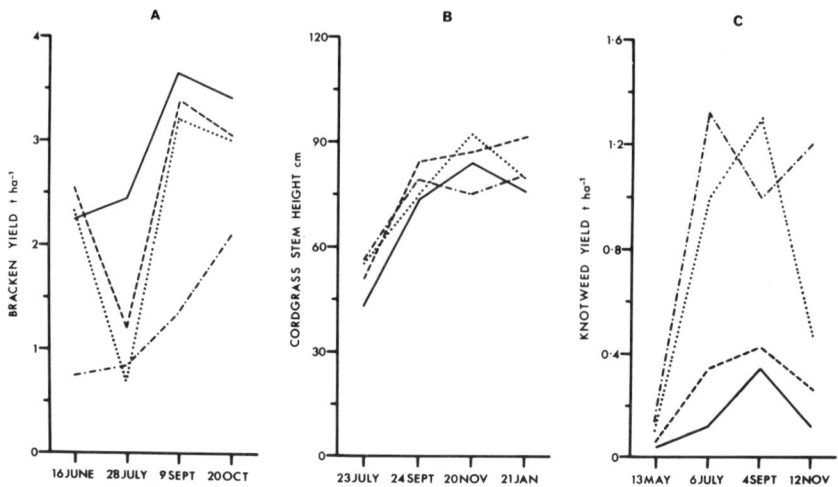

Fig. 3. Effect of the level of NPK fertilizer application on bracken yield (A) cordgrass stem height (B) and Japanese knotweed yield (C) after 3 years of annual harvesting on the same dates —— 0gm^{-2}; — — 50gm^{-2}; .... 100gm^{-2}; -.-. 200gm^{-2} (100gm^{-2} = 1t ha^{-1}).

required after perhaps 4 years of harvesting whereas longer periods of harvesting of cordgrass may be possible. The optimum rotations of harvesting and resting need to be determined.

It is difficult to predict how harvesting will affect subsequent yields of a dedicated energy crop established as a monoculture. However, the resilience of Japanese knotweed to harvesting annually for each of the 3 years since planting (Figure 4C) suggests optimism.

2.5 Conversion of natural vegetation to biofuels

Two basic scenarios have been developed for the conversion of natural energy crops to biofuels. Species harvested in summer could be anaerobically digested to give methane as the biomass would have a high water content, although it is also possible to convert wet biomass into low grade heat energy (Have, 1982). Trials on anaerobic digestion by Stafford (1981) at Cardiff University have yielded efficiencies of conversion ranging from 35 to 57% (Table II). However these modest efficiencies were only obtained after 100 days retention of a finely macerated feedstock. Tentative costings of biogas from the anaerobic digestion of bracken suggest an overall price of £11 per GJ compared with a range of £2.94 to £9.85 GJ for natural gas and propane (Callaghan et al. 1982). Anaerobic digestion of bracken is unlikely, therefore, to be justified

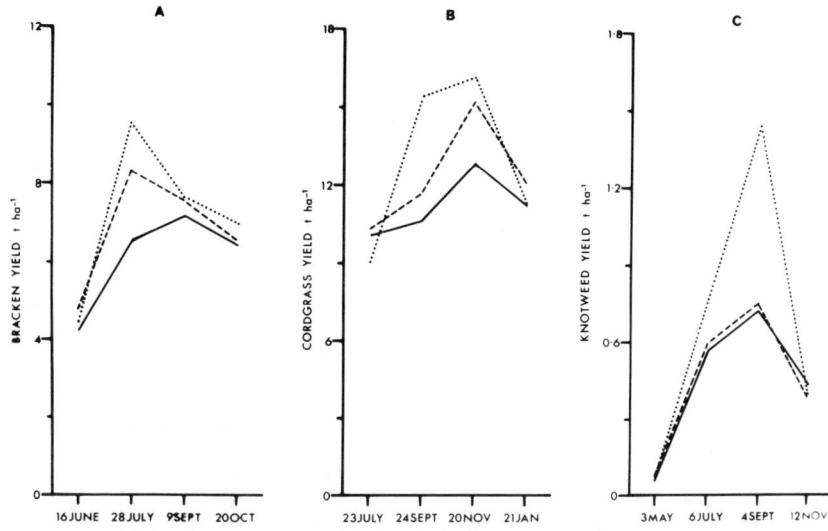

Fig. 4. Effects of repeated harvesting on yields of Bracken (A), Cordgrass (B) and Knotweed (C). Mean yields for the first year of harvesting (....), first 2 years of annual harvesting (-----) and first 3 years of annual harvesting on the same dates (———).

particularly as it can be dried in the field and used for thermal conversion. Other species such as Janapese knotweed have high yields in summer and high biogas recoveries which may justify anaerobic digestion. The possibilities of using digestor residues for animal feed or organic fertilizer, and extracting protein from chopped leaves prior to digestion will have a major effect on the profitability of biogas production (Wilkins *et al*. 1977) but the removal of natural vegetation in mid-summer will have the most harmful effects on subsequent yields.

Table II. The anaerobic digestion of Natural Energy crops. (From Stafford 1981).

	Bracken	Cordgrass	Japanese knotweed
biogas yield ($m^3 t^{-1}$ volatile solids)	400	380	550
efficiency of conversion (%)	38.4	35.3	57.3

Thermal conversion methods can be applied to biomass harvested in summer and dried as hay or senescent biomass harvested in autumn. Such material can be compacted cheaply in to conventional bales (98kg m^{-3} or into pellets (800-1120kg m^{-3}). Once compacted, particularly at the higher densities, the biomass becomes cheaper to transport, handle and store and is resistant to

decomposition and wetting. Biomass in this form is extremely versatile
and can be burnt directly in domestic or farm stoves and boilers, or can
be treated in larger conversion units such as gasifires to produce a range
of fuels and chemicals.

The direct combustion of pellets of bracken would give an energy cost of
£2.03 per GJ which is cheaper than coal to a domestic user whereas the cost
of methanol from the gasification of pellets, at £7.05 per GJ, is cheaper
than post- tax motor spirit (Callaghan et al. 1982). Equipment is already
available for the harvesting and compaction of weed species such as
bracken and cordgrass, and pellets of bracken have been used to fuel a
domestic central heating system.

3. CONCLUSION

This study was based on previous knowledge of the potential to derive
biofuels from productive and extensive natural vegetation, but it addressed
itself to the uncertainties of reactions of natural vegetation to
management and harvesting.

The results presented in this paper confirm that bracken and cordgrass
could be maintained as renewable sources of energy on underutilized land
covering more than $3000km^2$ of rural Britain by adopting specific management
techniques. Tentative costings of biofuels from bracken suggest that they
are already competitive with costs of some traditional fuels and the
technological feasibility of using bracken biofuels has been demonstrated.
However, more technical assessments are required to optimize the harvesting,
processing and long term management of natural vegetation.

New monocultures established from productive weed species represent a
longer term scenario. It has been demonstrated that a monoculture of fast
growing perennial weeds can be established easily, but several years are
required for sufficient stand thickening to permit sustained annual harvests.
The difficulty of obtaining data quickly from such new energy crops should
not detract from the importance of investigating the vast but relatively
unknown potential of uncultivated species, particularly when protein
extraction is considered as an adjunct to energy production.

Although the impact of biofuels may not exceed 10% of the gross primary
energy demand in Great Britain, this contribution could save £1 billion
in oil imports (Lawson and Callaghan in press). However, the greatest
impact will be in rural and isolated areas where biofuel production from
underutilized land could represent a new source of income for generally
poor and subsidized communities. It is important that the gap between
research and development on the one hand, and commercial exploitation
is bridged by a demonstration project.

4. REFERENCES

CALLAGHAN, T.V., MILLAR, A., POWELL, D. & LAWSON, G.J. (1978). Carbon as
    a renewable energy resource in the U.K. - Conceptual approach. Report
    to the U.K. Department of Energy from the Institute of Terrestrial
    Ecology, Cambridge. 165pp.

CALLAGHAN, T.V., SCOTT, R. & WHITTAKER, H.A. (1981). The yield, development and chemical composition of some fast-growing indigenous naturalised British plant species in relation to management as energy crops. Report to the UK Department of Energy from the Institute of Terrestrial Ecology, Cambridge. pp 178.

CALLAGHAN, T.V., SCOTT, R. & LAWSON, G.J. (1982). Biofuel production from natural vegetation in Great Britain. In: Energy from Biomass, Proceedings of EC Contractor's Meeting, Brussels, 5-7 May 1982. p 30-36, D. Reidel, Dordrecht, Holland.

CALLAGHAN, T.V., SCOTT, R. LAWSON, G.J. & PROCTOR, A.M. (1983). Reports on the energy production potential of bracken (a) Japanese knotweed (b) and cordgrass (c). Reports to the UK Department of Energy from the Institute of Terrestrial Ecology, Cambridge.

HAVE, H. (1982). Heat energy from animal waste by combined drying combustion and heat recovery. In: Energy from Biomass, Proceedings of EC Contractor's Meeting, Brussels, 5-7 May. p 170-175, D. Reidel, Dordrecht, Holland.

LAWSON, G.J. CALLAGHAN, T.V. & SCOTT, R. (1980). Natural vegetation as a renewable energy resource in the UK. Report to the UK Department of Energy from the Institute of Terrestrial Ecology, Cambridge. pp 177.

LAWSON, G.J. & CALLAGHAN, T.V. (in press). Primary productivity and the prospects for biofuels in the UK. Int. J. Biometeorol.

STAFFORD, D.A. & HUGHES, D.E. (1981). Fermentation to biogas using agricultural residues and energy crops. In: Energy from Biomass - 1st EC Conference, Palz, W., Chartier, P. & Hall, D.O. (eds.), p 406-410. Applied Science, London.

WILKINS, R.J., HEATH, S.B., ROBERTS, W.P. & FOXELL, P.R. (1977). A theoretical economic analysis of systems of green crop fractionation. In: Green Crop Fractionation, Wilkins, R.J. (ed.). p 131-142. Occasional symposium no. 9 of the British Grassland Society.

GROWING CATCH CROPS FOR FUEL

Authors	:	S.P. CARRUTHERS, A. WITHERS
Contract Number	:	ESE-R-034-81-UK (H)
Duration	:	36 months 1 July 1980 - 30 June 1983
Total Budget	:	£111.301
Head of Project	:	Professor C.R.W. Spedding, University of Reading
Contractor	:	University of Reading
Address	:	Department of Agriculture and Horticulture, University of Reading, Earley Gate, READING, Berkshire, RG6 2AT, U.K.

Summary

The technical potential for growing catch crops for fuel in the UK depends on land availability, crop yield and conversion efficiency. An estimate based on 1979 crop areas, yield data from field experiments and conversion efficiencies supplied by University College, Cardiff places UK potential at 72-162 PJ biogas per year. Assessment of economic potential suggests that biogas can be produced from catch crops at costs competitive with conventional fuel prices. Key determinants of profitability include: availability of machinery and labour and the extent to which digester residues can replace fertilizer inputs. Potential further depends on how practicable growing catch crops for fuel is on UK farms; various farm factors, including size, type, location, machinery and labour availability, capital, income, energy needs and farmers' attitudes, will constrain or promote the realization of potential. This study has stressed technical aspects (particularly in relation to crop production) and provided a broad assessment of economic potential; future studies need to stress the implementation on farms of the whole system, in particular the integration of anaerobic digestion into farming practice.

1. INTRODUCTION

1.1 Background

Catch crops are grown between the harvest and sowing of main crops. Traditionally used to extend the grazing season for ruminant livestock the practice represents an opportunity to provide a feedstock for anaerobic digestion to produce biogas. Growing catch crops for fuel would not displace food production nor require major changes in the structure of agriculture, but would fit into current farming practices, use existing farm resources (including land which may otherwise be unused) and reduce farm energy costs or increase farm income.

Early studies identified this opportunity and its attractions as a potential biofuel source for both UK and the EC and a more thorough assessment of potential was considered necessary.

1.2 Objectives

The objectives of this study were to assess the technical potential (ie how much fuel?), economic potential (ie how profitable?), and overall prospects for implementation (ie how practical?) of growing catch crops for fuel in the UK, and to consider the management of the system (ie how to grow the crop, convert to fuel and integrate these and other intermediate processes?)

Technical potential depends on land availability, crop yield and efficiency of conversion to biogas; economic potential on feedstock production and conversion costs, the value of the products and the profitability of alternative uses of the same resources. In practice catch crops will be grown on farms and the extent to which the practice is implemented will depend not only on technical and economic potential but also on various farm factors which will constrain or promote the realization of potential.

This study has placed emphasis on technical aspects and in particular on crop yield and four field experiments have been carried out. Liaison with University College, Cardiff has enabled the anaerobic digestion of catch crop material to be investigated.

2. TECHNICAL POTENTIAL

2.1 Land availability

Land is the essential resource for growing catch crops (and hence producing biogas), and its availability in terms of area (how much?), date (when?), duration (how long?), and location (where?) is a prime determinant of technical potential (Figure I). Land availability depends on current crop rotations and is, hence, subject to change. Catch fuel crops could be grown after tillage crops or temporary grass, but are likely, in the short term, to be excluded from farms with livestock and hence restricted to the 'cropping' regions of the UK (Figure I).

2.2 Crop yield

The parameters of land availability (ie date, duration and location) will influence crop yield; date and duration determine the sowing and harvest dates of the catch crop while location will determine its environment (which will not vary from year to year). In addition, crop yield will be affected by various management factors. And understanding of the effects of sowing, and harvest date and environmental factors enables yield to be predicted for a given situation, while understanding management effects enables it to be optimized. This project has gone towards meeting these objects.

Figure II illustrates the effect on yield of sowing and harvest date and

Figure I : Land availability

General effects

Main crop parameter	Land availability parameter	Catch crop parameter	Additional constraints
Area	Area	Area	Environment
Harvest date	Date	Sowing date	Farm factors
			Method of harvest
			Speed of harvest
Sowing date	Duration	Harvest date	Method of catch crop establishment
			Autumn cultivations
Distribution	Location		Wet autumn and heavy soils

Date of availability. General pattern

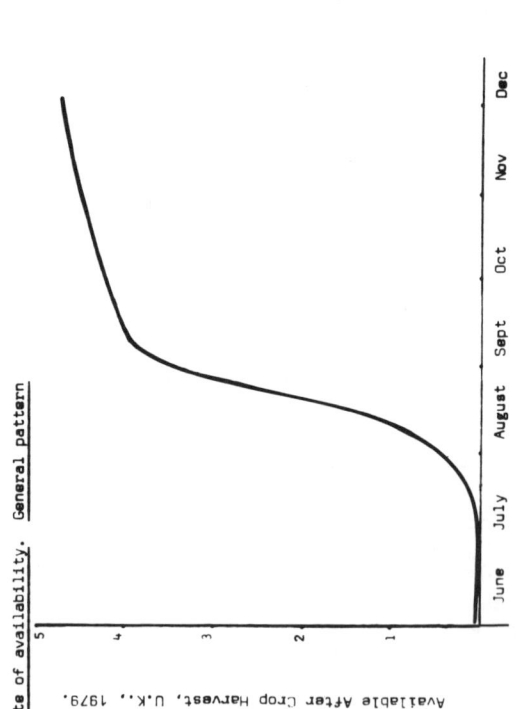

Estimated Accumulated Land Becoming Available After Crop Harvest, U.K., 1979.

Location of available land. Dominant farming systems in the U.K.

Cropping
Horticulture
Dairying and livestock rearing
General mixed farming

Source: Adapted from (1)

Figure II: Crop yield. Results of field experiments

General pattern of crop growth

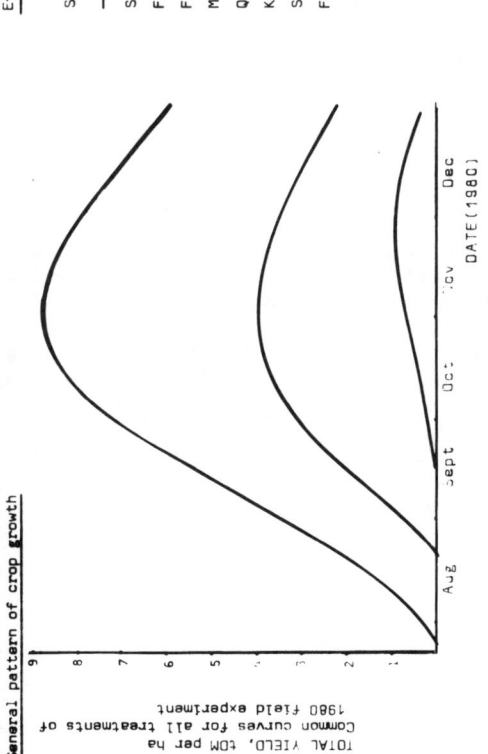

Effect of species. Comparison of species grown in field experiments 1982.

Species	Yield	Speed of growth	Speed of senescence	Frost resistance
Stubble turnip	::::	:::	:::	:::
Fodder radish	::::	:::	:::	:::
Forage rape	::::	::	.	:::
Mustard	::::	::::	:::	::
Quinoa	:::	:::	:::	::
Kale	::	.	.	::::
Sterile brome	::	:::	:::	::::
Forage pea	.	:::	::::	.

Effect of seed rate

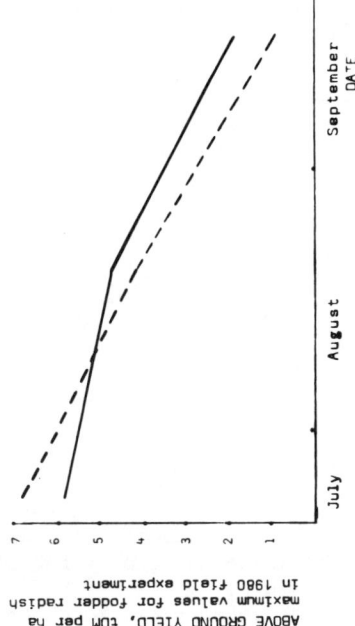

Effect of sowing date

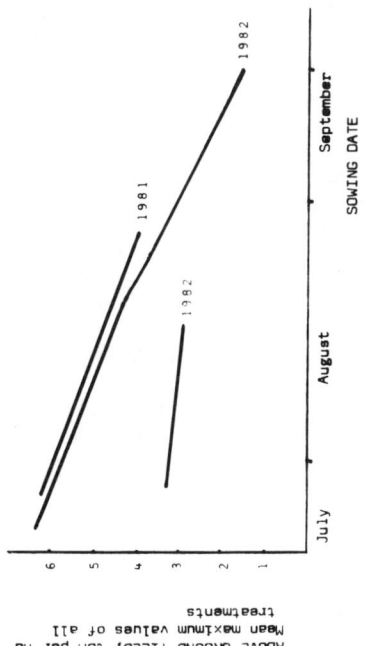

of two important management factors (seed rate and species) while Figure III illustrates the effects of location. Yields of up to 6 tDM per ha in southern regions can be provided from late-July sowings declining to about 1.5 t DM per ha; maximum yield is attained in 70-110 days depending on species.

2.3 Conversion efficiency

Studies at Cardiff suggest that net biogas yields (after allowing for digester heating) of 4-9 GJ per tDM are possible. Maceration seems likely to be essential for continuous digester processes; the batch process may require less pre-processing, cost less and be more suitable for a seasonal supply of feedstock as is provided by catch crops.

2.4 National assessment

Putting together an estimate of land availability based on 1979 crop areas, yield data in relation to sowing and harvest dates from field experiments and the above conversion efficiency, produced an estimate of the potential catch fuel crop production of the UK tillage area amounting to 72-162 PJ biogas per year (equivalent to 1 - 2% 1981 UK primary energy consumption).

3. ECONOMIC POTENTIAL

Economic analyses suggest that biogas could be produced from catch crops at a cost competitive with conventional fuel prices (Figure IV). Critical determinants of profitability include:
- crop yield;
- whether appropriate machinery and labour are available on the farm;
- the extent to which digester residues can replace fertilizer inputs;
- the siting and scale of the digestion processes.

Farmers with livestock are unlikely to grow catch crops for fuel at current energy and animal feed prices. If catch crops were substituted for bought in feed barley (similar in feed value) the equivalent value of the catch crop as animal feed would be approximately £10-67 per GJ biogas equivalent. Fuel price increases would change this situation.

4. IMPLEMENTATION

The practicability of growing catch crops for fuel on UK farms depends, further, on certain farm factors including: size, type (rotations practised, livestock, other enterprises), location (physical effects, energy supplies and transport effects), machinery and labour available, capital, income, energy consumption and needs and farmers' attitudes. These factors will determine the extent to which technical and economic potential can be realized, and how easily the catch fuel crop system can be integrated into current practice.

This latter aspect is of particular importance. While catch cropping is an established and largely acceptable farming practice anaerobic digestion is, to farmers, a new and largely unproven technology. The adoption of anaerobic digestion is, therefore, likely to be the major constraint on the implementation of catch fuel cropping.

5. MANAGEMENT

Important aspects of managing the catch crop include: choice of establishment method (ie aerial sowing, direct drilling, minimum cultivations, ploughing); choice of species and variety; seed rate; fertilizer input; weed, pest and disease control; moisture conservation and harvest date. The optimum management regime depends on the nature of the specific opportunity, but

Figure III : Crop yield. Effect of location. Yield location study.

Yields of stubble turnip at different sites in GB.

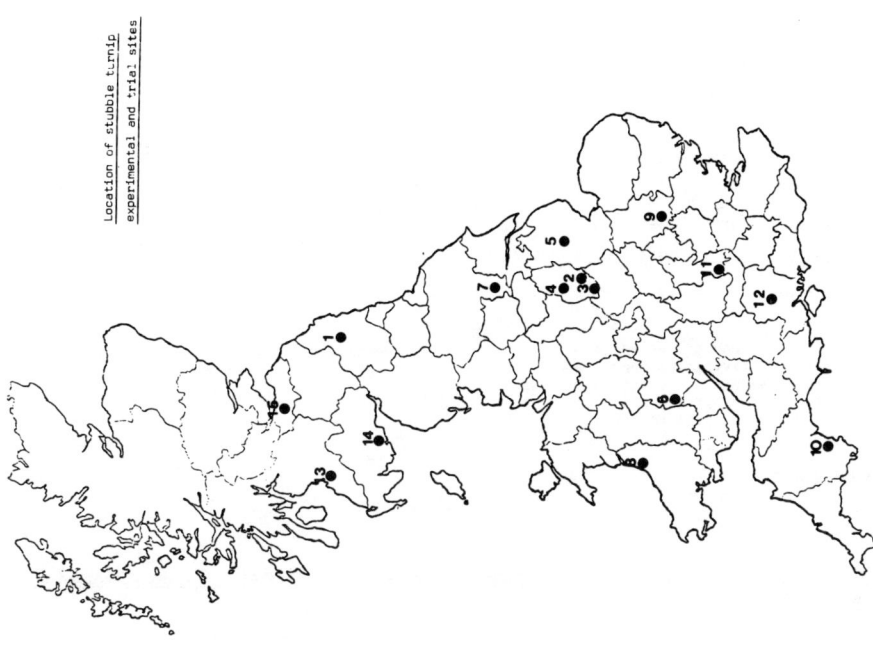

Location of stubble turnip experimental and trial sites

Site	Mean[1]	Range[1]	No. of varieties[2]	No. of years[3]
1	6.07	5.84 - 6.34	4	1
2	5.72	4.67 - 6.79	2	2
3	3.89	3.41 - 4.75	11	2
4	2.50	1.87 - 3.15	8	1
5	1.76	1.47 - 2.14	8	1
6	3.14	2.06 - 4.53	11	2
7	3.16	1.53 - 4.44	11	3
8	2.07	1.91 - 2.36	9	1
9	3.67	2.73 - 5.05	11	2
10	3.11	2.39 - 4.08	9	1
11	4.20	3.28 - 5.20	2	3
12	3.68	1.87 - 4.77	11	3
13	2.65	1.39 - 3.81	17	2
14	3.33	0.39 - 5.34	6	3
15	2.29	2.02 - 2.56	2	1
All sites	3.42	0.39 - 6.79	24	6

[1] Yields of tops after 90 - 150 days over a number of varieties and years at each site.
Adjusted to uniform sowing date of 2 August using a factor derived from field experiments.

[2] Complete range of varieties was : Arca, Civasto-R, Croppa, Debra, Gelria, Jobandi, Jobe, Labra, Marco, Nobitter, Polybra, Ponda, Siloga, Taronda, Tigra, Trofee, Vobra, Imperial Green Globe, Champion Green Top, Cyclon, Typhon, Appin, Rekord, Cobra.

[3] Complete range of years was 1975 - 1980.

Figure IV : Economic potential

Feedstock production costs

	Cost range[1] £ per tonne[2]
Variable costs	0.7 - 20.9
Variable costs plus fuel	1.8 - 70.9
Farmer's costs	15.4 - 92.9
Contractor's charges	24.0 - 103.8

[1] Exact cost determined by: establishment method, species, fertilizer input (whether digester residues used), harvesting method, whether stored, storage method, storage time, previous crop, soil conditions and farm circumstances.

[2] Assuming a harvested yield of 4 tDM per ha.

Conversion process costs

	£ per year	£ per GJ biogas
Digestion equipment	5600[1]	
Running costs	3500[2]	
Total	9100	2.9[3]

[1] 200m³ digester costing £35000 amortised over 20 years at 15% interest

[2] 10% capital cost

[3] Assuming net gas yield of 9 GJ per day for 350 days per year

Cost of producing biogas from catch crops compared with conventional fuel prices using 2 example systems

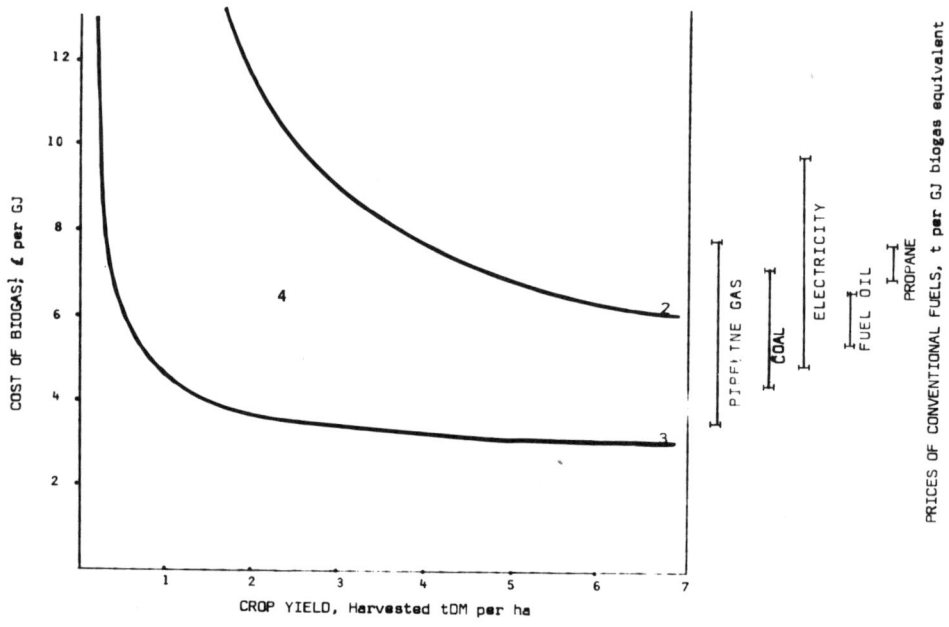

[1] Assuming 9 GJ biogas per t DM, conversion cost of £2.9 per GJ biogas

[2] Assuming stubble turnip established by farmer using minimum cultivations, compound fertilizer applied, harvested by a contractor using a forage wagon and stored as silage for 6 months in earth walled silo.

[3] Assuming stubble turnip established by farmer using minimum cultivations, digester residues replacing compound fertilizer, harvested by farmer using a forage wagon and fed directly to macerator or digester.

[4] Actual costs likely to fall within this range.

this study has provided much information to enable those decisions to be made.

6. CONCLUSIONS AND RECOMMENDATIONS

The broad UK potential of catch fuel crops appears promising both in terms of national production and profitability, but it is clear that actual potential and practicability, will depend very strongly on individual farm circumstances. The determinants of land availability, crop yield, costs and profitability, and the management of the crop production component of the system are understood. The stress of this study has been technical; future work should stress socio-economic aspects and the effect of farm and regional factors on the realization of potential. As the adoption of anaerobic digestion is a key determinant of catch fuel crop implementation a logical next stage would be to consider catch cropping in the context of a broader study of markets for anaerobic digestion. This would need to be farm based and use model farms and farm survey data, and would enable regional trends to be extracted.

Related to this is the need for further information on the full scale digestion of plant materials, the value of digester residues and the costs of the digestion process.

Catch crops, by definition, are opportunity crops and, unlike trees or dedicated energy crops, do not represent a long-term commitment of land or other resources. As such they are unlikely to merit big investment in conversion equipment, but could provide an appreciable and attractive source of feedstock as part of an integrated energy supply system.

7. REFERENCE

1. Edwards, A. and Rogers, A. (1974). Agricultural Resources. Faber and Faber, London.

JOINT RESEARCH ON ARUNDO DONAX AS AN ENERGY CROP

Authors   : M. ARNOUX, A. DAVENEL, M. LONG

Contract number   : ESE-R-042-F

Duration   : 36 months   1 May 1981 - 30 April 1984

Total budget   : F.F. 1.198.600   CEC contribution : F.F. 461.000

Head of the project   : M. ARNOUX, I.N.R.A., Montpellier-France

Contractor   : Institut National de la Recherche Agronomique

Address   : 149 rue de Grenelle   75341 Paris Cédex 07

Subcontractors   : CEMAGREF - Montpellier
                         CNARBRL - Nimes

Summary

This research afforded a possibility of improving the previous management and process of Arundo donax when cropped as a raw material for pulp and paper. A new larger rows spacing cropping system (screen type) allows an other simultaneous crop, in the same field, increases the real dry matter yield of the reed and finally improves the energy and financial balance in saving inputs. Moreover the mechanisation of the crop and the processing of the biomass harvested are now being improved by :
    - a new and simpler harvesting technique allowing a preliminary natural drying of the stalks in the field and avoiding a further deshydratation ;
    - an original stripping system of the chopped stalks associated to a granulation system of the grinded leaves ;
    - a new energetic transformation of the biomass by gasification instead of the previous direct combustion.
A study of the areas suitable for Arundo donax in south of France is now being achieved showing that 25.000 hectares could be utilised for this crop. Finally, the price of the therm for every kind of cropping system and process is now being estimated.

1. RESEARCH TO IMPROVE THE MANAGEMENT OF THE CROP

- I.N.R.A. -

Arundo donax is a wild mediterranean giant reed, which has been previously studied as a raw material for pulp and paper and which was then planted on a 2 meters rows spacing basis.

This design was decided to get as soon as possible a high dry matter yield : 20 tons/ha/year, i.e. 8 T.O.E. in terms of energy. However, according to the colonisation of the soil by the rhizomes (0,25 m./year in average) a competition for light, nutrients, water... occurs in the field, become more and more severe year after year and requires a heavy quantity of inputs (fertilisers, water...) which are expensive in terms of energy and money. Moreover, a high percentage of the field remains unoccupied during several years without any return to the farmer.

The hypothesis was done that Arundo donax could be crop as screens, i.e. large rows spacing (from 7 to 25 meters) and that an other annual plant could be simultaneously cultivate between the lines of reed. In this new design the intercalated crop become the main crop requiring its own fertilizers, irrigation, etc... and the reed, utilising this neighbouring, is consequently only planted and harvested without any appropriate inputs.

It is then only necessary to limit the growing of the rhizomes every 4 or 5 years, to avoid a complete colonisation of the field. In this design, the yield of the reed can be calculated on the real part of the field colonised by the rhizomes without any loss of acreage.

An experiment was designed in 1980 to check this hypothesis and to investigate different rows spacing from 2 meters (standard) to 25 meters.

As Arundo donax is a perennial plant, 5 years are at least necessary to get valid conclusions.

Nevertheless an example of the first results is given in the following tables :

1.1. STANDARD CROP : 2 METERS ROWS SPACING

	Average width of one row (meter)	% of the field A. colonised by reed B. vacant	Dry matter yield tons/ha	Inputs/ha (every year)
1rst year	0,25	A. 12,5 B. 87,5	40 |- 5,0 0	N : 250 K
2nd year	0,50	A. 25 B. 75	62 |- 15,7 0	P_2O_5 : 50 K K_2O : 260 K
3rd year	0,75	A. 37 B. 63	60 |- 22,5 0	Irrigation Herbicides

In this design, an other crop is impossible between the rows and the yield must be estimated on the total acreage of the field whatever the importance of the vacant part of the field (column 2-B).

1.2. EXAMPLE OF A LARGE ROWS SPACING : 7 METERS

	Average width of one row (meter)	% of the field A.colonised by reed B. vacant	Dry matter yield tons/ha		Inputs	
			reed only (1)	reed + another crop (2)	reed only (1)	reed + another crop
1rst year	0,30	A. 4.... B. 96....	50 0 ⊢ 2	50 + corn : 6	like above (standard) every year	reed : 0 normal (every year) for the other crop
2nd year	0,75	A. 10.... B. 90....	60 0 ⊢ 6	60 + corn : 5,5		
3rd year	1,15	A. 17.... B. 83....	67 0 ⊢ 11,5	67 + sunflower 2,8		

(1) In fact this possibility must be completely excluded.

(2) The yield of reed is estimated on the real acreage covered by the reed, i.e. the width of the row multiplied by its length/ha.

At least for the present the large rows spacing design is very attractive : the dry matter harvested in 3 years is of 11,5 tons (against 42,5 tons for the standard design) but without any inputs and plus a normal yield for the intercalated crop. Moreover, the yield of the reed, on the basis of the real acreage of the field colonised, appears to be of 59 tons/year in average, i.e. about 20 T.O.E./year.

These results must be confirmed during the next years. The screen influence on the intercalated crop and the best width of one row will be investigated as well.

Nevertheless, this new technique, probably not adapted to a large industrial production, seems very promising in terms of energy and money for small farms interested in producing energy for their own consumption.

2. RESEARCH ABOUT NATURAL DRYING AND HARVESTING MECHANIZATION OF ARUNDO DONAX

- CEMAGREF -

2.1. Harvesting method with artificial drying (standard)

Arundo donax has to be harvested between November 15th and March 15th to not endanger its growth. Then, the moisture content of Arundo donax varies from 50 to 55 per cent. Consequently, Arundo donax cannot be stocked without avoiding its fermentation and its damage. Moreover its burning or gasification are difficult. To stabilize this biomass it is necessary to reduce the moisture content to 15-20 per cent, like all agricultural product.

Consequently, the standard method for harvesting and processing Arundo donax is the following :
- harvest in the form of chips by a 250 H.P. self-propelled chopper equipped with the specific CEMAGREF harvesting head ;
- transport to the deshydratation unit ;
- deshydratation utilising a part of the biomass harvested as combustible ;
- separation, grinding and granulation of leaves.

On the basis of an average production of 20 tons of dry matter per hectare, i.e. a 8 T.O.E. potentiality, this method recover only 6.5 to 6.9 T.O.E. Morever, it needs a very expensive invest : the deshydratation unit.

Therefore CEMAGREF studied during last winter, in collaboration with GRAND MANUSCLAT farm, different systems capable to reduce in the field the moisture content of the harvested biomass (mainly the stems).

2.2. Tests of different methods inducing a natural drying of Arundo donax in the field

1rst method : conditionning in the form of chips threw down on the soil by the self propelled chopper equipped with the specific header.

2nd method : cutting Arundo donax with a mower cutterbar.

3rd method : conditionning biomass by a mower-conditionner equipped with fluted rubber rolls and conventionnal cutterbar or, better, rotory cutters.

The next graph indicates the data got during tests realized in January 1983. The last 3rd method appears specially well-adapted since a dryness over 80 per cent has been obtained in 15 days.

Later on, only the harvesting methods of biomass conditionned by a mower-conditionner (3rd method) will be set out.

During tests, the climatic conditions were propitious : we cannot ensure today that the natural drying will be always so easy and fast. Nevertheless we observed that after a rain, the produce dried again very quickly.

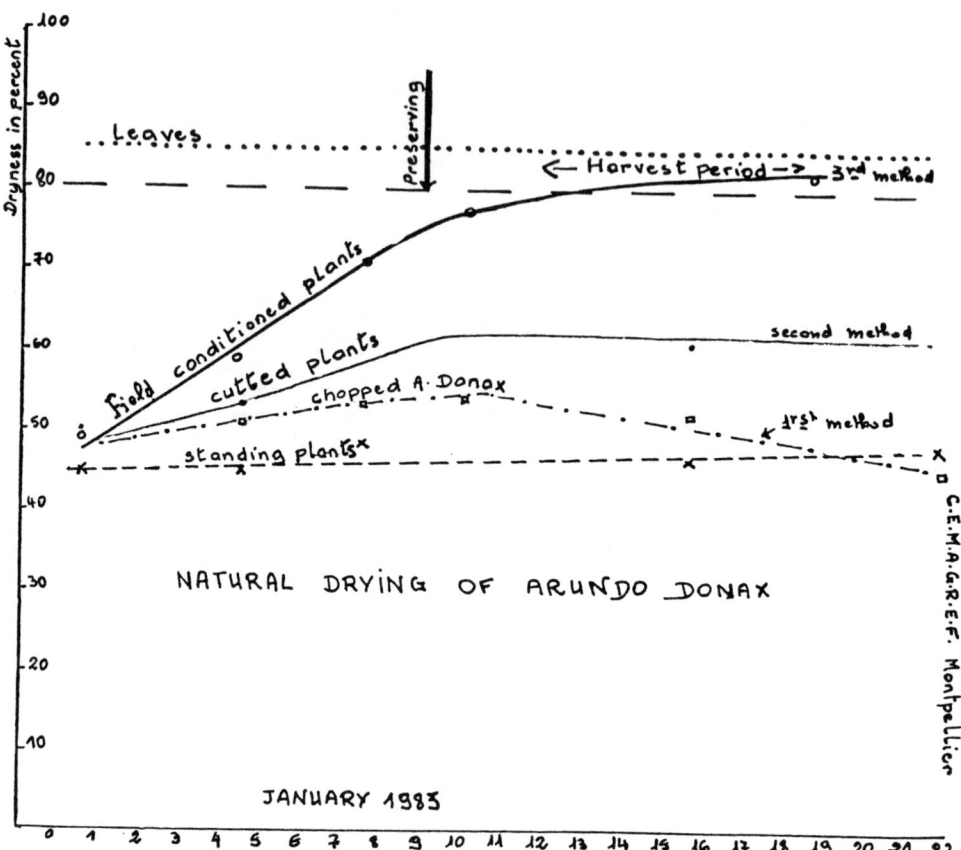

2.3. Handling of the field conditioned Arundo donax

Among the methods tested, three yields interesting results :
- a selfpropelled chopper equipped with the specific header or with a narrow common header (for grain and seed) ;
- a flail-type field chopper pulled by a 90-100 HP tractor ;
- a big rectangular baler (VICON).

These three methods give comparable costs. But the last one needs an additional chopping process. The second one requires more man-power but less specific invest : moreover, it almost eliminates all losses on the ground.

These encouraging results must be included in the economic approach work of the "Compagnies d'Aménagement", in taking into account the new conditioning process in the field and in excluding artificial drying cost (except grinding and granulation of the leaves).

The natural drying of Arundo donax avoids the expensive invest of a deshydratation unit, increases the usable energy between 20 to 25 per cent and reduces as much the production costs.

3. ECONOMICAL STUDIES   - C.N.A.R.B.R.L. -

The "Compagnie d'Aménagement de la Région du Bas Rhône Languedoc" in cooperation with the "Société du Canal de Provence" studied the areas suitable for cropping Arundo donax in the mediterranean part of France and estimated the average price of the therm produced versus the different cropping systems.

3.1. Areas suitable for Arundo donax plantations

The studies took into account, in "Languedoc" and "Provence", the topography and the qualities of the soils, the present cultures in these areas, to be replaced by Arundo donax, the farm structures, the possible markets in the different zones...

It finally appears that 25.000 hectares are suitable for an Arundo donax production in the Rhône delta : 3.500 ha are located in the west part of the "small Rhône" and 21.500 ha are located inside the delta and in the east part of the "main Rhône" as indicated in the following map :

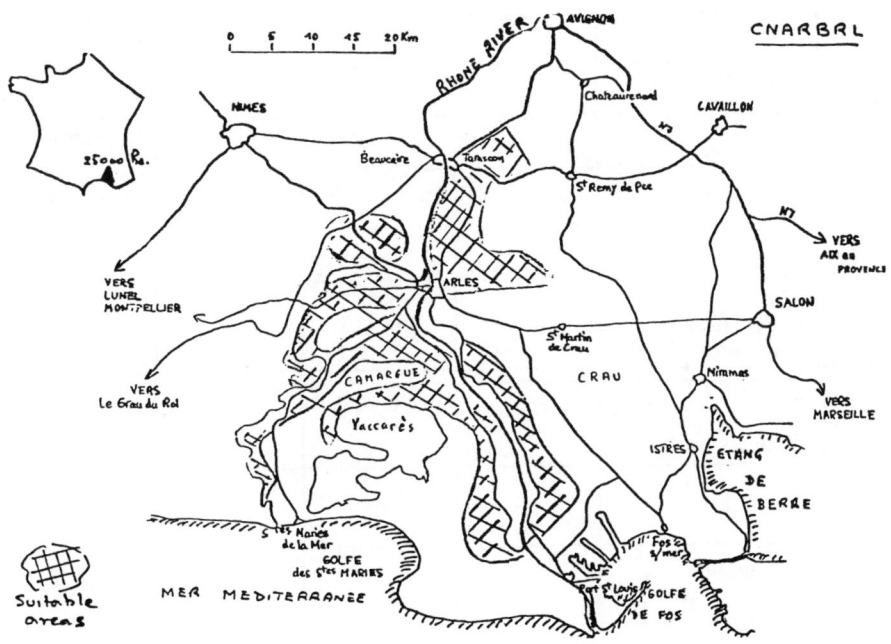

These 25.000 hectares could yield about 150.000 T.O.E. every year while the potential market appears to be very high : more than 200.000 T.O.E./year, for different purposes : greenhouses, hospitals, houses and local factories...

3.2. Study of the selling price of one therm

The following table condenses the main data and indicates what could be therm price delivered to the user :

**SELLING PRICE OF THE ARUNDO DONAX THERM
NATURALLY DRIED IN FIELD AND DELIVERED USER**

Hypothesis : - Standard crop : 20 tons/ha - (dry matter)
             - Large rows spacing : 30 tons/ha - (d.m.)

Irrigation price ha/year	300 F.F.	1.000 F.F.	1.600 F.F.	
Cropping system	\multicolumn{3}{c}{STANDARD CROP (2 meters)}	LARGE ROWS SPACING
Installation cost/year (divided by 11 years at 12 %)	1.300 F.F.	1.400 F.F.	1.500 F.F.	650 F.F.
annual costs — maintenance	890 F.F.	1.500 F.F.	2.100 F.F.	
annual costs — conditioning	\multicolumn{3}{c}{365 F.F.}	365 F.F.
annual costs — harvest and transport	\multicolumn{3}{c}{2.100 F.F.}	2.100 F.F.
Total annual costs	4.655 F.F.	5.365 F.F.	6.065 F.F.	3.115 F.F.
Loss of profit margin of replaced crop	\multicolumn{3}{c}{2.900 F.F.}	
Price at profit threshold : - one ton of dry matter - one therm (without taxes)	377 / 0,092	413 / 0,100	448 / 0,109	200 / 0,049
Price of therm to user (taxes included) chips with leaves	0,126	0,135	0,146	0,075
Chips only	0,150	0,160	0,170	0,100
Granulated leaves only	0,182	0,192	0,203	0,132

Actual price of one therm in France (April 1983) :
- industrial fuel oil : 0,162
- distillate fuel oil : 0,280

C.E.M.A.G.R.E.F.
C.A.R.B.R.L.

The large rows spacing system associated to a natural drying gives the best result because the price of the therm can easily compete with the oil therm.

The price of the therm coming from a standard crop associated to a natural drying is also attractive : but the thermic yield of Arundo donax is not so high than oil and this biomass is not so easy to handle. Nevertheless, it could be interesting when compared to the distillate fuel oil therm.

Finally, at least for the present, the new large rows spacing system associated to natural drying of the stalks in the field which have been investigated in the present research are the only one technic enabling a competitive selling price for the Arundo donax therm.

Moreover, these new techniques are now adapted to small farms (smaller and cheaper equipments, lower inputs...) and could be useful in affording the farmers a possibility of producing their own energy.

QUALITY AND QUANTITY OF LATEX WHICH CAN BE PRODUCED FROM NATURAL VEGETATION IN GREECE

Authors   : N. S. MARGARIS and D. VOKOU

Contract number : ESE-R-064-GR(B)

Duration   : 30 months, 1 January 1982-30 June 1984

Total budget   : DRA 9.817.500   CEC contribution: DRA 2.865.000

Head of project : Prof. N. S. Margaris, University of Thessaloniki

Contractor   : University of Thessaloniki (Laboratory of Ecology)

Address   : Laboratory of Ecology
                         University of Thessaloniki
                         Univ. Post Box 119
                         Thessaloniki-GREECE

Summary

Euphorbiaceae is one of the major latex producing families very promising in terms of exploitation as energy sources. Research in the Greek areas proved that *Euphorbia* species are very numerous with high contribution in biomass terms in many of them. Provided are data concerning growth characteristics of *E. dendroides* and *E. acanthothamnos* to answer the question of the feasibility of harvesting. *E. helioscopia* is proved to be a very promising species with many ecotypes. Its occurrence and extremely increased growth in olive, almonds, and pear plantations makes it a very important species needing further research to evaluate the possibility of combined cultivation with the above mentioned trees. It is estimated the oil production that such plantations may yield.

1.1 Introduction

Hydrocarbons producing plants of the *Euphorbiaceae* family have been proposed to be used as energy crops (1, 2, 3). Problems related refer to agricultural land competition for food production (4) as well as to protection of "marginal" lands (5), since these non-cultivated and usually semi-arid areas are generally considered as the appropriate areas for cultivation of plants like *Euphorbia*, jojoba and guayule. The number of *Euphorbia* species dealt with up to now in experimental projects are not more than twenty in spite of the fact that it is a very rich genus represented only in Greece by more than 60 species many of which dominate in the natural systems. We have started therefore this project aiming: (a) to identify the species and ecotypes occurring in the natural systems of Greece, and estimate their contribution in them and (b) to evaluate possibilities for alternative solutions of cultivation of selected species.

1.2 *Euphorbiaceae* in natural ecosystems

Euphorbia dendroides is the dominant plant in more than 60 km^2 in Galaxidi. The same is true in about 20 km^2 in Xirosterni area in the island of Crete. In the last report we provided data dealing with its biomass contribution in some natural systems of Greece as well as with the correlation of plant age and biomass (6). We further estimated the age distribution of *E. dendroides* in Galaxidi area. This plant grows up to forty years; there were estimated 40 plants per ha in the age class 36-40 years and the data included in Figure 1 show a normal age pyramid. Because of its high germination-there were approximately 4000 seedlings per ha -harvesting of *E. dendroides* will not necessitate subsequent planting. The solid residue yield is about 7% on dry weight basis, a similar value observed also by Calvin's group in California (2). Its major component is cycloartenol.

Unfortunately, in 90% of the cases studied, *E. dendroides* does not grow again from the same roots after harvesting. However, some of the harvested plants do have resprouting behaviour. That means that we possibly have available a genetic stock for possible development of a resprouting race (7) of *E. dendroides* which in consequence will be able to recover after harvesting.

Euphorbia acanthothamnos has been chosen to be studied because of its presence as a major component of natural ecosystems in the more arid places of Greece (8, 9, 10).

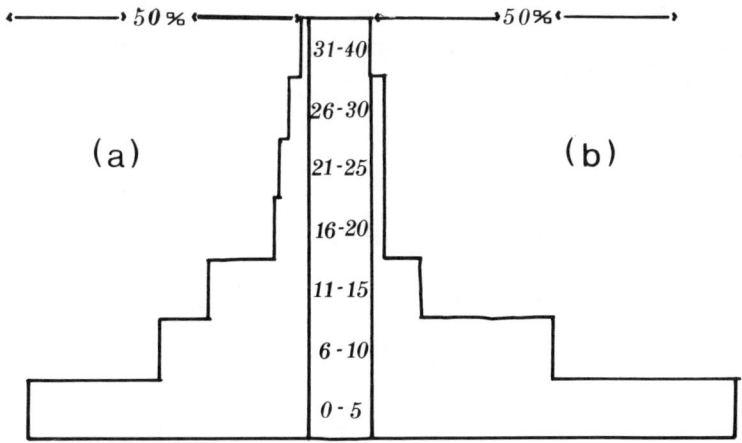

Figure 1. Age distribution of *E. dendroides*(a) and *E. acanthothamnos*(b) grown in Natural Ecosystems of Galaxidi and Attica, respectively.

Data on biomass contribution has been shown in our last report. Figure 2 shows the correlation of age to maximum diameted and biomass. Data dealing with biomass distribution in different heights of the plant and ages are presented in Figure 3. We consider these data necessary for the evaluation of the possibilities of a future harvesting in the field. Since *E. acanthothamnos* recovers after harvesting both by resprouting as well as from seeds (11) this plant can be considered very promising.

1.3 Screening of *Euphorbiaceae* all over Greece

Table 1 contains a catalog of *Euphorbia* species collected from different areas of Greece. The aim of this work is to evaluate the different species as well as the different ecotypes of the same species growing under different soil and climatic conditions. It can be seen that *Euphorbiaceae* are found all over Greece under all climatic types and altitudes. This screening provided some data of special importance concerning *Euphorbia helioscopia* that we will analyze in detail.

Table 1. *Euphorbiaceae* species collected by us in different places of Greece.

Species	Places they were found	Comments
E. dendroides	Galaxidi, Simi, Crete	Abundant in Natural Ecosystems
E. acanthothamnos	Attica and S. Crete	Abundant in Natural Ecosystems
E. characias	All over Greece	Very promising
E. helioscopia	Amphissa, Stylida, Mt. Pelion	Abundant in olive and fruit plantations; of high interest
E. cyparissias	Axioupolis	Promising
E. oblongata	Agras	No evaluation
E. salicifolia	Agras	No evaluation
E. peplus	Skopelos, Mt. Pelion	
E. deflexa	Xourihti	Seems promising
E. barrelieri	Prespa	No evaluation
E. myrsinites	Mt. Pelion, Gravia	Promising for high elevations
E. brittingeri	Labinou	No evaluation
E. paralias	South Crete	Promising; it grows on sand
E. terracina	Mouressi	No evaluation
E. pubescens	Milies	No evaluation

1.4 The *Euphorbia helioscopia* case

We found this species distributed all over Greece, occurring abundantly in olive plantations. Since huge areas of Greece are covered by olive groves we consider interesting to check the possibility of cultivation of this plant in them. For this reason we studied the growth characteristics of *E. helioscopia* growing in olive groves. Usually this plant reaches not more than 10 cm with only 0.1 g of dry weight per plant and each individual has only one stem. It was a surprise to us to realize that in the extended areas of olives in Amphissa we found that *E. helioscopia* growing under olive trees

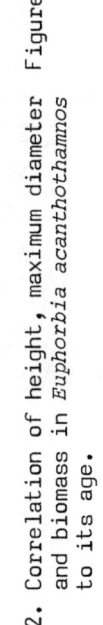

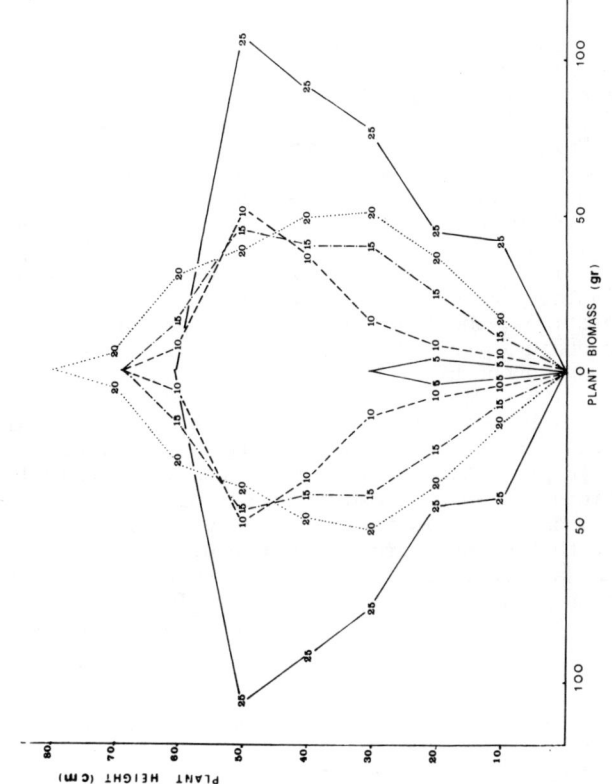

Figure 2. Correlation of height, maximum diameter and biomass in *Euphorbia acanthothamnos* to its age.

Figure 3. Graphic representation of *Euphorbia acanthothamnos* biomass at different heights of plants aged from 5 to 25 years.

attains a dry weight ten times more (1.1 g per plant) due to the fertilization of the olives. Continuing our research we found an ecotype that develops more than three stems per plant and a total biomass more than 6 g per plant when it grows close to fertilized trees. That means almost 60 times more than the normal biomass production. We continued our observations in the extended olive plantations of Stylida (second in area in Greece) and other representative areas (Attica, Volos etc.). In all cases we found *E. helioscopia* growing normally under the olive trees. We considered this finding very interesting because the combined cultivation of olive trees and *E. helioscopia* can give a solution not only to the energy problem but also to the lack of land available for cultivation.

Due to the above findings we extended our investigations in areas given to fruit trees cultivation. *E. helioscopia* was also present in almond as well as in pear plantations, where we found an ecotype more than 70 cm high and with dry biomass approximately 20 g per plant!

If we consider that about 8% of the dry biomass is "oil" the cultivation of the latter ecotype with a density of about 100 plants per m^2 will provide more than 10 barrels per ha per year. This value is about two times less than the proposed production in California by Calvin. However, by simple breeding processes this production can be significantly increased. According to the above estimations the idea of experimental cultivation of *E. helioscopia* in olive, almond, and pear plantations seems very promising. It should be taken into account that by such procedure we avoid the problems arising because of the lack of agricultural land to be given to energy crops. On the other hand if pilot projects prove to be successful, the extended areas of olive and fruit plantations may provide the necessary land for parallel energy crops.

1.5 References
1. Calvin M. (1978). Interdisciplinary Science Reviews $\underline{3}$:233.
2. Calvin M. (1979). Bio-Science 29:533.
3. Calvin M. (1980). Naturwissenschaften $\underline{67}$:525.
4. Ehr lich PR et al. (1977). Ecoscience, Freeman, San Francisco.
5. Goodin JR and Northington DK. (1979). Arid Land Plant Resources, International Center for Arid and Semi-Arid Land Studies, Texas Tech. University, Lubbock, Texas.
6. Margaris NS. (1982). In: Energy from Biomass (Series E, Volume 3, G. Grassi and W. Palz, Eds.), D. Reidel Publishing Co., Netherlands.
7. Kumar A and Joshi B.(1983).In: Energy from Biomass, 2nd E.C.Conference (A. Strub et al., Eds.) Applied Science Publisher, London.
8. Margaris NS. (1975). Berich. Schwiz. Bot. Ges., $\underline{85}$:96.
9. Margaris NS. (1976). J. Biogeography $\underline{3}$:249.
10. Diamantopoulos J and Margaris NS. (1981). Flora, 315.
11. Arianoutsou M and Margaris NS. (1981). Biol. Ecol. Medit., $\underline{8}$:119.

FORESTRY – WOOD WASTES

An experimental study of short rotation forestry for energy

Coppiced trees as energy crops

Improvement of forest trees for short term biomass production

Utilization of coppice forests biomass for fuel and other industrial uses

The production of energy from short rotation forestry

Short rotation forestry harvester chipper

Design and building of a forestry wastes harvester

Harvesting before the fire for energy; Mediterranean-type ecosystems in Greece – Costs and benefits

AN EXPERIMENTAL STUDY OF SHORT ROTATION FORESTRY FOR ENERGY

Author   :   C P MITCHELL

Contract Number :   ESE-R-016-UK(N)

Duration   :   36 months   1 July 1980 - 30 June 1983

Total Budget   :   £112,000   CEC contribution : £50,400

Head of Project :   Professor J D Matthews

Contractor   :   Aberdeen University, Forestry Department

Address   :   Aberdeen University
                     Forestry Department
                     St Machar Drive
                     Aberdeen   AB9 2UU, UK

Summary

   The work done in the period of a contract with the CEC is reported. The programme has 3 sections; assessment of above-ground biomass produced by forest trees, establishment of trials of short rotation forest energy plantations and preparation of specifications for equipment and systems suitable for harvesting forest biomass for energy. All 3 sections have been completed.
   The first section resulted in a complete data set of above-ground biomass production for ten tree species growing on different site classes and a set of management tables for conifers which can be used to predict biomass production from stand basal area.
   Trial short rotation forest energy plantations have been established on 3 site types (marginal agricultural land, scrub and young woodland) in 4 geographic areas in Britain (north-east Scotland, the Scottish borders, southern and south-west England). The logistics and costs involved in establishing plantations on small areas of land have been recorded.
   Machines and systems for harvesting forest residues, scrub woodland, single stem and coppice energy plantations have been studied. The importance of integrated harvesting systems has been recognised. A small-scale coppice harvester has been developed for harvesting willows.

An Experimental Study of Short Rotation Forestry for Energy

Introduction

Short rotation forestry is a method for growing trees to provide a feedstock for various energy markets. Such markets may include domestic and industrial combustion, or production of pipeline gas or methanol via thermochemical processing (1). Short rotation forestry systems can be either coppice or single stem trees.

Coppice production in Britain tends to be restricted to fertile lowland and sheltered sites where productivities of up to 20 dt/ha/yr on cutting cycles of 5 years can be expected. The availability of this type of land is limited so single stem short rotation forestry systems were designed for rotations of 12 to 20 years on less fertile sites in the lowlands and uplands of Britain. Productivities of up to 12 dt/ha/yr can be expected using whole tree harvesting (2).

It was estimated in a feasibility study (3) that an additional 1.5 Mha of land is potentially available for growing trees for energy in Britain. Of this 0.5 Mha comprises small areas of unused or under-used land such as field margins, rough grazing, unmanaged scrub in the lowlands and 1 Mha of unused or under productive marginal agricultural land in tne uplands.

The present study was undertaken to provide further information to enable a more realistic assessment of the feasibility of the production of wood for energy. The aims were to provide empirical data on yield, the logistics and costs of establishing forest energy plantations and the ways in which they can be harvested.

The programme is in three sections:
1 - determining production of above-ground biomass from plantations of single-stemmed trees up to 28 years old;
2 - establish trials of short rotation forest energy plantations on representative sites in Britain;
3 - prepare specifications for equipment and systems suitable for harvesting forest biomass for energy.

Section 1 Estimates of Biomass Production

Possible sources of wood for energy are the produce of whole tree harvesting of early thinnings and forest residues. Empirical data on the above-ground production of young forest crops was not available. The main aim of this section was to provide reliable estimates of above-ground production of biomass, and its partitioning, for young crops of ten species growing in existing plantations on several site classes in Britain. In addition, the data generated can be used to determine the likely production levels for single stem, short rotation forestry crops.

The objectives were:
1 - to tabulate, for each site class, total above-ground biomass per tree and its partitioning between stem, branches and foliage;
2 - to derive mathematical relations between total biomass and its components and some easily measured tree parameters for single trees (the Single Tree Model);
3 - to derive a mathematical relation between stand dry matter production and some stand mensurational parameter (the Stand Model);
4 - to produce a set of stand biomass tables to aid management of forests for energy.

Ten tree species were studied (alder, birch, Nothofagus, sycamore, Corsican pine, Scots pine, Douglas fir, hybrid larch, Sitka spruce and Western hemlock) on a total of 27 sites. These encompassed a range in age of 12 to 28 years and Yield Class of 4 to 28.

This section was completed in 1981. The data for each site have been published (4) as well as the single tree and stand models (5, 6). The predicted biomass yields and partitioning of components produced for each species and site using the single tree models are shown in Table I.

Table I Predicted Biomass Yields (t/ha) and Partitioning (%)

Species	Age (yrs)	Yield Class (m³/ha/yr)	Total Wt (t/ha)	Stem* (%)	Branch (%)	Foliage (%)
Alder	12	6	26	50	32	18
	18	10	74	75	21	4
	22	14	210	79	17	4
Birch	25	6	100	73	23	7
	20	9	99	69	24	7
Nothofagus	13	17	75	74	23	3
	24	11	145	85	13	2
	17	14	102	76	21	3
Sycamore	28	4	97	79	18	3
	20	10	114	75	21	4
Corsican pine	19	12	86	53	30	17
	20	16	136	61	27	12
Douglas fir	14	20	86	47	24	28
	18	12	68	42	29	29
	21	16	126	62	24	14
Hybrid larch	13	10	36	43	45	12
	20	10	111	65	30	5
	20	15	156	68	28	4
Scots pine	14	14	43	45	32	23
	18	8	60	51	25	23
	19	14	109	56	27	17
Sitka spruce	14	26	109	51	26	23
	22	14	105	55	30	14
	20	26	209	64	24	12
Western hemlock	15	28	174	68	21	11
	17	16	58	56	25	19
	18	20	109	59	24	16

* To 5 cm top diameter

In general, broadleaved trees had a greater proportion of biomass in the stemwood than did conifers (74% and 55% respectively). The conifers had more biomass in the branches and foliage (26% and 19% respectively) than the broadleaves (21% and 5% respectively).

Stand Tables were produced for the pines and the other conifers (6). These can be used by forest managers to determine yields of forest biomass for energy. As long as the stand to be measured falls within the range covered by the table (species, age, Yield Class and stocking) the yield of biomass can be read off when the basal area is known. An example of how the management tables might be used to determine the quantities of biomass from whole tree harvesting of thinnings and the various options for marketing the produce has been discussed (7).

Section 2 Establishment of Trial Forest Energy Plantations

The object of this section is to ascertain accurate figures for costs of operations and gain an understanding of the logistic problems of establishing forest energy plantations on small areas of land.

The plantations have been established on three site types - marginal agricultural land, scrub woodland and to replace young plantations. The latter was included so that problems associated with a second rotation

energy plantation could be studied. The ten species studied in Section 1
above are being tested in comparative replicated trials situated in four
geographic regions of Britain (Table II). At each site 2 replicate plots
of each of the species were planted at 1 x 1 m spacing; the work was done
by forestry contractors.

Table II   Location of Trial Forest Biomass Plantations

Site	Region	Land Use	Area (ha)	Planting Year
1 Craibstone	North-east Scotland	Marginal agriculture	0.2	1981
2 Banchory	North-east Scotland	Scrub woodland	0.6	1981
3 Aldroughty	North-east Scotland	Young woodland	1.0	1981
4 Marlefield	Scottish Borders	Marginal agriculture	2.0	1982
5 Kilham	Scottish Borders	Scrub woodland	2.0	1982
6 Witney	Southern England	Marginal agriculture	1.0	1982/3
7 Tar Wood	Southern England	Scrub woodland	1.0	1982/3
8 Longridge	Southern England	Young woodland	1.0	1982
9 Crowcombe	South-west England	Marginal agriculture	1.0	1982
10 Queenhill	South-west England	Scrub woodland	1.0	1982
11 Holmington	South-west England	Young woodland	1.0	1983

North-east Scotland
1 Craibstone: A small area of agricultural land between two good fields
was marginal, because it was too steep for tractor operations. It was not
possible to plough the site prior to planting. The site was already
fenced against stock and rabbits. Open-rooted plants were direct planted
in March 1981. Extensive hand weeding and the use of a polythene mulch
were necessary to control profuse growth of grasses in the summer following
planting. In the winter months a chemical herbicide (Clanex) was applied.
2 Banchory: An area originally covered with birch scrub was cleared and
sold as firewood in February 1981. The site was ploughed at 1 m spacing
and fenced against rabbits. Open-rooted plants from cold storage were
direct planted in May. No weed control has been necessary so far although
a hand weeding may be required this year. Some damage was done to the
trees by hares during the winter.
3 Aldroughty: This site had a cover of a young coniferous crop which was
felled and sold for pulp and palletwood. Residues were removed completely
from the site before planting with open-rooted stock from cold store in
April 1981. The site was fenced against rabbits and roe deer. Despite
much hand weeding and the use of chemical herbicides regrowth of sycamore
stools and seeded-in birch is a problem. In addition wild raspberries are
growing profusely with the result that the site will need to be cleaned
this year.
   The logistics and costs of operations for the three sites above have
been reported fully elsewhere (8).
Scottish Borders
4 Marlefield: An area of agricultural land was ploughed to 1 m spacing
and planted with open-rooted plants in February 1982. The site was fenced
against stock and rabbits. Plant survival appears good and a hand weeding
was done last year.
5 Kilham: An area of scrub woodland was cleared to produce firewood in
January 1982. The site was ploughed to 1 m spacing before planting with
open-rooted plants in February. The site was fenced against stock and
rabbits. This is a fairly exposed site with shallow soil so there were
some plant losses due to drying out. A hand weeding was done last year.
Southern England
6 Witney: An area of marginal agricultural land was covered with black
polythene mulch prior to planting with open-rooted stock last spring. Due

to the extreme dryness in 1982 some planting was left until 1983, otherwise the larger plants would have died. The polythene mulch was removed this year because it was damaged during the winter and was no longer effective as a weed control. The remaining plants were planted in April 1983.

7 Tar Wood: This site of old mixed broadleaved trees was cleared during the winter of 1981 to produce firewood and fenced against rabbits. As with site 6 it was not prudent to plant all the plots in 1982 and some were planted in 1983. As with the Aldroughty site extensive control of old coppice, seedling growth and brambles will be necessary.

On sites 6 and 7 the Forestry Commission have also established two of their coppice trials (ESE-R-017 UK).

8. Longridge Wood: A young plantation of European larch was felled and sold for pulp and firewood. The residues were cleared from the site which was fenced against rabbits. Planting with open-rooted stock was done in April 1982.

South-west England

9 Crowcombe: An area of marginal agricultural land which had been invaded by bracken. Attempts to clear the bracken using Asulox met with little success. The site was fenced against rabbits. Open-rooted plants treated with an anti-mammal smear were planted in January 1982. Bracken growth was profuse in the summer necessitating two hand weedings. Plant survival, however, was excellent. Some deer damage was noted during the winter so plants were re-treated with anti-mammal smear.

10 Queenhill Wood: An area of scrub woodland which was cleared for firewood and fenced against rabbits. Open-rooted plants were planted in March 1982 and treated with anti-mammal smear. Despite last years drought plant survival was excellent. The site needed to be weeded once last year and a hand weeding may be necessary this year. The plants have been re-treated with anti-mammal smear.

11 Holmington Wood: An area of young birch woodland was cleared to produce firewood in January 1983. Planting was planned for February but had to be delayed until May (the plants being kept in a cold store) because of the extremely wet spring. The site was fenced against rabbits and the plants treated with anti-mammal smear.

The total costs of establishment up to May 1983 and the breakdown by operation are given in Table 3. It is quite clear that the scale of operation has a significant effect on cost of establishment. Costs of establishing plantations on old woodland sites are particularly expensive. The problems of establishing energy plantations on forest sites have been discussed by Mitchell and Puccioni-Agnoletti (9).

Weeding operations were expensive due in part to the close spacing of the young trees. The use of chemicals had to be closely supervised and hand weeding to be done carefully. Reducing initial stocking to 5000/ha may reduce this component without unduly affecting yield of biomass.

Costs of operations can be significantly reduced if the landowner can plant the land himself using surplus labour from on the farm.

Clearly, costs of establishment are only one side of the economic equation; yields and hence revenues are the other. In this short study yield figures from the plantations will not be obtained but if the yields postulated earlier are achieved then most operations will be viable. The 'breakeven' economic position of growing three species at the Banchory site on a 20 year rotation and using a 3% discount rate are shown in Table 4. For alder, which is expected to have a productivity of 15 t/ha/yr on this site, wood for fuel would need to be worth £25/t for the operation to break even. However, for Nothofagus growing at a similar rate the value of wood would need to be £34/t to break even.

Table III   Total Cost of Establishment[1] and Operations (%)

Site	1	2	3	4	5	6	7	8	9	10
Operation										
Clearance	-	4	13	-	11	4	36	23	9	12
Ploughing	-	11	-	3	2	-	-	-	-	-
Fencing	-	23	33	21	25	21	21	17	24	25
Plants & planting	26	50	37	65	51	24	34	47	44	46
Beating-up	8	12	10	-	-	-	-	10	-	-
Weeding	68^2	-	8	12	12	50^2	-	3	19	13
Protection	-	-	-	-	-	-	-	-	4	4
Total/ha (£)	5264	3067	3650	2072	1997	4082	4087	4645	5185	5240

1 - to May 1983
2 - includes polythene mulch

Table IV Banchory: Value of Wood Required to Break Even at 3% Discount Rate

Productivity (t/ha/yr)	Discounted Costs[1] (£)	Break Even Revenue (£)	Break Even Value (£/t)
Nothofagus			
5	5117	9241	92
10	5450	9843	49
15	5616	10142	34
20	6003	10841	27
Alder			
5	3602	6505	65
10	3935	7107	36
15	4101	7406	25
20	4488	8105	20
Sitka spruce			
5	3344	6039	60
10	3677	6641	33
15	3843	6940	23
20	4230	7639	19

1 - assumes 20 year rotation and includes harvesting costs

The economic viability of such production systems is very dependent on the yield and value of the produce. Realistic values for the product need to be established before landowners will enter into a commitment of growing trees for energy.

This study has highlighted the need for in-depth studies of the attitudes of farmers to energy forests in addition to the availability of land, capital and labour.

Section 3 Harvesting Forest Biomass for Energy

The purpose of this section is to examine existing machines and systems for suitability for harvesting forest biomass for energy and prepare specifications for possible new equipment.

Forest biomass for energy will be available from the following sources: scrub woodland, forest residues, thinnings, coppice and single stem energy plantations. The many ways in which they can be harvested have been discussed previously (1, 10, 11).

In order that materials handling can be simplified, the biomass should be supplied to the centralised conversion plant as chips. There are two main options for the supply of forest biomass - 'hot' and 'cold' systems (1). In a hot system all the components in the harvesting, chipping and transport chain are closely linked and provide material directly to the user. This system is preferred as it eliminates problems associated with storage. If storage is necessary then a 'cold' supply system is required. Here it is recommended that the biomass be stored as whole trees, bundles or bales either in the forest or at some intermediate point.

For scrub woodland and single stem energy plantations there are harvesting machines and systems which are suitable for use with little adaptation (12), but there is scope for developing equipment which is better suited to the small-scale of operations which are envisaged in Europe.

Harvesting of forest residues is more of a problem. On clearfell sites the solutions are relatively straight-forward although barely economic at present. In the case of thinnings it is becoming clear that the only way to handle residues is by whole tree harvesting. Equipment is available or under development in North America or Scandinavia but is likely to be too capital intensive for most European operations. One way in which returns to the owner might be increased is by use of integrated harvesting systems in which a conventional timber product and an energy product are harvested in 'one pass'. Such systems appear attractive on paper, particularly when considering the very great potential small roundwood resource, but must be shown to be financially viable in the forest. *Practical studies of such systems should be given high priority as it is likely that they will provide most of the biomass for energy in the short to medium future, particularly in north-west Europe.*

In order to complement the work with the Bord na Mona harvester (13) the development of a tractor-mounted machine for harvesting 2 to 3 year old coppice on disadvantaged farmland was sub-contracted to Loughry College, Northern Ireland. The harvester is mounted on the three point linkage of a 44 kw agricultural tractor and cuts the willow with a circular saw. The cut sticks are then tied into bundles weighing 30 to 40 kg. Progress on the design, building and testing of the 1981/82 build of the machine has been reported (14). During the winter the machine has been redesigned to reduce its weight and improve the handling and bundling of the cut material. Trials with a test rig suggested that the new design will be suitable and the machine is currently being rebuilt. The 1982/83 build will be tested in the field later this year. *With this build a significant step forward has been achieved and that, following trials, the next stage will be a pre-production prototype.*

Conclusions

The original programme of work has been accomplished. The significant results are:
- a set of forest biomass management tables has been produced;
- 11 trial plantations have been established and their costs recorded;
- the importance of integrated harvesting systems has been identified;
- a coppice harvester suitable for commercial development is now available.

Significant steps have been taken yet we are still some way from demonstrating the economic viability of forest biomass production. One area of work which should have high priority is that of systems studies to match availability of the resource with the supply needs of the end users (15). Much of the information produced so far could be used to produce realistic figures for incorporation in such a study.

References
1. Mitchell, C P & Pearce, M L (1983). Feedstocks and Characteristics. Proc 1st European Workshop on Thermochemical Processing of Biomass. In press.
2. Mitchell, C P (1980). Forest Biomass for Energy in the United Kingdom. In 'Forestry Energy'. Proc IEA/IUFRO Congress, Sweden. Ed J E Mattson & P O Nilsson. pp 52-58.
3. King, G H (1981). An assessment of forest energy. EUR 7550 EN.
4. Mitchell, C P; Matthews, J D; Proe, M F & MacBrayne, C G (1981). Biomass Yields of Forest Trees. Tech Rep to ETSU. ETSU-B1081a.
5. Mitchell, C P (1981). An Experimental Study of Short Rotation Forestry for Energy. In 'Energy from Biomass Vol 1'. Reidel pp 30-34.
6. Mitchell, C P; Proe, M F & MacBrayne, C G (1981). Biomass Tables for Young Conifer Stands in Britain. Kyoto Biomass Studies pp 45-50.
7. Mitchell, C P; Matthews, J D; MacBrayne, C G & Proe, M F (1981). Determination of Yield of Biomass from Whole Tree Harvesting of Early Thinnings in Britain. In 'Energy from Biomass'. Ed W Palz et al. Appl Sci Pub pp 181-186.
8. Mitchell, C P & Matthews, J D (1983). A Study of Short Rotation Forestry for Energy in Britain. In 'Energy from Biomass' Ed A Strub et al. Appl Sci Pub pp 181-185.
9. Mitchell, C P & Puccioni-Agnoletti, M C (1983). Forest Energy Plantations on Forest Sites. IEA/FE Pub. NE 1983: 1 pp 28.
10. Mitchell, C P (1981). Harvesting Systems for Small Trees in Britain. In 'Harvesting Small Timber' FPRS Proc, P-81-32, pp 95-97.
11. Mitchell, C P & Säll, H O (1982). Biomass Harvesting, Transport and Storage - Needs, Specifications and Mechanization. IEA/FE PGB/PGC Meeting Oslo. In press.
12. Scaramuzzi, G (1982). Utilisation of Coppice Forest Biomass for Fuel and Other Industrial Uses. In 'Energy from Biomass Vol 3' Reidel pp 71-76.
13. Keville, B J (1982). Short Rotation Forestry Harvester Chipper. In 'Energy from Biomass Vol 3' Reidel pp 83-89.
14. McLain, H D (1983). The Development of a Harvester (patent pending) for 2-3 year old Willow Coppice. In 'Energy from Biomass' Ed A Strub et al. Appl Sci Pub pp 225-229.
15. Mitchell, C P & Miles, T R Jr (1983). Biomass Supply and Pretreatment - Workshop Report. Proc 1st European Workshop on Thermochemical Processing of Biomass. In press.

COPPICED TREES AS ENERGY CROPS

Author	:	M.L. PEARCE
Contract number	:	ES-E-R-017-UK(N)
Duration	:	36 months July 1980 - June 1983
Total budget	:	£59,000 CEC contribution : £23,600 (40% of total)
Head of Project	:	A.J. Grayson, Director Research
Researcher	:	M.L. Pearce, Research Chief Forester
Contractor	:	Forestry Commission, Research and Development Division
Address	:	Westonbirt Arboretum Tetbury Glos England

Summary

This terminal report describes the successful establishment of a series of field experiments designed to provide data from which a series of production models can be created. The experiments form the basis of a prognosis which suggests that some tree species grown on a short rotation coppice system can provide economically viable feedstock for energy conversion. The four year programme has been marginally sufficient to bring only a small part of one experiment into production. This early data is encouraging and suggests that further intensive assessment is required before reliable models can be formulated. The physical and practical elements of the contract have been fulfilled and it can be confidently predicted that the experiments will provide for the long term nature of the project.

1. Introduction
It is necessary to reiterate the objective of this contract in order that its purpose in the wider nature of the research programme (CEC. Solar Energy Series E) can be explained. Annex 1 of the contract calls for the establishment of a series of regional field experiments and at this time (June 1983 - the end of the contract) this objective has been successfully achieved. However, this is only an intermediate stage to the longer term objective of generating from these experiments, data which will provide the basis of production models. The level of production (tonnes d.m. ha^{-1} yr^{-1}) for a chosen species grown on a coppice system in a given set of edaphic and environmental circumstances, will be the subject of many further years of assessment. The trees, and in some cases where coppicing has already been achieved, the stools, are growing successfully, and this status will be maintained with periodic harvesting and assessment in accord with the variable rotation element of the experiment plan. Protection against predation by any organism will be sought and attempts will be made to monitor any factors which affect production levels.

2. Current situation
Table 1 of the report contained in Vol.3. (Proceedings Brussels 5 - 7 May 1982) lists each of the experiments and gives details of their composition. In this report of the current status, each experiment will be discussed under its' title only.

2.1. Sheffield Park (Sussex)1/81.
The two introduced species on this rehabilitated woodland site have not grown exceptionally vigorously during the last two growing seasons. A decision was taken at the end of 1982, not to coppice (the initial harvest following establishment) until after one further growing season, in order to benefit the root systems. The comparative control treatment is to encourage growth from the stools of the existing woodland species which were cut over in the initial site clearance, and this has been successful. As expected the competition from natural weed growth has had to be controlled with the use of herbicide application. Hand held sprayers with 'Arborgaurd' crop protectors were used to apply Glyphosate during the first two growing seasons.

2.2 Mepal (Cambs) 1/81.
Our experience gained from the use of Nothofagus spp in forest conditions of similar environment, encouraged the inclusion of the species in this experiment. It must be noted however that this fenland site on an arable farm is extremely exposed, and despite the suitable nature of the soil, this tree species failed to establish itself satisfactorily due to low winter and spring temperatures. The inclusion of a small subsiduary plot of Eucalyptus spp suffered the same fate, but was expected. The establishment of the remaining two species has been slow but successful and coppicing has been delayed until the end of the 1983 growing season. Annual weed growth has had to be controlled with herbicide as reported for the previous experiment, but on this site the weed species are not seriously competing with the planted tree crop.

2.3 L.A.R.S. (Avon) 1/81.
This is by far the most optimal site and the most likely to produce the highest levels of biomass. Growth to date has been sufficiently vigorous to enable coppicing to take place after the initial growing season and some biomass production levels to be assessed at the end of the second growing season. This data will be discussed later, but in order to

obtain it, the experimental design was modified slightly. The original design called for two rotation intervals, two years between harvests and four years between harvests. On a single replication (i.e. 6 plots out of a total of 36) the four year harvesting cycle has been reduced temporarily to 4 x 1 year harvesting cycle, which will enable us to measure the environmental effect upon production. The two year cycles have not been interrupted and at the end of 1985 we will have collected four sets of 1 year data and two sets of 2 year data. From this it should be possible to extrapolate the shape and slope of the production curves for ultra-short rotation lengths. Remembering that on this site all species were coppiced after the first growing season, it is interesting to observe that in the following year (1yr shoot on 2 yr stool), all species completely colonised the site at the $1m^2$ spacing amd almost so at the wider $2m^2$ spacing. Herbicide was used during the first growing season, but not during the second.

2.4 Alice Holt (Hants) 337/82.

The site chosen for this experiment had been previously grazed by cattle for many years and the soil condition of the top 30-40 cms was semi-anaerobic. Shallow agricultural ploughing was employed to ameliorate this condition and will in the long term be benificial, but initial establishment of the planted crop has been positive but slow. Weed control has been achieved with herbicide and the initial coppice cut will be made at the end of 1983.

2.5 Witney (Oxon) 1/82.

Establishment has been slow, largely due to excessive vegetation re-growth resulting from the initial clearance of the woodland condition. Herbicide control was partially effective and will be repeated during 1983, following replacement planting of initial losses. In twelve of the 48 plots, regrowth from the stools of the cut-over woodland crop has been designated as the fourth species treatment and will be assessed to measure its' production level against the three other species. The initial coppice cut will be made at the end of 1983.

2.6 Witney (Oxon) 2/82.

This experiment has established well and growth has been good, but again, a modification in the design has been necessary from experience gained during the first growing season. The site was expected to produce heavy weed competition and be liable also to summer soil moisture deficit. To counteract both conditions, all plots were covered with 500 gauge ($125m^u$) black polythene sheet and the plants introduced through cross slits made in the sheet. Throughout the following growing season, damage occured to the sheeting through wind penetrating the planting slits and extending these openings to major rifts and tears. Without a permanent labour presence on the site it was impossible to effect repairs and by the end of the season the integrity of the treatment was destroyed. Experience suggested that the treatment was impractical to maintain and the polythene removed entirely before the commencement of the 1983 growing season. The initial coppice cut will be achieved at the end of 1983.

2.7 Wentwood (Gwent) 6/82 Ext:

The conversion of this older experiment and the establishment of conditions suitable for the initial coppice cut at the end of 1983 are complete. Production data from these 20 yr old stools will provide valuable comparisons with the other six experiments.

3. Results.

There are few parameters which can be empirically assessed at this stage to determine the success or otherwise of this contract. In general terms the four year project to establish the set of experiments has been successful. Only one experiment has grown sufficiently vigorously to have achieved two coppice cuts during the 4 years and this will be discussed under 'production'. Firstly, an objective evaluation will be made of the establishment and maintenance phases during the contractual period.

3.1 Establishment

An adequate coverage of the broad environmental zones which might be suitable for a coppice fuelwood system has been achieved. (Figure 1). Using a crude zoning of precipitation - wet, through moist to dry, coupled with temperature - cold, through warm to hot, then each experiment can be said to be sited on one of the combinations (with minor exceptions). Despite the variety of ownership of the sites, adequate access and co-operation has been secured, not only for the period of this contract, but for the longer time necessary to collect production data.

The planting stock used to establish the experiments varied according to species, Populus and Salix with 30cm hardwood cuttings, Nothofagus and Alnus with rooted transplants and Eucalyptus with rooted seedlings containerised in Japanese paper pots. All three types of plant are accepted as common practice and did not contribute to any losses following planting. Survival was generally high throughout all the experiments with highest losses being attributed to delayed planting - these being subsequently replaced the following spring. All sites have been protected against rabbits, either by being sited in existing compounds or by erecting an independent fence around the experiment. Only one site (Sheffield Park) was thought to be immediately vulnerable to deer and the fence extended to provide protection and the progress of the remaining experiments will be kept under surveillance. The polythene mulch used during the first growing season at Witney 2/82, did attract large populations of voles and although no damage was sustained to the planted crop - this may have proved a problem if the polythene had been retained.

3.2 Maintenance

The most important aspect of the establishment phase is to maintain optimum growing conditions for the introduced plants as far as is practically possible.

As it is not possible to modify the environment to any large extent, concentration on the elimination of any competing vegetation has been considered important. The polythene mulch used at Witney 2/82 was unfortunately not practical to maintain, but theoretically it does have the advantage that as well as eliminating competing vegetation, it prevents evaporation of soil moisture until such time as the planted stock has colonised the site completely. (Trials are currently being conducted with this treatment independently of this contract).

The alternative in these seven experiments is to control competing vegetation with herbicide to a point in time (2 - 3 yrs) when the planted crop has shaded out the site completely. Every effort is made to protect the trees from toxic chemicals and to date this treatment is successful and will be continued.

3.3 Production

The only hard data available has resulted from the good growth at L.A.R.S. 1/81 and by virtue of the modification to the rotation treatment.

Table 1 shows the fresh weight values for 1 year coppice shoots on two year stools. Remembering that this is both very early data, unreplicated and from a single experimental site - there are however, some interesting indications of the variability of individual stools and the effect of crop spacing. The analysis of dry matter production is not yet complete, but a 'rule of thumb' estimate can be made by applying a factor of 0.5 to the figures in Table I_1 - i.e. Populus at $1m^2$ spacing has produced approximately 7 tonnes (dry) $ha^{-1} yr^{-1}$ for the first year of coppicing on 2 yr old stools.

4. Discussion

It should be evident that the contractual obligation has been fulfilled to all practical intent, but as envisaged from the outset, the data which will build the production models is only just beginning to emerge. The first results are encouraging in that they confirm the need to test crop spacing as a factor of species choice, the indication that production levels are likely to be dependent both upon spacing and rotation length, and that production levels can be expected to reach the figures suggested in earlier prognosis.

It is interesting to note the wide range of individual stool fresh weights from the early L.A.R.S. 1/81 data, yet indicating an effect on all three species from crop spacing. It is dangerous of course, to extrapolate from individual stool production to a per hectare production and with the accumulation of subsequent data the extrapolation will be achieved proportionally on a unit area basis. However, despite the heterogenity that can be expected in almost any crop of trees, it is encouraging that individual stools can produce very high production levels of up to 6kg/ stool in a single growing season. It should be the purpose of this and similar projects, to identify the factors limiting this level of production in large scale plantations grown as an energy feedstock.

The experiments reported on here, are designed to provide many of the answers to the outstanding problems of such systems - Table 2 sets out the current status of each one and should indicate also their suitability to provide valuable data in the future.

Figure I.
Environmental distribution of experiments

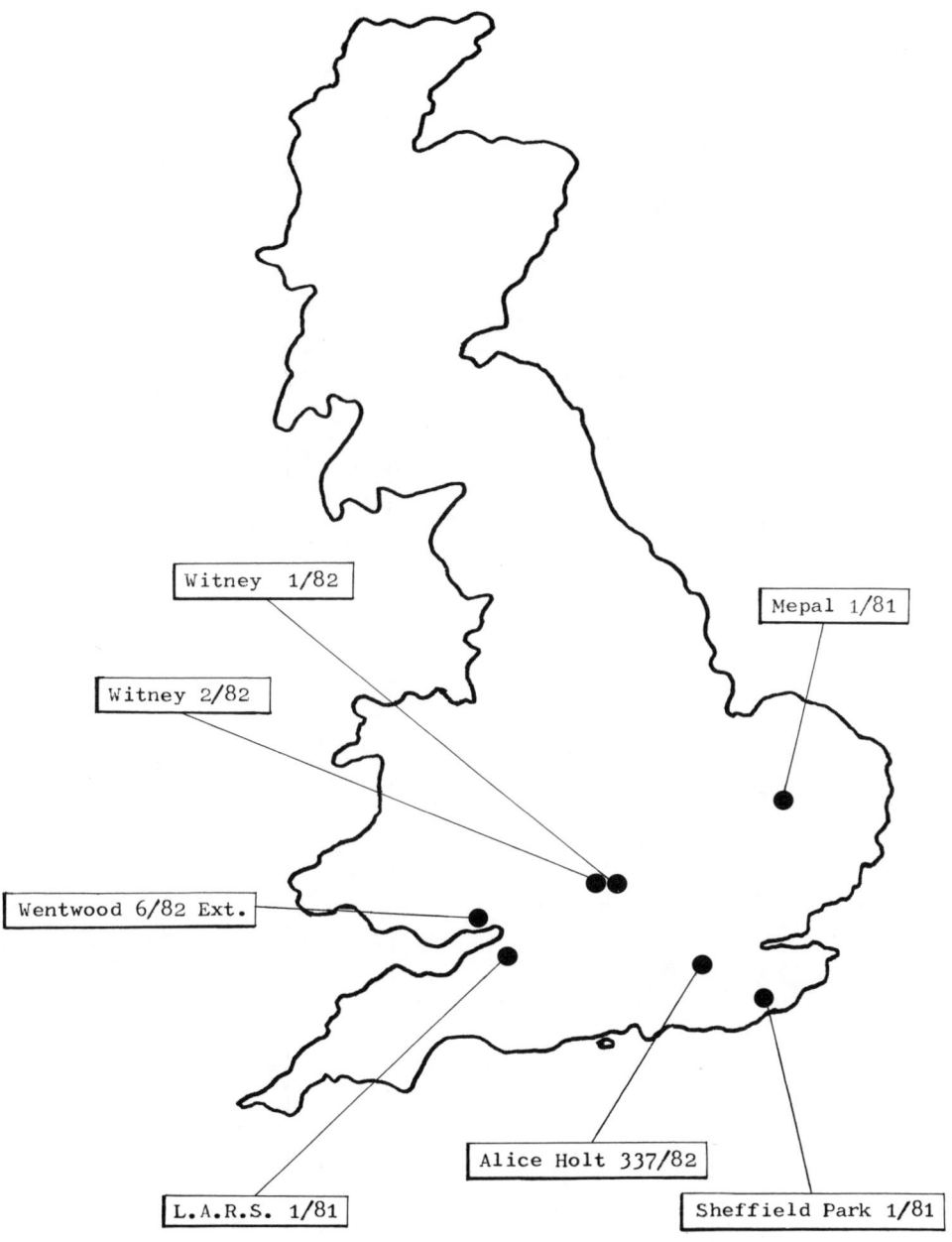

L.A.R.S. 1/81. Fresh weights. 1yr coppice on 2yr stools.   Table I.

Species	2m² spacing (2500 stools/ha)			1m² spacing (10000 stools/ha)		
	Plot means kg/stool	Plot range kg/stool	Extrapolated tonnes/ha/yr	Plot means kg/stool	Plot range kg/stool	Extrapolated tonnes/ha/yr
Populus 'RAP'	4.13	2.33-6.30	10.3	1.52	0.78-2.64	15.2
Salix 'AG'	3.33	1.87-4.73	8.3	1.12	0.18-2.34	11.2
Eucalyptus archerii	1.25	0.08-4.90	3.1	0.83	0.02-3.25	8.3

Table 2.

Description	Environment		Previous Land Use			Species						Spacing		Rotation			Replications	Nelder plot	Date of initial Coppicing
	Rainfall	Temperature	Woodland	Agriculture	Stream-side Agriculture	Populus	Salix	Nothofagus	Alnus	Eucalyptus	Natural	2500 pa	10000 pa	1 yr	2 yr	4 yr			
Sheffield Park 1/81	Low	High	x			x	x				x	x			x	x	2		Winter 83/84
Mepal 1/81	Low	Low			x	x	x					x	x		x	x	3		Winter 83/84
L.A.R.S. 1/81	High	Mod		x		x	x	x		x		x	x	x	x	x	3	x	Winter 81/82
Alice Holt 337/82	Mod	High	x			x	x	x	x	x	x	x	x		x	x	3	x	Winter 83/84
Witney 1/82	Mod	Mod		x		x			x	x		x	x		x	x	3		Winter 83/84
Witney 2/82	Mod	Mod			x	x			x	x		x	x		x	x	2		Winter 83/84
Wentwood 6/82 Ext.	High	Low				x									x	x	3	x	Winter 83/84

IMPROVEMENT OF FOREST TREES FOR
SHORT TERM BIOMASS PRODUCTION

by

Eric TEISSIER du CROS

Contractor number : ESE-R-018-F
Duration : 36 months ; 1 july 1980 to 30 june 1983
Total budget : FF 1,000,000 (INRA, AFME and CEC)
CEC contribution : FF 460,000
Head of project : Eric TEISSIER du CROS
Contractor : Institut National de la Recherche Agronomique
Station d'Amélioration des Arbres Forestiers
Ardon
45160 OLIVET
France

Summary
This programme aims at selecting and improving highly productive forest tree species for short term biomass production. During the period of the contract 13 field trials involving 4 softwoods : Larix eurolepis, Cryptomeria japonica, Picea sitchensis and Pinus contorta ; and 9 hardwoods : Alnus glutinosa, A.rubra, A.cordata, Populus trichocarpa, P.canescens, the Belgian poplar clones "UNAL" and "BEAUPRE", Robinia pseudo-acacia and Quercus borealis, have been established with planting densities ranging from 1100 to 6000 trees per hectare. Collections of provenances in the natural ranges of 4 mains Alder species will enable comparing their performances. A 7-meters-high greenhouse now enables large-scale interspecific crosses for search of hybrid vigor. Seedbed inoculation of Frankia, the nodulating bacteria of Alders, seems to increase height growth of 1 and 2-years-old seedlings. This programme has to be developed in the following directions : increasing of the number of Genera to be tested, wider sampling of the potential biomass production areas in France on which to establish trials, more accurate study of the Alder-Frankia symbiosis to produce a fast adaptation and quick early growth of seelings.

1. INTRODUCTION

In France, different research and development programs deal with short term forest biomass production. The first one is lead by AFOCEL, mostly for fiber production. The Genera involved are mostly Poplars and Eucalypts but also Redwoods and other Conifers (1). The second one concerns the evaluation of the production of existing coppices, methods to increase it and the sylviculture of newly established high density stands. It is run by the Laboratory of Sylviculture of INRA (Orleans)(2)(3). The third one will be initiated in 1984 and will aim at demonstrating newly selected forest tree material in large close-spaced stands, at testing harvestingdevices and at striking economic balances. It is sponsored by the Agence pour la Maitrise de l'Energie (AFME) and will be run by CEMAGREF (§). The fourth one deals with selection and creation of forest tree material for short-term biomass production. It is run by the Forest Tree Improvement laboratory of INRA (Orleans) with the help, for field trials establishment,of the Biomass Technical Division (INRA-Orléans).Its funding is shared by AFME and CEC. Its objectives, research routes and 3-year-practical realisations will be developed here after.

2. OBJECTIVES OF THE PROGRAMMES

Objectives have been fully developed in previous reports (7)(6) and are shortly recalled here.
France has only small amounts of left agricultural land or peat-bogs, but it has large quantities of low grade forest sites on hydromorphic soils, covered with low producing coppices. Species chosen for biomass production in replacement of existing ones, will have to be adapted to such sites. Testing sites should be representative of land available.

3. RESEARCH ROUTES

Because of the particularities of the sites and of the production goals, it was felt that a new and powerfull selection and breeding programme had to be initiated. Different routes are followed at the same time.
- choice of the most vigorous genotypes derived from classical improvement programmes : Poplar,Larch, Sitka spruce, Sugi.
- partial adaptation of recently started improvement programmes : American red oak, Yellow poplar

---

(§) Centre National du Machinisme Agricole, du Génie Rural, des Eaux et des Forêts.

Species	Types of populations	Year	Plantation densities (trees per ha)	Number of rotations (age of coppicing not yet defined)	Region	Number of trees	Observations
1 *Larix eurolepis* *Cryptomeria japonica*	clone mixture clone mixture	1982	1100, 1600, 2200 and 4400 1100 and 4400	1	Western slopes of Massif Central	2500 740	
2 Sitka Spruce	49 clones	1983	2500 and 5000	1	Bretagne	1200	
3 Black Alder	12 provenances	1982	2000 and 4000	2	Sologne	2700	
4 Black Alder Red Alder	12 provenances 2 provenances	1982	2000 and 4000	2	Orléans State Forest	2870	
5 Black Alder Cordate Alder Red Alder	3 provenances 7 provenances 2 provenances	1983	2000 and 4000	2	Rhône river valley	2900	Agricultural land
6 Black Alder Grey Poplar Contorta Pine	4 provenances 4 clones 4 provenances	1983	2200	1	Orléans State Forest	1200	Hydromorphic soil Mixed and pure plots
7 Black Alder Poplar "UNAL"	1 provenance 1 clone	1980	3300	1	Sologne	290	Mixed and pure plots
8 Blackcotton-wood	36 clones	1982 1983	2000 and 4000 4000	2	Sologne Gatinais	3800 3800	Forest Land Agricultural Land
9 Poplar UNAL Poplar BEAUPRE Blackcotton-wood	3 clones	1983	2000, 3000 4000, 5000	3	Perche	5000	
10 Black Locust	4 provenances	1983	3000 and 6000	1	Orléans State Forest	1350	
11 Black Locust Cordate Alder Poplar "UNAL"	1 provenance 1 provenance 1 clone	1983	3000	1	Gatinais	1536	Mixed and pure plots
12 Cordate Alder Black Alder	2 provenances 4 provenances	1983	1500, 3000 4500	3	Sologne	3150	
13 American Red Oak	9 provenances	1983	3000	1	Sologne	540	Compared to wide spacings (1500/ha)

TABLE II. Biomass field trials established during the period of the contract

Figure 3 : isolation bags for hybridization of Alders

Figure 2 : pollination chambers used previously

Figure 1 : the "large" hybridization greenhouse

-initiation of completely new improvement programmes : Alders, Black locust
- interspecific crosses with special attention to hybrid vigor : Alders, Poplars
- study of factors affecting the juvenile growth, particularly with Alders and Frankia, their nitrogen fixing endophyte.

4. SITUATION IN MAY 1983

4.1 Collection of species and provenances

This concerns Alders (Table I) and has been realised with the help of the International Energy Agency and different research institutes of Poland, West Germany, Belgium, Italy, USA and Canada.

Species	Types of populations	Origin	Observations
Cordate Alder A.cordata	3 provenances 10 provenances 41 provenances	artificial Italy Corsica (France)	Sites located, seeds to be collected fall 1983
Black Alder A.glutinosa White Alder A.rhombifolia Red Alder A.rubra Sitka Alder A.sinuata A.cremastogyne A.hirsuta A.japonica A.nepalensis A.trabeculosa	24 provenances 1 provenance 87 provenances 1 provenance 1 provenance 1 provenance 1 provenance 1 provenance 1 provenance	Europe California Canada, USA Oregon (USA) China China China China China	

4.2 Fiel trials (Table II)

Thirteen trials have been established. They concern 12 species and 10 sites ranging from good and well drained to poor and hydromorphic forest or agricultural land. They cover roughly 11 hectares. Survival for 1982 tests was more than 99 percent for Alders and Poplars. It was 85 percent for Larch and Sugi.

4.3 Controlled crosses

A 318-square-meters-large 7-meters-high green-house, partly funded by CEC, and built in 1981 and 1982, now enables wide scale interpecific crosses with Alders (figure 1). A collection of container, grafted or self rooted, flowering trees is being realised. 1981 crosses showed a 3 weeks-period of receptivity of Black Alder female flowers in greenhouse conditions (8). 1981 and 1982 interspecific crosses produced large amounts of seeds but doubt was cast upon their hybrid status because of the amount of seed obtained from unpollinated flowers. The isolation technics which had shown to be reliable for poplars (pollination chambers-Figure 2) were not sufficient for Alders. The 1983 hybridization compaign is giving emphasis to isolation of flowers from undesirable pollen (Figure 3). The species involved in this campaign are A.glutinosa, A.incana, A.cordata, A.rubra, A.inokumae and A.hirsuta. Hopefully, our first interspecific Alder seeds will be sown in Spring 1984.

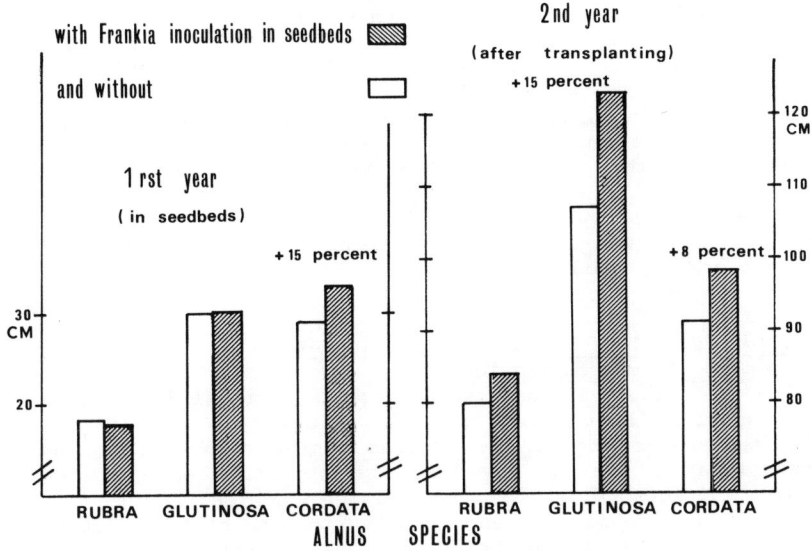

Figure 4. Inoculation of Frankia
may increase height growth of Alder seedlings

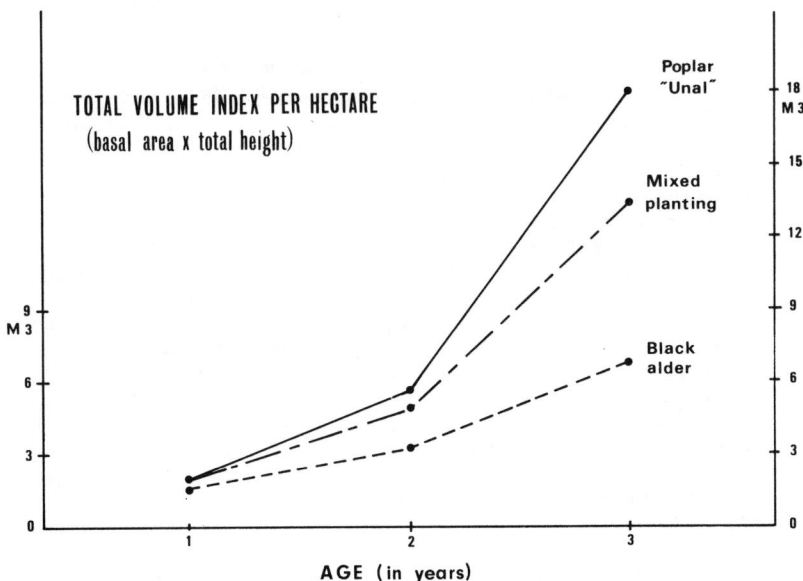

Figure 5. Black Alder, a nitrogene fixing species may be competing
the Poplar "UNAL" instead of cooperating

4.4 Alder-frankia interaction
In the previous report (8), significant effect of Frankia inoculation in seed beds was shown on 1-year-old seedlings of Alnus cordata. One year later after transplanting in classical nursery conditions, and without any new inoculation, the favorable effect is still present on A.cordata (+8 percent) and appears on A.glutinosa (+15 percent), (Figure 4). A similar experiment layed out in 1982, and involving 3 provenances instead of 1, for the same 3 species, shows no superiority of the artificial inoculation in comparison with natural inoculation. This variation between years suggests that more work has to be done in this field. In 1983, we intend to use 2 Frankia strains, one of F.glutinosae and one of F.rubrae. A complete laboratory screening of the Alder x Frankia interaction involving more than 100 Frankia strains and 3 Alders species is now initiated by soil microbiologists of the Lyon University in cooperation with our Laboratory.

4.5 Volume growth of Poplar x Alder close-spaced pure and mixed plantings
Trial nr 7 in Table II has been planted in Spring 1980 at 1.5 x 2.5 m spacing. It consists of pure 72 sq.m. plots of a Black Alder provenance and of the Belgian Poplar clone "UNAL", and mixed plots of the two species. The objective was to study possible effects of a Nitrogene fixing species on the total production of the mixed stand compared to pure stands. Figure 5 shows an increasing volume-per-hectare superiority of pure Poplar plots compared to mixed plots. A literature survey suggests that the effectiveness of such species mixtures might be higher or might occur earlier with closer spacings. (4) (5).

5. CONCLUSION
This paper has reported progress of a 3-year EC research programme aiming at selecting and improving different forest tree species to produce biomass. As it had to be initiatedwith collections of base populations and producing of seedlings for field tests, the older experiments haven't had more that one year vegetation yet.
Any accurate result from these tests will have to be waited for till the first harvest. None destructive yield estimations which will give rankings of the different genotypes will nevertheless be made before harvest with help of production tables worked out by other contractors and by the INRA Laboratory of Sylviculture. In addition to field tests this programme needed development in two directions : (a) interspecific hybridizations within the Genera Alnus, Populus and Larix (b) practical use of the nitrogene fixing Alder x Frankia symbiosis. For part of this work, help was asked for from two Laboratories of the Lyon University. The first one deals with the detection of interspecific hybridization barriers in Poplars, the second one with a systematic screening of Frankia strains on different Alder populations.
After this initiation period, this programme definitely needs development in at least two directions :

(a) increase of the number of forest species which may have a high biomass production potential ;
(b) more systematic sampling of the regions where forest biomass could be produced and establishment of field tests in representative sites.

REFERENCES

1) AFOCEL, 1982.
Culture de biomasse ligneuse. Taillis à courte rotation.
Paris-214p.
2) AUCLAIR D. and CABANETTES A., 1981.
Methods for the estimation of above-ground biomass and biomass production in the classical coppice, and first results.
Energy from biomass. 1rst conference. Brighton. 216-221.
3) AUCLAIR D. and CABANETTES A. 1983.
Method for the estimation of biomass in coppice and coppice with standards, by conversion from "large timber" volume to total biomass.
Energy from biomass. 2nd E.C. Conference. Berlin 1982-222-225.
4) BORDERS B.E., 1980.
First year growth of two Populus clones grown in mixture with Alnus glutinosa. Masters degree thesis. Iowa State University. Ames. Iowa. 44p.
5) DE BELL D.S, RADWAN M.A., 1978.
Growth and nitrogene relations of coppiced Blackcotton-wood and Red Alder in pure and mixed plantings. Forestry Science Laboratory. PNWFRES. USDA For. Service. Olympia. Whashington.
(6) (7) (8) TEISSIER du CROS E., 1980, 1981 and 1982.
Selection and creation of forest tree material for biomass production. Establishment of short rotation coppices. CEC. Projet E. Energy from biomass. Coordination meetings of Contractors.
Amsterdam, 5p., Copenhagen, 4p. and Brussels 5p.

UTILIZATION OF COPPICE FORESTS BIOMASS FOR FUEL AND OTHER INDUSTRIAL USES

Author   :G. SCARAMUZZI

Contract number : ESE-R-035-I(S)

Duration   :36 months   1 July 1980 - 30 June 1983

Total budget   :Lit.519.500.000 CEC contribution: Lit.259,750,000

Head of project :Prof. G. Scaramuzzi, ENCC/SAF-Centro di Sperimentazione
                    Agricola e Forestale, Roma

Contractor   :Ente Nazionale Cellulosa e Carta

Address   :V.le Regina Margherita, 262
                    00198 ROMA

Summary

The project is part of a larger one promoted by the Italian Agency for Cellulose and Paper (E.N.C.C.) for an enhanced utilization of coppice forests in Italy. It concerns a Turkey oak (Quercus cerris) stand in Calabria. Stand characteristics and results of harvesting trials, first after-felling observations and technological investigations are reported. Machinery damage to stumps was greatly reduced (1.7% of high-damaged stumps) by sprouts concentration by skyline. Sprouts weight appeared as a major factor affecting harvesting yield. Harvesting costs varied from 27,795 to 39,502 It. liras/t with different stand and terrain characteristics. Because of the high transport costs, the use of current coppice forests biomass is restricted within a very short distance from the harvesting area. Industrial trials for fiberboard and paper pulp production from coppice whole-tree chipped biomass proved the possibility of its use in mixture with current raw material up to 75% for fiberboard production and up to 35-50% for corrugated paper pulp production. The biomass calorific value of the main species of italian coppice-forests was assessed, while an estimate of the expectable availability of coppice biomass for energy uses and an analysis of the conditions required for its economic use as an energy source are under completion.

1. Outline of the project

The background and objectives of the contract-project, which is part of a larger project of the Italian Agency for Cellulose and Paper (E.N.C.C.) were illustrated in the reports submitted to the previous coordination meetings.

The contract-project, concerning an experimental area in Calabria, was expected to cover the following aspects:
- silvicultural aspects, including assessments on site and forest cover and on machinery effects upon stumps and standards vegetative conditions, natural regeneration, stability and evolution of soil and vegetation;
- harvesting aspects, including felling, skidding and chipping trials of the obtainable stock;
- technological aspects, including assessments on the use of coppice biomass for energy, on the ground of its technological properties and of limitations to its industrial use.

2. Results obtained

2.1. The experimental area

As indicated in previous reports, the experimental area is located in the district of Cerva, Catanzaro (Calabria), within the Regional Forest Property, and is representative of widespread oak formations in southern Italy for which the conversion into high forest – even if feasible – does not prove convenient. In the case of the area under study, its non-convenience was mostly depending on the stand structure (presence of a large number of small- and very small-sized sprouts), which would have brought about conversion costs so high as to find in any case no counterbalance from a hypothetical increase of value of the resulting high stand.

The area is located at 800-1120 m a.s.l., with a south-west prevailing aspect and a slope gradient exceeding 25% up to 100%. Soil morphology is moved to rather rough, with rock outcrops even considerable.

Soils are of the acid-brown kind, more or less deep and developed according to physiographic position, slope gradient, aspect and coverage degree. They are usually open-textured and fairly provided with organic matter.

Mean annual rainfall is 1200-1300 mm, with a summer minimum of 85-90 mm. The Turkey oak (Q.cerris) prevails in the experimental stand, being associated with Italian oak (Q.farnetto), evergreen oak (Q.ilex), pubescent oak (Q.pubescens), European hophornbeam (Ostrya carpinifolia) and Italian maple (Acer opalus).

2.2. Stand characteristics

Sampling areas, each representing approximately one hectare, were used for stand investigations. The following characteristics were assessed (referred to net forest surface): age, about 30 years; stem density, 5,623/ha (133 standards, 599 seed-originated trees, 4,981 sprouts); stump density, 3,619/ha; basal area, 42 m^2/ha; standing volume (cormometric volume up to 3 cm-diameter), 230 m^3/ha; standing biomass, 292 t/ha fresh weight and 160 t/ha oven-dry weight; annual mean volume increment, 7.7 m^3/ha; sprouts b.h. mean diameter, 9 cm; sprouts mean height, 9 m.

The experimental area presents, therefore, a considerable standing volume, thanks to both the high stand density and stem height occurring, that are indices of the site fertility.

The following characteristics of the wood stock were also ascertained: weight distribution by diameter, 76% over 6 cm, 13% between 6-3 cm, 11% less than 3 cm; green specific gravity, 1,129 Kg/m^3; basic density (oven-dry weight/green volume), 655 kg/m^3; bark per-cent, 21.6% by volume, 19.3% by weight; moisture (referred to green conditions), 42%.

2.3 Harvesting trials

The whole-tree chipping system was adopted, with the following scheme of work:
- felling by motor-saw and piling up of whole sprouts
- concentration by skidding (on short distances) or by skyline (on long distances)
- hauling by skidder
- chipping of whole sprouts at the landing.

The whole-tree chipping system allows a greater amount of biomass to be harvested, which proved in this case over 30% higher than that obtainable by the traditional system, by which only material up to 6 cm diameter is utilized.

Within the harvesting operations carried out, three areas were selected, showing different stand density, stem size and terrain characteristics (area A: 6,186 stems/ha, mean stem weight 27 kg, slope gradient 25-30%, soil morphology moderately rough; area B: 5,088 stems/ha, mean stem weight 47 kg, slope gradient 25-60%, soil morphology rough; area C: 3,023 stems/ha, mean stem weight 80 kg, slope gradient 60-100%, soil morphology rough).

Harvesting operations being now completed, the overall main results achieved are summarized hereunder.

Felling and piling. This operation should keep into account the requirements of the machinery to be employed in the concentration and hauling phases. To this end, it is necessary for the sprouts to be felled in such a way as to have their basis up-slope, then bundles of plants being formed to permit a higher working yield by the machinery to be subsequently used. Two workers are employed in this operation, i.e. a chain-saw operator and an assistant, who provides for sprouts piling up.

The output recorded for the three different areas as well as percent working times of the various phases proved as follows:

	area A	area B	area C
- output, t/h	1.87	2.50	4.86
- percent times of various working phases:			
. felling	22.4	19.7	32.0
. moving	23.7	26.9	18.0
. piling	7.0	12.4	3.8
. stumps and undergrowth clearing	8.0	5.3	8.6
. motor-saw supplying and blade sharpening	20.6	15.4	16.8
. unprod. times and halts	18.3	20.3	20.8
	100 -	100 -	100 -

The higher output recorded for areas B and C are to be mostly attributed to the higher mean weight of sprouts, largely making up for the effects due to the lower stand density.

Concentration. Apart from the border strip, where it was carried out by tractor, the concentration of sprouts to the skid border was performed by means of two different types of skyline, a 'K 3000' Koller and a 'Mini Urus' Hinteregger, both self-powered but with a different steel spar height (7 m and 4.7 m, respectively).

Besides the skyline-operator, two workers are needed for hooking to the carriage the tree bundles arranged along the concentration line (both skylines have an automatic carriage with a capacity up to 1 t). A length of the cable way up to 350 m was used on slope of between 25 and 100%.

The working yield recorded for the different areas and for the two skyline types is set down hereunder:

		area A	area B	area C
Koller:	average distance, m	72.60	35.50	53.50
	average time, min	3'98	3'79	4'42
	average load, kg	175	239	266
	yield, t/h	2.64	3.78	3.61
	average distance, m	75.50	52.20	62.50
	average time, min	4'00	4'09	4'00
	average load, kg	206	255	217
	yield, t/h	3.09	3.74	3.26
mean:	average distance, m	74.05	44.00	58.00
	average time, min	3'99	3'94	4'21
	average load, kg	190	247	241
	yield, t/h	2.86	3.76	3.43

The above data show the highest average yield for area B, which is ascribable to the combined effect of a lower concentration distance and the greater sprout weight, with a resulting increase of the average load per run. On the other hand, the particularly high sprout weight for area C, because of greater piling difficulties, kept the load per run at the same level as for area B and therefore, due to the longer concentration distance, a lower yield was recorded here, intermediate between those of the other two areas.

No significant yield difference was recorded between the two skyline types.

Hauling. This working phase refers to the transferring of sprouts bundles, along the hauling track, from the concentration point to the chipping-landing.

It was carried out by means of a 'Timberjack 255 E' 75-kw tractor, provided with an 'Esco' grapple and a 70 m-rope dragging winch.

The yield recorded for the different areas is set down below:

	area A	area B	area C
- average hauling distance, m	150 -	240 -	350 -
- average hauling time, min	7'12	5'70	8'57
- average load, t	1.73	1.64	1.60
- yield, t/h	14.50	17.30	11.20

The higher yield for area B is to be mostly attributed to the greater sprout weight, resulting in a reduction of loading and unloading times (area A, 4'27; area B, 3'25); the even lower time recorded for area C (2'82) was here counterbalanced by the higher hauling time caused by the greater distance and, even more, by the high slope gradient of the terrain.

Chipping. For this operation a 'Morbark 12' 190 HP-chipping machine was used, equipped with a 2-blade disk.

Yield resulted 4.47 (A), 5.30 (B) and 7.0 (C) t/h for the three areas respectively, the increasing yield being ascribable to the increasing size of sprouts.

Harvesting costs. A summary of the yields recorded for the various harvesting phases in the three areas is given below (t/h):

	area A	area B	area C
- felling and piling	1.87	2.50	4.86
- concentration	2.86	3.76	3.43
- hauling	14.50	17.30	11.20
- chipping	4.47	5.30	7.04

Percent times of the various working phases proved as follows:

	area A	area B	area C
- felling and piling	45	44	28
- concentration	30	29	40
- hauling	6	6	12
- chipping	19	21	20
	100	100	100

A 55.5% of the total harvesting time was covered for areas A and B by the three phases with a higher mechanization (concentration, hauling and chipping) as against 44.5% represented by the single phase of felling and piling. Their weight increased to approximately 70% for area C because of the higher concentration and hauling times recorded.

Single operations and total harvesting costs for the different areas proved as follows[1] (It. Liras/t).

	area A	area B	area C
- felling and piling	8,900	6,700	3,400
- concentration	12,350	9,350	10,300
- hauling	2,700	2,250	3,500
- chipping	9,400	7,950	5,950
- loading on lorry	700	700	700
- total harvesting cost	34,050	26,950	23,870
- forest road construction	300	300	300
- miscellaneous expenses, 15%	5,152	4,087	3,625
	39,502	31,337	27,795

---

(1) For the calculation of machine operational costs the following simplified relation suggested by the British Forestry Commission was used: hourly costs = 3 x machine cost/life time of machine.

A 30% financial support of the machine cost is obtainable from the State for Southern Italy, which cuts down harvesting costs of about 15%.

2.4 After-felling observations

As indicated in previous reports, some permanent sampling areas were selected in the harvested part of the stand, being representative of different conditions of concentration systems and soil fertility, with a view to evaluating machinery damage to stumps and standards and its effects on the resumption of vegetative activity and on wood increments, soil and vegetation.

The first observations carried out proved more considerable injuries to stumps when concentrating by tractor (border strip) than when operating by skyline (14% of seriously damaged stumps vis-à-vis 1.7%). Ten to fifteen per cent of the standards left standing were high-damaged and so eliminated.

2.5 Investigations on the use of coppice biomass

Industrial trials of fibreboard and pulp production from the harvested coppice biomass were performed within the framework of the larger E.N.C.C. project. They proved the possibility of utilizing whole-tree chipped material in mixture with current raw material, up to 75% for fibreboard production and up to 35-50% for corrugated paper pulp production.

Biomass calorific value for the main species of Italian coppice forests were assessed, resulting values being as folllows (kcal/kg):

	wood	bark	overall
Acer opalus (Italian maple)	4.685	4.226	4.641
Castanea sativa (Chestnut)	4.810	4.251	4.752
Ostrya carpinifolia (Eur.hophornbeam)	5.050	4.260	4.972
Quercus cerris (Turkey oak)	4.801	4.683	4.783
Quercus farnetto (Italian oak)	4.925	4.508	4.837
Quercus ilex (Evergreen oak)	4.481	4.115	4.438
Quercus pubescens (Pubescens oak)	4.773	4.307	4.660

An estimate of the expectable availability of coppice biomass for energy uses and an analysis of the conditions required for its economic utilization as an energy source are presently under completion.

2.6   Concluding remarks

Technical results of the project were very good: machinery damage to stumps was very low, while damage to standards left standing may be overcome leaving a slightly higher number of stems during felling stage and subsequently reducing it to the number required.

The harvesting system adopted proved also economically feasible. However, the high transport costs, under the present price conditions, restrict coppice biomass use within very short distances from its harvesting area.

Industrial trials performed showed the possibility of using oak coppice whole-tree chips in mixture with current raw material for fibreboard and corrugated paper pulp production. However, industrial use of coppice biomass appears to be greatly restricted by the geographic distribution of industries and consequently relevant opportunities for energy uses may occur. An estimate of its expectable availability will be included in the final report.

Scaling up and extension of investigation to different coppice species and harvesting operational conditions are needed to substantiate the results obtained.

THE PRODUCTION OF ENERGY FROM SHORT ROTATION FORESTRY

Authors	:	M. NEENAN & G. LYONS	
Contract Number	:	ESE-R-036-EIR H	
Duration	:	36 months	1 July 1980 - 30 June 1983
Total Budget	:	£Ir 398,177	CEC contribution £Ir 173,800
Head of Project	:	Dr. M. Neenan, Agricultural Institute, Oak Park, Carlow, Ireland.	
Contractor	:	An Foras Taluntais (translated - Agricultural Research Council)	
Address	:	19 Sandymount Avenue, Ballsbridge, Dublin, 4.	
Sub-Contractor	:	Forest & Wildlife Service 1, Leeson Lane, Dublin, 2.	

Summary

Some 25 field experiments were carried out and these were followed through with laboratory tests on dry matter, calorific values, and specific gravity.

The results have been evaluated in a specially developed systems analysis program. On the basis of 1983 costs, on improved blanket peat, mined out fen peat, or low grade farm land, short rotation forestry based on Salix or Populus coppice is cost effective. The critical factors identified are spacing at $1m^2$ or less, coppicing, in the first year followed by one heavy application of fertiliser, and adequate weed control in the first and second year. Hill land and virgin peat will produce biomass economically from coniferous species on a 7 to 10 year rotation. Little difference has been found in the calorific value of species. Dry matter content varies from 40-55% in the growing wood, and when stored as chips decreases rapidly to 30% and more slowly to 25%. A domestic boiler based on wood chips has been developed and is now undergoing final testing.

1.1 Introduction
The objective of the research was to optimise the various factors of production so that the energy produced would be competitive in price with oil or coal. The main costs in production are land, land development, planting material, labour costs of planting and harvesting, fertilisers, and pesticides.
The project consisted essentially of 5 parts:
(i)   Availability of suitable land
(ii)   Species by soil type interactions
(iii)   Silvicultural practices which increase yield or decrease costs
(iv)   Calorific properties of the fuel
(v)   Economic evaluation of the systems of production

2.1 Materials and Methods
On the basis of the F.A.O. Soils Map of Europe and the Irish Soil Survey, and taking into account the Disadvantaged Areas of the EEC, four major types of land were initially chosen as having possibilities for energy crops.
The final selection of sites was made in consultation with the Dept. of Forests & Fisheries who are subcontractors in this programme.
In addition, three other types of land of which large areas exist were investigated, mined out peatland (80,000 ha by the end of the century), improved alpine peatland, and low grade agricultural land [1].
The first major series of trials on this was undertaken in 1977 by the Forest and Wildlife Service. Four locations were chosen as being representative of these site types. These were:-
(1) Old Red Sandstone: Mountain and hill land; 33% of the total land area of the country and currently used for rough grazing.
(2) Surface-water gley: Wet mineral hill land; 21% of the land area, currently used for agriculture (cattle and dairying) at low-return level.
(3) Raised bog (Fen): Approximately 5.7% of the area. These are currently used mainly for industrial scale peat production.
(4) Deep blanket peat: Occurs mainly along the exposed western seaboard and at high elevations on the mountain ranges. Main use is hand peat production.
Afforestation would be an economic option for the first three categories of soils and, with soil improvement, for the deep blanket peat.

2.2 Species
The following species were used:- Alnus glutinosa, A.rubra, Betula pubescens, Castanea sativa, Populus trichocarpa, Populus hybrid Rap, Salix aquatica gigantea and Salix viminalis, Picea sitchensis and Pinus contorta.
Spacings used were those considered most appropriate for the species on these soils. Soil preparation methods have been described in an earlier report [2].

3.1 Results
Table I shows the above ground dry matter yields attained after five and six growing seasons respectively in situations where results are of economic significance [3].

TABLE I : Above ground dry matter (tonnes/hectare) after 5 and 6 growing seasons respectively (mean annual dry matter increments in brackets). In the case of conifers the yield includes the foliage. All data refer to single stems, not coppiced.

	Soil Type					
	O.R.S.		Gley		Raised Bog	
Species	5 yrs.	6 yrs.	5 yrs.	6 yrs.	5 yrs.	6 yrs.
Lodgepole pine	18.4 (3.7)	34.4 (5.7)			17.6 (3.5)	51.0 (8.5)
Sitka spruce	13.1 (2.6)	28.2 (4.7)			10.8 (2.2)	22.2 (3.7)
Alder (A.rubra)	F		6.7 (1.3)	12.8 (2.1)	F	
Poplar T (MB)	F		14.7 (2.9)	17.1 (2.8)	F	
Poplar TT 32	F		6.9 (1.4)	8.8 (1.5)	F	
Willow (S.aquatica gigantea)	F		8.2 (1.6)	7.8 (1.3)	F	

The results indicated that the performance of many species which has proven satisfactory on better quality sites was severely limited by nutrient status and/or adverse climatic factors such as elevation and/or exposure. The performance of the broadleaved species (which are also the coppicing species) on most of these sites has not been good.

Of the coniferous species, lodgepole pine (coastal) is clearly the best biomass producer, in these situations. Growth of the species in the sixth growing season has been good especially on the raised bog site. The mean annual overground biomass increment of 8.5 tonnes/hectare after six years is more than double the mean annual increment over the first five years (3.5 tonnes/ha/annum). It is considered that the maximum rate of increase may not yet have been attained, but a satisfactory yield should be reached by year 10. Picea sitchensis followed a somewhat similar pattern.

The techniques involved in these investigations have been successfully incorporated in the Biomass Demonstration Project.

3.2 Species soil type trials by An Foras Taluntais

These were carried out at 3 centres, two on peat and one on mineral soil. The latter was laid down in 1977 and covered a range of 13 species/clones. The results, which have already been reported elsewhere [4] have shown that yields of 15-18 tonnes ha^{-1}annum of dry matter can be obtained from coppice. In 1978 more extensive trials on species and growing techniques were laid down on a mined out fen peat site provided by Bord na Mona. In 1979 a further trial was laid down on improved western type of blanket peat. In all some 25 field trials were initiated. The more useful results obtained are given in Table II.

TABLE II: Main results from field trials. Yields are in oven dry tonnes per hectare per year equivalent. Unless otherwise indicated the planting material consisted of one year old rooted plants.

EXPERIMENT	BRIEF RESULT
Species trial on low grade farm land	Populus - 15-18 t ha^{-1}annum Salix spp. 15.3 do. Alnus spp. 3.4 do.
Planting density (coppice) on low grade farm land (m)	Populus @ 0.6x1.0 - 18.0 t Salix do - 4.9 t Populus @ 0.3x1.0 - 15.5 t Salix do - 8.3 t
Coppice yield as % of primary yield	Populus: 127-129%; Salix a.gigantea: 250%; Alnus cordata: 69%
% Establishment from cuttings on mined out fen peat	Salix a.gigantea 86-96% Populus sp. 72-91%
Coppice trial on mined out fen peat	Salix aquatica gigantea - 4-5 years 13-16 t ha^{-1}annum
Nelder trial with S.viminalis on mined out peat	Optimum spacing 0.9m^2 to 1.8m^2 per plant. Result complicated by weed infestation and leaf litter yield.
Spacing trial (randomised blocks)	1.089 m^2 - 1.49 m^2 - 4.0 t ha^{-1}annum at 3 year stage primary growth on worst area of mined out peat
Salix species x spacing trial on improved peat (Co. Mayo)	3.7 t ha^{-1}annum^{-1} primary yield 7.7 t ha^{-1}annum^{-1} 1st coppice yield from Salix aquatica gigantea
Plants surviving coppicing (various trials)	Salix 75-95%; Populus 99%.
Dry matter content of coppice wood	40-45% in winter
Fertiliser trial NPK factorial on Salix vittelina	$N_3P_3K_3$ N_3P_3 N_3 P_2 Response to nitrogen 160% of control

3.3 Weed Control

Many instances of poor yields can be attributed to ineffective weed control during the first two years of growth. Peatland normally requires up to twice the normal rate of soil applied herbicide. In many instances the tolerance of the woody species to these higher rates has not been verified. It has been found that Salix a.gigantea will tolerate Simazine, Atrazine, Oxadiazon, Propyzamide, Terbutylazine and Chlorthiamid [5]. The most economic treatment is Simazine at 3 kg ha^{-1} of a.i. followed by a directed application of Paraquat during the first and second years.

Juncus effusus, which is one of the more serious weeds, reduces yield by 20-23%. If this occurs in the first or second year the full potential yield will not be obtained. It can be controlled by Paraquat alone at 0.84 kg a.i. per ha in summer and 0.30 kg a.i. in winter.

3.4 Alnus species

Fertiliser nitrogen is a major cost. It may be more economic to obtain a moderate yield from a nitrogen fixing species such as Alnus rather than a high yield from a species such as Populus which has a high fertiliser requirement. A major investigation was initiated on the botanical and microbiological aspects of Alnus.

Four species were planted in a field trial on mined out peat in 1980 [1]. Recovery after coppicing was almost nil for Alnus rubra, and rather poor for A.cordata. An attempt is being made to vegetatively propagate by tissue culture those plants of A.rubra and A.cordata which will coppice. Selections of a shrub type A.incana are also being propagated in this way.

Studies on the propagation of Alnus glutinosa, A.incana, A.cordata and A.rubra by means of softwood cuttings and using 0.08% indole butyric acid has resulted in 50-80% rooting success. The use of smaller sized cuttings for the propagation of Populus is also being investigated [5].

A collection of 15 Alnus species and hybrids was established at Kinsealy. These include A.cordata, A.incana, A.glutinosa, A.rubra, A.japonica, A.rhombifolia, A.tenuifolia, A.orientalis, A.rugosa, A.subcordata, A.matsumurae, A. x spaethii, A. x cordinca, A.serrulata, A. hirsuta, A.viridis, A.nitida, and A.crispa [6].

Studies have been undertaken on the occurrence of nitrogen fixing organisms in Irish soils, and on the possibility of inoculating plants with highly effective nitrogen fixing strains of Frankia spp [7]. Some preliminary results are given in Table III.

TABLE III : Nitrogen fixation by Alnus and % occurrence of effective nodules on plants taken from different soils

	Kg ha^{-1}annum^{-1} N fixed	Occurrence of effective nodules	%
A. incana	212	Good mineral soil	50
A. glutinosa	120	Mined out peat (Clonsast)	42
A.cordata	115	Western blanket peat (Glenamoy)	25
A.rubra	105	Gley soil from Forestry trial site	16

3.5 Tissue culture

A number of systems were initiated for micropropagation of Populus, Alnus and Salix [6]. By culturing buds of Populus, multiple shoot formation was induced on Murashige & Skoog medium with benzyladenine (1.0 mg/l). An average of 5 shoot primordia was obtained per 6 mm long stem-internode. Production of multiple shoots from buds and internodes offers the potential for large scale build up of clones where original material is scarce.

Plants recovered from buds, stem explants and cell cultures were transferred to soil reaching 2.6m after two seasons. Procedures for generating genetic variation in Poplar from existing superior clones using gamma rays were initiated. Hybrids of P.tremula x P.tremuloides obtained from Prof. H. Muhs (Germany) were germinated in agar. Approximately 450 plants were obtained and will be ready for field trials at the end of 1983.

Procedures are being developed to clone individual trees which show good coppicing ability. Cultures were established from selections of the following material (a) 20 year old trees of A.rubra and A.incana, (b) 4 year old trees of A.incana grown at Clonsast, and (c) coppiced trees of A.rubra grown at Clonsast.

Buds of Salix aquatica gigantea were cultured. Conditions for multiple shoot formation remain to be established. A collection of native willows has being made, and these are being propagated by conventional procedures.

3.6 S.R.F. Fuel Analysis

Moisture content, specific gravity, calorific value, ash and % volatile constituents of tree components, as well as bulk density and particle size of chipped wood were determined. Some results are shown in Fig. 1.

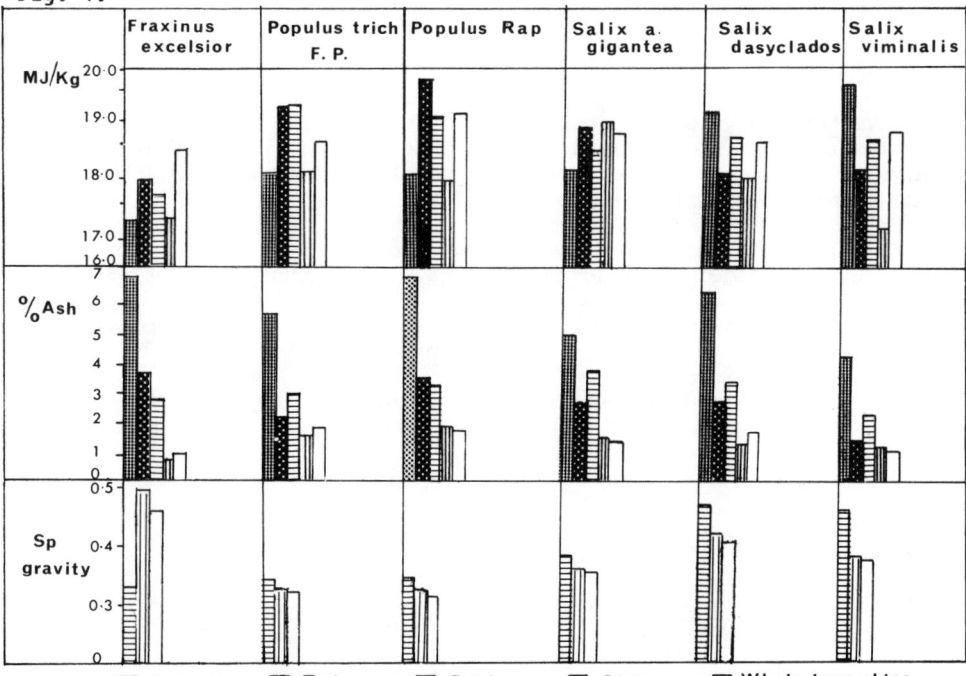

FIG.1: Calorific value, ash content and specific gravity (wet volume/dry weight basis) of 6 species of short rotation wood.

For specific gravity, ash content and calorific value tests, differences between species were statistically significant at the 0.001 level. Differences between components were also significant at this level, except in the variation of specific gravity, where differences were not significant for several species.

3.7 Moisture content and storage of chipped wood fuel

Moisture content can vary from about 30 to 45% depending on the species, age of trees and other factors.

As sufficient short rotation fuel was not available from experimental plots, material for a moderately sized storage experiment was freshly harvested coppice of young (8-10 years) ash (Fraxinus excelsior). Figure 2 illustrates the drying curves obtained from the mean of seven chip pile samples for three distinct storage seasons.

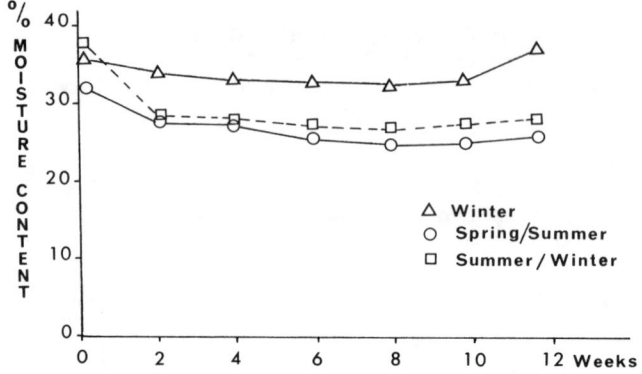

FIG.2 : Drying curves of chipped wood fuel, stored in the open.

3.8 Systems analysis

The results from the field experiments to date were used in a systems analysis for nine scenarios. Labour and other costs were as in May 1983. The results are given in Table IV.

TABLE IV : Estimated cost of producing wood fuel based on physical data obtained in field experiments, at discount rates of 4% and 8%.

Scenario	Species & planting method	Yield D.M. t ha^{-1}annum^{-1}	Cost per TOE ($) 4%	8%
1	Salix a.gigantea (0.6 x 0.6m) cuttings	16.2	68.17	85.21
2	Salix a. gigantea (1.0 x 1.0m) plants	13.0	68.62	84.04
3	Poplar Rap (1.0 x 1.0m) plants	4.0	153.52	205.75
4	Salix a.gigantea or Poplar Rap (1.0 x 1.0m) cuttings	12.0	50.23	55.78
5	Poplar plants (1.0 x 0.6m)	15.5	95.59	125.58
6	Poplar plants (1.0 x 0.3m)	18.0	137.31	188.78
7	Salix a.gigantea (1.0 x 1.0m)	12.0	94.18	122.33
8	Pinus contorta (2.0 x 0.3m)	8.5	148.72	163.77
9	Pinus contorta (0.8 x 0.8m)	5.7	223.11	249.33

These values should be regarded as tentative, because although many of the parameters such as yield have been established, some costs, as for example planting material, may have been overestimated. The price of a rooted plant was taken as 22p and that of cutting 6.5p. This compares with a cost of about 0.05p per plant where conifers are mass produced. If one attributes the same value to the cuttings obtained after one year, then the nett cost of planting material for s.r.f. would be nil.

4.1 Conclusions

The cost of production of all agricultural crops has escalated considerably over the past 7 years, as oil prices have become reflected in the various input costs. Nevertheless, most of the scenarios show the possibility of wood fuel being competitive with oil. On the basis of the experience gained in these investigations input costs can be greatly reduced and yields increased. In the case of the Forest & Wildlife trials soil preparation was more elaborate and more expensive than in commercial practice. This was necessary to ensure a positive result. In the short rotation experiments, inputs were also at higher rates than necessary, in many cases at excessive rates, or were of the wrong type (e.g. herbicides and fertilisers). New techniques such as the use of refrigerated material has enabled a high success rate to be achieved in establishing plantations from cuttings.

In these costings cuttings and rooted plants of the short rotation species are put in at an estimated cost and no allowance is made for revenue generated from cuttings obtained after the first harvest. In hindsight, it seems coppicing should begin at the end of the first year.

The drying tests show that moisture decreases fairly rapidly to about 30%, which is satisfactory for turf burning boilers. The calorific value of species and plant components is very similar which means that the species can be adapted to the soil type available.

References
[1] NEENAN, M. & G. LYONS (1980) (Editors). Production of Energy from Short Rotation Forestry. An Foras Taluntais, 19 Sandymount Ave., Ballsbridge, Dublin 4.
[2] McCARTHY, R. (1980). Production of energy from Short Rotation Forestry. p.29-44. Edited by M. Neenan & G. Lyons
[3] CONDON, L. (1983). Dept. Forests & Fisheries, 1 Leeson Lane, Dublin. Personal communication, May 1983.
[4] NEENAN, M. (1982). Short Rotation Forestry as a Source of Energy and Chemical Feedstock. 2nd EC Conference Energy from Biomass. Ed. A. Strub, P. Chartier & G. Schleser. Applied Science Publishers. p.142-146
[5] KELLY, J.C. (1983). Agricultural Institute, Kinsealy, Malahide Rd., Dublin 5. Personal communication, May 1983.
[6] DOUGLAS G. (1983). Agr.Institute, Kinsealy. Personal communication, May 1983
[7] O'NEILL, P. & P. MURPHY (1983). Agric. Institute, Johnstown Castle, Wexford. Personal communication, May 1983.

Acknowledgement
The authors wish to thank Bord na Mona for providing land for the experiments, Mr. J. Devereux and Mr. N. McNamara for technical assistance.

SHORT ROTATION FORESTRY HARVESTER CHIPPER

Author	:	Bernard J. Keville/ Edward J. Devenish
Contract Number	:	ES-E-R-019-EIR(N)
Duration	:	36 months 1.7.1980 1.7.1983
Total Budget	:	IR£278,350
CEC Contribution	:	IR£ 80,720
Head of Project	:	B. J. Keville Chief Design Engineer Bord na Mona.
Contractor	:	Bord na Mona
Address	:	Lower Baggot Street Dublin 2.
Sub-Contractor	:	Irish Sugar Co., Carlow, Ireland.

1.1 Summary

The Biomass Short Rotation Harvester which was designed and built in Ireland as part of the E.E.C. biomass programme has been undergoing a range of tests during the past year. The tests have been carried out on both single stem and coppice plantations. The machine was specified to produce billets approximately 150 mm long from material up to 85 mm dia. A schematic layout of this machine is shown in Figure 1.

The principle elements in the machine design are:
(a) Cutting and intake system
(b) Billeting mechanism
(c) Conveyors
(d) Tipping Hopper
(e) Tracks.

A full description of these elements has been given in a previous report (1).

Some of the planned tests could not be carried out fully because of the scarcity of suitable biomass material. This scarcity was due to the failure of the tree growing programme to achieve projected targets on cutaway peatland.

However, sufficient results have been obtained to indicate that a major re-assessment of the machine design and specification is required.

1.2 Introduction:
As part of the E.E.C. Biomass Programme, a Short Rotation Harvester was designed and built to cut, collect and billet single stem and multi-stem material. Preliminary tests were carried out on this machine during February and March 1982.

A decision was made at that time to undertake a more comprehensive series of tests during the remainder of 1982 and into 1983 when more biomass material would become available. The results of these tests are shown in paragraph 1.4.

In the course of this programme some modifications were carried out to the machine. The details of these modifications are shown in Figure III.

1.3 Test Programme:

Location	:	Cutaway peatland plantation at Clonsast, Co. Laois.
Test period	:	April 1982 to March 1983
Crop 1	:	Single stem poplar plantation, 3, 4 and 5 year olds with diameters up to 75 mm, row spacing of 2 m and tree spacing of 1.5 m.
Crop 2	:	Coppice Salix aquatica gigantea 3, and 4 year old plantation with stem diameters up to 50 mm, row spacing of 1.5 m and tree spacing of 1 m.
Tractor normal operating speed:		2.5 km/hr.
Tractor power	:	98 KW

1.4 Performance Tests
Single Stem

The harvester performed the cutting, felling, collection and conveying on the single stem poplars successfully at this speed. However, problems were experienced with the billeting process and these were as follows:-

(i) Incomplete billeting of the tree top section where small lateral branches occur.

(ii) The billeting shears losing their synchronisation.

The billeting mechanism is shown in Figures II and III and consists of two counter rotating square sectioned shafts with four adjustable chisel edged blades bolted in position. The shearing mechanism is set so that each revolution of the shafts produces four billets and the gap between matching chisel edges is set at 2 mm.

The billets are formed by a combined squeezing, wedging, shearing action.

When a tree top section with lateral branches passes into the billeting mechanism the shafts tend to deflect. Only the tree main stem is severed, but the lateral branches remain intact. These branches restrict the conveying of the billets to the harvester hopper. In addition they bind on the rotating elements of the harvester and seize up the mechanism.

When the two billeting shafts lose their synchronisation as shown in Figure IV the tree stems are not fully severed and therefore, cannot be conveyed to the hopper.

Coppice

When the harvester trials on the coppice or multi-stem plantations commenced the billeting problems were repeated but cutting and collecting problems were more serious. Figure V shows a schematic layout of the cutting and intake system.

When the harvester was operating within the dense plantation rows, the primary feed drums were unable to clear the felled material fast enough after it was cut. This resulted in a large volume of stems gathering at one point at the bottom of the feed drums and becoming entangled in the guards surrounding the cutting discs. The thin flexible lateral branches also tended to wrap tightly around the secondary feed drum and other rotating elements.

Since the processing elements of the machine are protected by slip clutches, no serious damage was caused and the machine had to be cleared manually. Modifications were carried out to the guards surrounding the cutting discs and were moderately successful in relieving the problem.

Further problems were incurred within the Salix plantations because the lateral branches of adjacent trees became entangled.

1.5 Modifications

Modifications were carried out to the cutting and intake arrangement and to the billeting mechanism. The safety guards surrounding the cutting blades were modified to prevent the tree stems from becoming entangled at the front of the machine. In addition the speed of the primary feed drums was increased to improve the flow of the cut material.

The billeting mechanism drive shafts were increased both in diameter and length. They were then geared together and the machine frame was stiffened to prevent deflection so that once synchronised and set, they would maintain their position relative to each other.

The steel chevrons on the secondary feed drum were also modified in an effort to prevent the thin lateral branches from binding on it.

When the harvesting trials re-commenced the modifications to the intake system had improved the cutting and collection of the material but the problems were not completely eliminated. The performance of the billeting mechanism was greatly improved but the thin lateral branches continued to escape between the cutting edges and were not severed by the billeting mechanism.

The binding of the thin branches on the secondary feed drum was much reduced.

1.6 Boiler Tests
Generally, the lateral branches have much smaller diameters than the main billets and due to the weakness in the billeting and flailing design, the lateral branches remain intact. Hence material will not readily flow by gravity to the boiler from an overhead bunker. Figure VI is a schematic diagram of a typical chain grate boiler and its associated gravity feed bunker. This is the accepted method of feeding Sod Peat fuel to boilers in Ireland. Because of the billet dimensions, screw conveyors of conventional design cannot be utilised for stoking boilers.

Much discussion has taken place with the Electricity Generating Company who are performing boiler feed trials using these billets. However, the trials to date have not been successful since the lateral branches breech and pack within the feed hopper.

1.7 Results and Conclusions
The Short Rotation Harvester is capable of harvesting and billeting within both single and multi-stem plantations with yields of 6 tonnes per hectare of dry matter with moderate success. However, further design modifications will be required to help alleviate the persistent problems of harvesting in coppice plantations and the billeting of tree stems with lateral branches.

Figure VII shows design proposals for these modifications. The proposed vertical taper augers positioned in front of the primary feed drums will spread the cut and felled material vertically to prevent it from gathering at the lowest point of the harvester and thereby regulate the feed into the machine.

The vertically positioned barrel cutting drum will separate the plantation's thin branches, from the branches in neighbouring rows by cutting a path through them.

The billeting mechanism would be replaced by a guillotine arrangement where the process material thickness will not be a limiting factor.

These modifications have not been carried out, since the dry matter yields to date fall short by about 50% of the viable targets on cutaway bog areas.

Since the boiler feed arrangement is by gravity from its associated bunker, the lateral branches hinder the natural flow and create breaching problems. Therefore, the lateral branches must be removed for successful boiler feeding.

The Electricity Supply Company has carried out large scale tests using wood chips as a fuel with good results and comparable with peat fuel from a gravity stoking and combustion view point.

1.8 Billet Storage
When the billeted material had been stored in the field after harvesting much fungi had developed and was visible on the billets and throughout the stock pile. This would suggest that sufficient natural air convection had not occurred within the stock pile. The storage period was of six months duration. However, there was no incidence of

fungi growth on the one month old stock piles. These results indicate that the billeted material does not facilitate solar drying to a large extent and would suggest that an alternative harvesting system should be considered. This view is also reinforced, if dense plantations with an economic viable yield target of 12 tonnes per hectare are to be taken into account, since the existing machine will not harvest plantations with a yield in excess of 6 tonnes per hectare of dry matter.

1.7 Recommendations

Because of the disappointing results obtained todate, a review of the harvester design and specification should be undertaken. Unless fairly major modifications are carried out, it is unlikely that reliable performance will be obtained or that an acceptable end product will be produced.

Since the utility company has expressed dissatisfaction with the billeted material both as a fuel and because of problems in boiler feed, it is recommended that the concept of billeting should be re-assessed in favour of a chipped end product.

The alternative harvesting process would then be as follows:-

(i)   Cut and fell the trees

(ii) Bundle the trees

(iii) Store until required

(iv) Transport to destination (Utility Company).

(v)   Chip and feed to boiler for combustion.

Reference

(i)   Energy from Biomass

      Proceedings of the E.C. Contractors' meeting held in Brussels, 5-7 May, 1982.

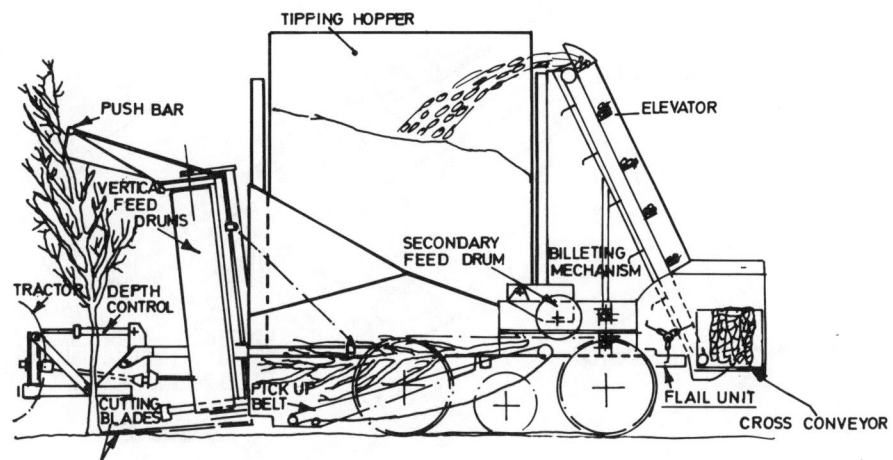

SCHEMATIC LAYOUT OF BIOMASS HARVESTER

FIGURE 1

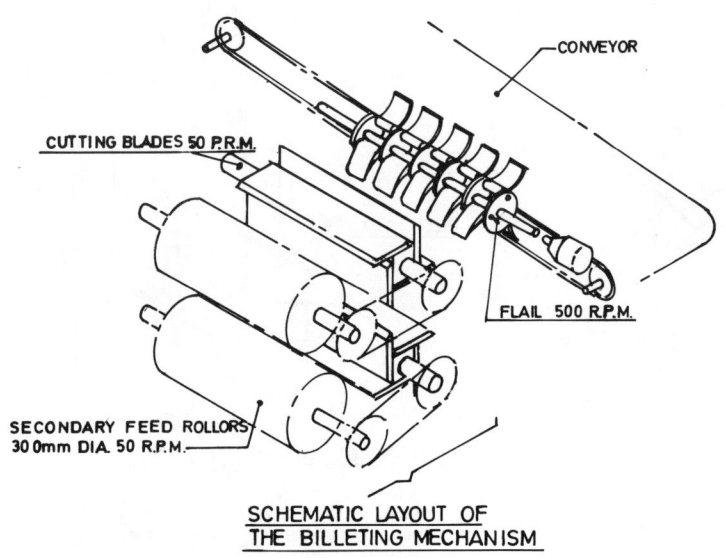

SCHEMATIC LAYOUT OF
THE BILLETING MECHANISM

FIGURE 2

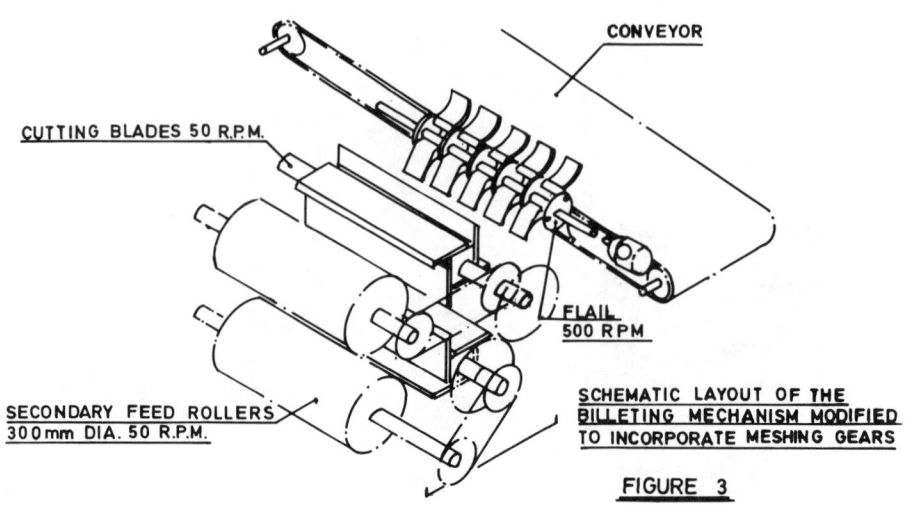

SCHEMATIC LAYOUT OF THE BILLETING MECHANISM MODIFIED TO INCORPORATE MESHING GEARS

FIGURE 3

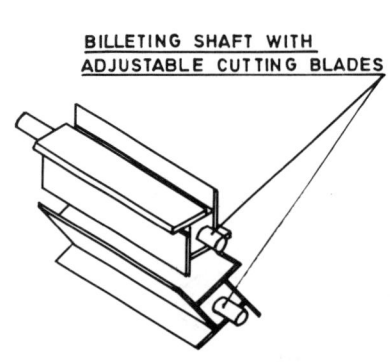

SCHEMATIC LAYOUT OF BILLETING CUTTING BLADES OUT OF SYNCHRONISATION

FIGURE 4

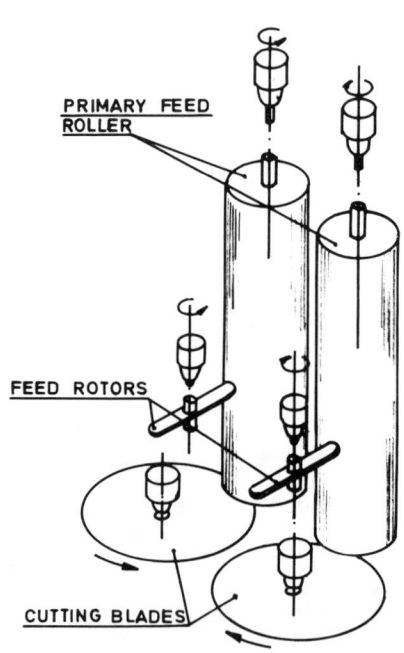

SCHEMATIC LAYOUT OF THE CUTTING AND INTAKE SYSTEM OF THE SHORT ROTATION HARVESTER

FIGURE 5

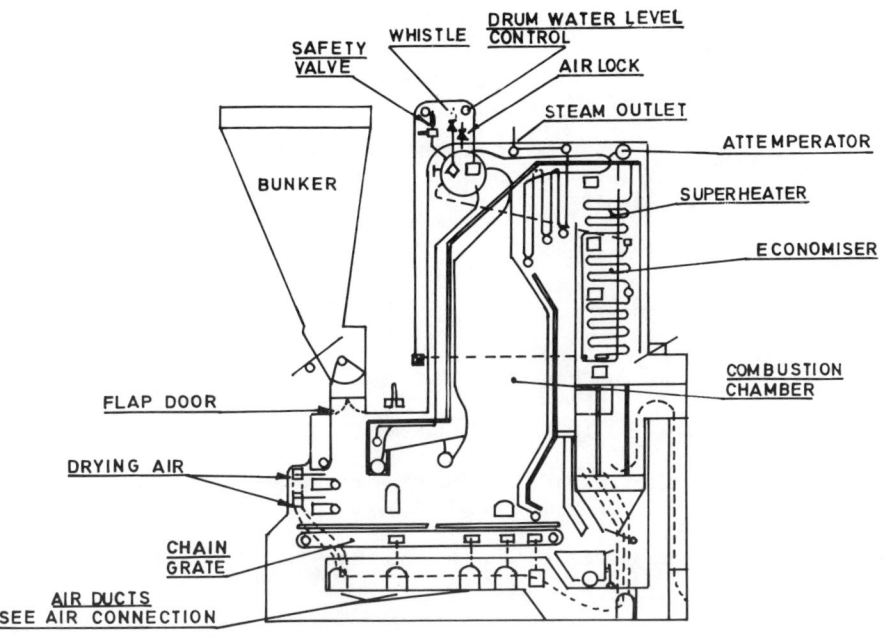

FIGURE 6

SECTION THROUGH A 5 MW TURF BURNING BOILER

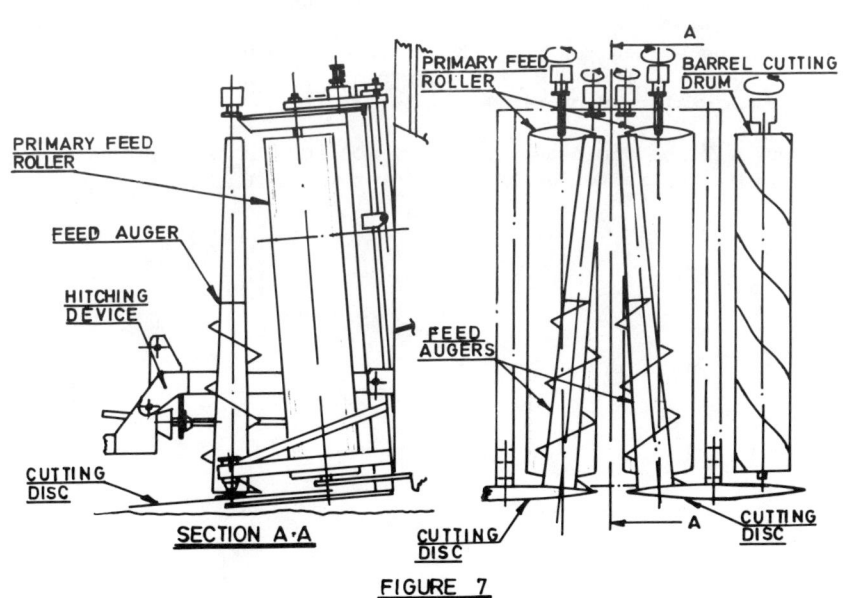

FIGURE 7

DESIGN AND BUILDING OF A FORESTRY WASTES HARVESTER

Contract number	: ESE-049-F
Duration	: 24 months
Total budget	: 100 000 units of account. ECC contribution : 50%
Head of project	: LUCAS Jean, Chief Engineer - rural engineering (GREF) at the CEMAGREF
Authors	: B. BONICELLI, P. BARBE, F. SEVILA
Contractor	: Centre National du Machinisme Agricole, du Génie Rural des Eaux et des Forêts.
Address	: CEMAGREF - Parc de Tourvoie B.P. 121 - 92164 ANTONY

Summary :

A polyvalent frame to harvest and clean under chips or pellets any kind of forest production or wastes has been determined. A choice has been made between two solutions : a cutting system mounted on a conveying arm out of special tractor and a horizontal drum included at the front of 6 wheels suspended chip and carry tractor; it has a 100 KW engine, a slope correcting suspension and a hydro-mecanic transmission which powers and steers the unit. Its 4 m3 container can be discharged as high as 2.5 m in commercial road carriers.

Its cost will be approximately of 80 000 ECU; it will produce chips for 80 ECU/TOE compare to 130 ECU/TOE for fire wood in France.

The purpose of this project is to develop a very polyvalent
frame to be used in forest to clean young plantation, to harvest
coppices and to collect harvest wastes.

It has to be :

. of a reduced size to move easily between lines of growing trees
  and seed bearers (1.6 m wide; 2.8 m high)

. easily manouvrable to avoid standing trees on any kind of ground
  (front slope 70 %, lateral slope 25%, 7 tons maximum weigh)

. a carrier to have the chip and carry operation on the same circulating frame (2 tons of load)

. able to self unload into commercial road container (2.5 m above
  ground)

. able to be driven as rapidly as possible on roads from one plot
  to the other

. able to condition the product into chips or pieces compatible
  with the use of an automatic feeding of a domestic boiler.

With these constraints we end up with two configurations :

1/ Felling and chipping tool mounted on a conveying arm :

The motive power frame substructure is composed of an hydraulic
unit (a 80 KW-diesel engine and a hydrostatic transmission) with
continuous speed variations toward two powering and steering axles
attached on a simplified frame containing the power unit. A turret
includes the driving post, the tool-handling arm, the wood pieces
container.

The structure is similar to a for driving wheels tractor and has
several advantages compared to existing machines such as : large
versatility of applications and motions in a forest environment.

The design of this structure has been an important task in the
first part of the project. Although it lead us to a very evoluted
4 powering and steering wheels unit of great interest, the choice
of the main components such as the driving axles in the available
production has determined far too big sizes to meet our constraints.

2/ Horizontal drum included in the machine

The frame is based on a non directive wheel suspension drive set.
The motion of 6 wheels is achieved by a chain transmission and two
hydraulic chain wheel drive separately fed by variable flow pumps.
The overall suspension is designed in order to :

. absorb high mechanical constraints

. secure stability through an automatic correction of slope (lateral and frontal)

The frame is made out of two horizontal beams. Each one has three holes for the upper axle of the "leg" supporting each wheel. At the bottom of the frame is mounted a vertical fork which substain the container when it is discharged.

Following schemes describe this equipment we have decided to built and which is under its testing procedure.

Its cost will be approximately of 80 000 Ecu. It is designed to produce on the average 25 tons of humid product per day.

This gives us a probable price of 80 ECU/TOE for the harvested wood, compare to 130 ECU/TOE for fire wood in France.

DETAIL OF THE FRAME OF THE HARVESTER

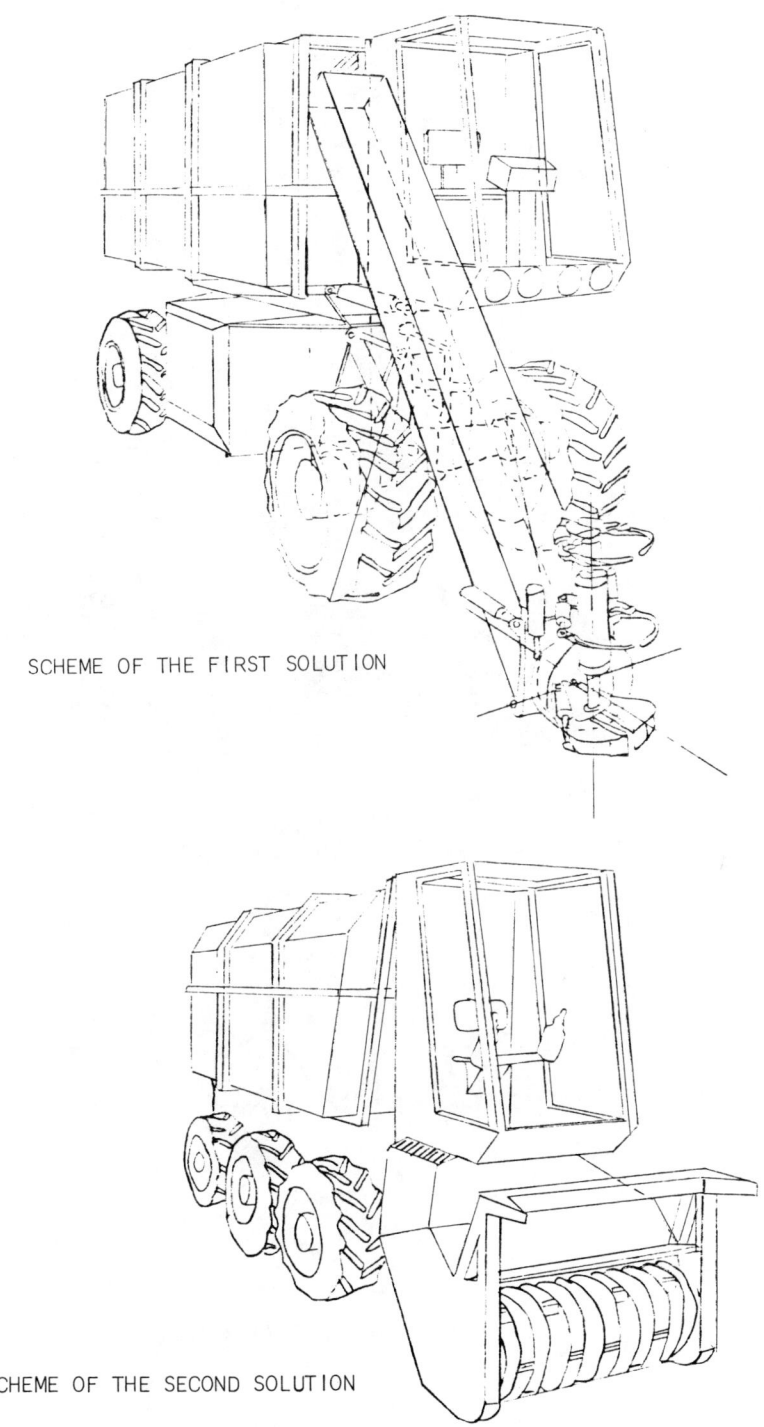

SCHEME OF THE FIRST SOLUTION

SCHEME OF THE SECOND SOLUTION

DETAIL OF THE DIFFERENT FUNCTIONS OF THE HARVESTER

DETAIL OF THE HYDRO MECANIC TRANSMISSION

DETAIL OF THE HYDROSTATIC PUMPS MOUNTING

HARVESTING BEFORE THE FIRE FOR ENERGY; MEDITERRANEAN-TYPE ECOSYSTEMS IN GREECE. COSTS AND BENEFITS

Authors : N. S. MARGARIS, M. ARIANOUTSOU and S. PARASKEVOPOULOS

Contract number : ESE-R-063-GR(B)

Duration : 30 months, 1 January 1982-30 June 1984

Total budget : DRA 11.463.000   CEC Contribution: DRA 2.865.000

Head of project : Prof. N. S. Margaris, University of Thessaloniki

Contractor : University of Thessaloniki (Laboratory of Ecology)

Address : Laboratory of Ecology
University of Thessaloniki
Univ. Post Box 119
Thessaloniki-GREECE

Summary

Harvesting natural mediterranean-type ecosystems as a managemental policy to provide energy and organics and at the same time face the fire problem demands first answer to questions dealing with its ecological soundness. It was found that the system recovers very well after harvesting, in a similar way to that after a fire with a significantly increased net productivity. The losses of nutrients due to biomass removal are compensated in a ten-year period when nitrogen is concerned, whereas the same is not valid for phosphorus. Because of the herbaceous bloom in the first post-harvesting years and the resprouting behaviour of the woody plants the soil subsystem continues to be replenished in carbon and energy. The diversity of the system increases and new habitats and niches are created. It is concluded that such a managemental policy is both profitable and ecologically sound.

1.1 Introduction
It seems that energy from biomass is not the panacea for solving energy and environmental problems. Without any doubt utilization of agricultural and animal residues for energy production may be used to some degree to face the energy and environmental crisis; on the other hand proposals dealing with energy crops as well as with cultivation of "marginal" lands are not free of ecological problems. Referring to energy crops, the questions arising concern competition for agricultural land, and all the problems relating to conversion of natural to man-made fragile systems such as the need and impacts of pesticides, insecticides and fertilizers use. On the other hand utilization of marginal lands for the cultivation of species like jojoba and guayule (1) is not necessarily acceptable from the ecological point of view. The word "marginal" has any sense only from the economic point of view since such lands, mainly in the semi-arid regions of the world are often very important because they represent an invaluable pool of genetic information.

Despite the above mentioned shortcomings of biomass utilization, we believe that there could be found solutions ecologically sound. Natural ecosystems have the advantages of covering extended areas of the Earth being highly resilient at the same time, able to absorb perturbations. By following basic ecological theory the limits of their resilience could be estimated, and if so we could further proceed to a profitable exploitation without degradation.

On that basis we started this project; systems proposed for exploitation are the mediterranean-type ecosystems of Greece (2, 3, 4). These systems are attacked by frequent fires with resulting loss of valuable energy and organics. Since they are adapted to regenerate after fire the basic idea of this project lies to the possibility of fire substitution by harvesting. If so, fire hazards are minimized while energy and organics are gained and exploited.

The questions formulating the research skeleton are:
- can harvesting substitute fire?
- which plants can recover?
- is the system's diversity maintained?
- is fertilization necessary in consequence of nutrients removal by harvesting?

1.2 Recovery after harvesting
Figure 1 contains data representing the recovery of dominant plants after harvesting. Diagrams A, B and C concern harvestings in spring, summer and winter. In all cases woody plants recover quite well. Of course during the stress period (summer) the recovery mechanism does not function but after the first autumn rains it starts again and results taken remind the data from mediterranean-type ecosystems recovery after fire (5, 6). We must also point out that even herbaceous plants, absent in the maquis system before harvesting are present afterwards.

As we stated in the last report (7) we chose an area cleared in the past from the National Electric Corporation for installation of high voltage lines. We have already mentioned that recovering of these systems is quite satisfactory. These areas were harvested 18 and 8 years ago and we proceeded to a new harvesting two years ago. Data presented in Figure 1 -Diagram D- show that the dominant plants recover in a way resembling that of stands harvested in the past in terms of species composition and quantity of plant production. Another point of interest is the fact that not only woody plants but also many others such as herbaceous and geophytes, coming always after fire, appear as well (Figure 2). Among the geophytes *Asphodelus microcarpus* dominates. Therefore, it is obvious that these systems react very positively after harvesting without any sign of degradation.

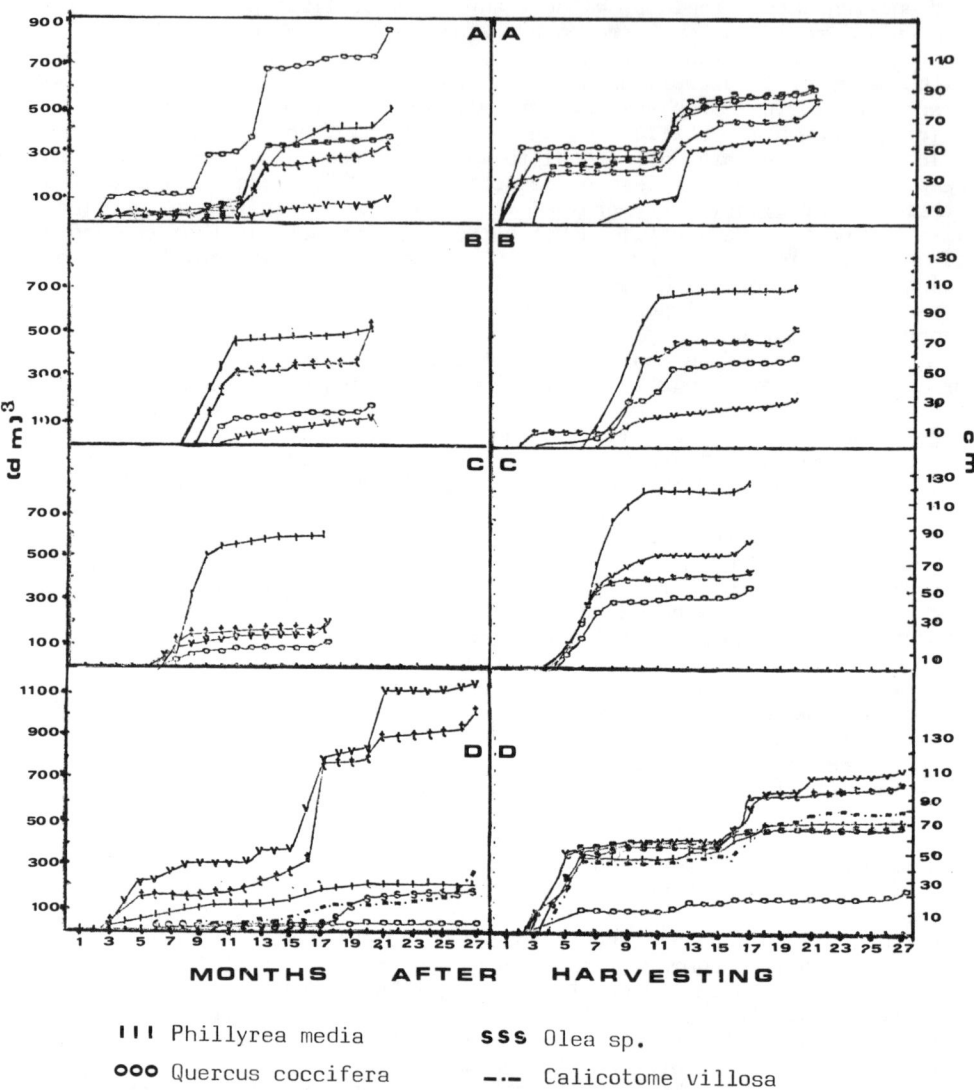

Figure 1. Recovery of dominant maquis shrubs after harvesting in different seasons (A, Spring; B, Summer; C, Autumn) is very quick. Same is true (D) after the successive harvesting before 18, 8 and 2 years.

Figure 2. This photo shows one year recovery of a harvested system.

Before proceeding to exploitation of maquis ecosystems more data are required deriving from other typical areas, as well. For this reason we started experimental harvestings in Mt. Pelion. In that area harvestings have been carried on by the local population every 15-20 years providing therefore means of comparison. The species composition of maquis systems before harvesting consists of evergreen sclerophyllous shrubs like *Arbutus unedo, Quercus coccifera, Quercus ilex, Erica arborea, Phillyrea media Spartium junceum* etc. Though the species composition of all the harvested areas was not identical, recovery was observed in all cases, and the productivity attained during the first year was more than 14 tons.ha^{-1}.

1.3 Productivity characteristics

When a maquis system is aged more that 20 years, the number of the component species is relatively low. A striking feature is the absence of herbaceous plants below the shrub canopy (Figure 3). The biomass of the system remains almost unchanged and all photosynthetic production is directed to maintaining the biological activity. After harvesting productivity is very high. If we compare the green to the non-green aboveground biomass in Mt. Pelion, one year after harvesting, the ratio is 1:1 while before it was 1:20. It is obvious that through harvesting maquis systems are brought back to the early successional stages what means serious increase of net productivity. The theoretical model we presented last year on maquis productivity (7) seems to be valid under field conditions.

1.4 Nutrient losses

It was stated in the last report that carefull estimation of nutrient losses due to biomass removal with harvesting should be elaborated. At least 100 kg.ha^{-1} of nitrogen and 10 kg.ha^{-1} of phosphorus are removed and must enter the system during the next ten years before a new harvesting to keep balance.

For this reason in 1982 we installed raingauges in Stavros experimental site, from which we collected the rain water in order to estimate the quantities of nutrients entering in the system through precipitation. In a one year period (1982-1983) 5 kg.ha^{-1} nitrogen have entered in the system;

Figure 3. Unharvested maquis aged at least 33 years are almost free from herbaceous vegetation. This system is waiting for the next fire while the biomass is almost stable.

consequently we can predict that half of the quantity removed by harvesting will be back in a ten-year period.

At the same time, since the presence of Papilionaceae species the first post-harvesting year was great enough, we tried to estimate the quantity entering the system as a result of atmospheric nitrogen fixation by bacterial nodules in their roots. According to our data 3.2 and 0.7 $g.m^{-2}$ are fixed the first and second year, respectively. If these values are added to those referring to precipitation it can be easily seen that more than 80% of the removed nitrogen comes back in a ten-year period. If we further include values of nitrogen fixation by woody shrubs such as *Spartium junceum* and *Calicotome villosa*, occurring in high number after harvesting, we might eventually conclude that there will be no problem of nitrogen shortage deriving from this type of management.

Concerning with phosphorus the situation is not so promising as with nitrogen. From a quantity of about 10 $kg.ha^{-1}$ removed, only 0.3 $kg.ha^{-1}$ came back through precipitation in the first year (1982-1983). That means that with rotating harvesting every ten years only 30% of phosphorus returns through precipitation. If we consider also that our soils are somewhat poor in this element it seems that fertilization with phosphorus will be needed. More data should be collected in the near future in this matter.

1.5 Soil metabolism

Since the energy and/or carbon offer to the soil subsystem might be seriously decreased because of the removal of the above ground biomass it was pointed out in our last report the need to estimate litter fall. It seems after all that such a problem does not exist. The bloom of herbaceous plants appearing during the first two years after harvesting and the resprouting shrubs like *Arbutus unedo*, *Erica arborea*, *Cistus* sp., *Calicotome villosa* etc., provide enough material replenishing the soil.

1.6 General remarks

Looking to the harvested systems we can conclude that maquis systems do not suffer from this kind of management. On the contrary, they show an active succession, and the number of plant species after harvesting is relatively high. During the spring of 1982 and 1983 herbaceous plants were in full bloom. The peak coincides with increased numbers of insects, which is reasonable since many of these plants are insect pollinated. It means that by this kind of management we make a complex environment where diverse organisms have the chance to find habitats and niches (Figure 4).

Figure 4. Harvested maquis, from the National Electric Corporation, 8 and 18 years ago, are recovering very well since not only all the plants are coming back but also birds are present.

It should be also taken into account that by such a managemental policy there is no flammable material in the system and therefore there is no more any need for expences for protection from fire. According to our calculations the biomass available for exploitation is high enough and the combination of all estimations and results make us propose this sort of natural systems exploitation which does not harm them but on the contrary efficiently protects them.

1.6 References
1. Goodin JR and Northington DK. (1979). Arid Land Plant Resources. International Center for Arid and Semi-Arid Land Studies, Texas Tech. University, Lubbock, Texas.
2. Margaris NS. (1979). In: Biological and Sociological Basis for a Rational Use of Forest Resources for Energy and Organics (S. Boyce, Ed.), USDA/ Forest Service, Southeastern Forest Exp. Stn., Asheville, N.C.
3. Margaris NS. (1981). Biomass, $\underline{1}$:159.
4. Margaris NS. (1983). International Journal of Solar Energy (in press).
5. Biswell HH. (1974). In: Fires and Ecosystems (T.T. Kozlowski and C.E. Ahlgren, Eds.), Academic Press, N.Y.
6. Conrad CE and Oechel WC (Tech. Coordinators). (1981) Dynamics and Management of Mediterranean-Type Ecosystems, USDA/Forest Service, Pasific Sothwest Forest and Range Experimental Station, Berkeley.
7. Margaris NS. (1982). In: Energy from Biomass (Series E, Volume 3, G. Grassi and W. Palz, Eds.) D. Reidel Publishing Co., Netherlands.

ALGAE

Development of a production size system for the mass culture of marine microalgae

Exploitation of lagoon macroalgae for biogas production - Preparatory study for a comprehensive project to exploit the algae in the lagoon of Venice to produce energy and improve the environment

Biomass from offshore sea areas

Biomass from offshore areas

Solar biotechnology study and development of tubular solar receptors for controlled production of photosynthetic cellular biomass for methane production and specific exocellular biomass

Culture of a hydrocarbon producing alga, Botryococcus braunii, at pilot level

Renewable hydrocarbon production by cultivation of the green alga Botryococcus braunii - Investigation of the factors affecting hydrocarbon production

Methane production by anaerobic digestion of algae, I. Pilot plant biomethanation of cultivated marine algae Tetraselmis for energy production in southern Italy

Methane production by anaerobic digestion of algae, II. Production of algae

DEVELOPMENT OF A PRODUCTION SIZE SYSTEM FOR THE MASS CULTURE OF MARINE MICROALGAE

Authors	: R.MATERASSI, M.R.TREDICI, F.MILICIA, C.SILI, E.PELOSI, M.VINCENZINI, G. TORZILLO, W.BALLONI, G.FLORENZANO, K.WAGENER
Contract number	: ESE-P-021-D
Duration	: 24 months 1 January 82 - 31 December 83
Total budget	: 320.000 DM CEC contribution:160.000 DM
Head of Project	: Prof.K.Wagener
Subcontractor	: Centro di Studio dei Microrganismi Autotrofi del CNR
Director	: Prof. Gino Florenzano
Address	: Piazzale delle Cascine, 27 50144 FIRENZE, Italy

Summary

Based on the previous work carried out in the small cultivation units installed in the "Palazzo" farm at Lamezia Terme (Calabria), a pilot plant for the outdoor mass culture of the green microalga Tetraselmis tetrathele in fertilized seawater was realized. The plant includes five cultivation units (total area 560 sq metres), a digester of 1 cubic metre working volume and ancillary equipments for harvesting the biomass (by sedimentation) and seawater supply. The cultivation units are provided with a new mixing device (mixing board) requiring a very low energy input.

Researches carried out in the pilot plant have demonstrated the feasibility of the MCL process as a biofuel production system. The knowledges acquired are adequate for scaling-up the process to a full scale production plant.

Since MCL is a versatile production system, in warm regions where the yield of biomass is very high it can find applications, besides energy production, in other fields, including aquaculture, wastewater reclamation, production of feeds and chemicals.

I. Introduction
   Within the Project E of the Solar Energy R & D Programme this research Centre initiated, from 1978, a research on the mass culture of marine planktonic microalgae in southern Italy, with the aim of developing the MCL process and to verify its feasibility.
   A number of species of marine microalgae and cyanobacteria were isolated from the S. Eufemia gulf and their growth properties in laboratory and outdoor culture determined. Among the strains tested, Tetraselmis tetrathele was choosen for further study, owing to its high productivity, good competitive ability towards extraneous algal species and a relatively high settling rate that made possible to harvest the biomass by sedimentation.
   Using six small paddle wheel ponds (PW ponds), the culture technique for T. tetrathele was set up and the influence of some important parameters (dilution rate of the culture, nutrient regime, etc.) on the yield was determined.
   The yield of biomass obtained in the period 1979-1981 was remarkably high, ranging from 58 tons of dry biomass ha^{-1} $year^{-1}$ in 1979 to 65 tons in 1981 (1-3). The data obtained were utilized in scaling up the process to a pilot level.
   The principles informing the process are the continuous operation of the algal culture under natural climatic conditions, the use of mineral compounds as a source of nutrients and the realization of a minimum energy input for cultivation and harvesting.
   The construction of the pilot plant started in the autunn of 1981 and was completed in the spring of 1982. The plant was installed in proximity of the sea. The area has been kindly placed at our disposal by the calabrian department of the Cassa per il Mezzogiorno. The same organization helped us in providing for the electric energy and fresh-seawater.
   The culture units are provided with a new tipe of mixing device, the mixing board, devised by Wagener (4) for minimizing the energy requirement of the culture. The ancillary equipments include the harvesting system and the seawater supply system. Moreover, a digester of I cubic meter working volume has been provided by the Biotechnology Unit of the Catholic University of Louvain.

2. Materials and methods
   2.1. Description of the culture units
   Five cultivation ponds have been built. Each pond is composed of two identical independent sectors. Three ponds are 20 metres long and 4 metres wide each (80 sq metres of illuminated surface) while the other two are 20 metres long and 8 metres wi

Figure 1 (left) - General view of the pilot plant for outdoor culture of <u>Tetraselmis tetrathele</u> showing 4 of the 5 culture units and the digester on the left side.
Figure 2 (right) - Particular of the mixing device.

de each (160 sq metres of illuminated surface). The mixing device has been realized by the Aachen University group. It consists of a board placed vertically across the pond. Between the board and the bottom of the pond a slit is left, across which the culture suspension flows in the direction opposite to the movement of the board. The speed of the mixing board was adjusted to 10 cm.sec.$^{-1}$.

The seawater is supplied by a submersible pump placed into the sea and connected to the pilot plant by a 2 inches pipe.

2.2. <u>Harvesting procedure</u>

In the evening the required volume of culture suspension is transferred from the culture ponds into a sedimentation pond. The algal cells sediment over night so that in the subsequent morning the surnatant liquid can be eliminated. The concentration factor reached in the sedimentation pond ranges between 20 and 30. The concentrated algal suspension is transferred into a decanter designed and built by the group of proff. Nyns and Naveau. From the bottom of the decanter after few hours 30-50 litres of algal slurry, having the concentration in volatile solids required by the digester, are withdrawn. The residual suspension is returned to the sedimentation pond. The algal slurry exceeding the needs of the digester is sun dryed. The data on the digester performance are referred by the Belgian workers (this volume).

2.3. <u>Management of the algal culture</u>

The cultures of <u>Tetraselmis tetrathele</u> were run on a semi-continuous regime as described earlyer (2). The volume of culture suspension drawn every day from the ponds was adjusted in order to keep constant the biomass concentration. The culture removed was replaced with an equal volume of seawater. Carbon, nitrogen and phosphate were supplyed daily as sodium bicarbonate, urea ad potassium dihydrogen phosphate respectively, in the amount required by the growing algal population. Nutrients were dissolved in a small amount of fresh water and added continuously from 8.00 a.m to 13.00. Periodically the concentration of ammonia nitrogen and orthophosphate in the culture solution was checked.

2.4. Analytical methods

Ammonia nitrogen was determined colorimetrically after steam distillation with the Nessler reagent. Orthophosphate concentration was estimated by complexing orthophosphate with molybdenum blue according to the method of Murphy and Riley (5) and evaluating the absorption of the resulting solution at 750 nm.

The biomass concentration in the culture was determined with the membrane filter method.

Total solids, volatile solids, minerals and ashes content in the suspension introduced into the digester were determined with the procedures described by Nyns and Naveau (6).

3. Results

3.1. Productivity of Tetraselmis tetrathele in open ponds stirred with the mixing board.

The data on the productivity of Tetraselmis tetrathele in 80 sq metres ponds are reported in table 1. The mean productivity for the period 1 June 82 - 20 May 83 (355 days) has been 14,2 g of dry weight sq metre^{-1} day^{-1}, corresponding to a yield of 51 tons ha^{-1} year^{-1}. This value is about 20% lower than the yields obtained in previous years in small "raceway" ponds (2,8 sq metres each) installed in the "Palazzo" farm at Lamezia. The difference is not great if we consider the great diversity between the two experimental installations attributable, firstly, to the different stirring system and pond dimensions. In evaluating these data it must be considered also that this is the first long run with an entirely new cultivation system, so that improvements are expected either from a better knowledge of the physiology of algal growth in board stirred ponds (B ponds) or from improvements in the culture technique.

In the period November-February the yields in B ponds were equivalent to the yields previously reported for the PW ponds. In March and April productivity in B ponds were considerably lower than in PW ponds. These lower yields were not attributable to the adverse effect of temperature, since in B ponds the temperature profile is better than in PW ponds due to a lower heat loss by evaporation. In this period in B ponds a massive growth of pennate diatoms occurred (see later). The control measures

Table 1 - Yields of outdoor mass culture of Tetraselmis tetrathele in 80 sq metres board stirred ponds (Lamezia Terme - Calabria).

month	mean temperature (°C) air min.	air max.	culture min.	culture max.	total solar radiation Kcal m^{-2} day^{-1}	yield g m^{-2} day^{-1}
1982						
June	16,0	27,4	19,5	32,5	6.900	23
July	18,7	30,4	23,4	35,6	7.030	32
August	18,3	29,2	23,4	35,0	6.280	26
September	16,5	28,3	21,7	30,9	4.680	16
October	13,2	22,8	17,7	23,9	3.100	11
November	8,6	18,5	12,4	18,4	2.500	8
December	5,3	14,8	8,7	12,9	2.050	5
1983						
January	5,2	14,0	8,8	13,5	2.200	6
February	4,6	13,0	6,7	13,0	2.750	5
March	7,2	16,0	10,6	18,0	3.570	9
April	8,6	20,0	13,5	24,7	5.100	11
May	11,5	21,8	17,9	30,0	6.240	18

interfered with the normal pond management causing considerable losses of Tetraselmis biomass. In September a marine strain of Chlorella sp. developed in the ponds. Correspondingly in this period the productivity was lower than expected.

As in the PW ponds, the highest productivity in B ponds was reached in July (32 g sq metre^{-1} day^{-1}). This value is close to the yields previously reported for this period.

During the day in B ponds a considerable accumulation of oxygen in the culture suspension occurs. This may adversely affect the yield, due to the enhancing effect of high oxygen tensions on photorespiration. An intermittent increase in the speed of the board could prevent oxygen accumulation.

In July 1982 the influence of the dilution rate on productivity was investigated. As shown in figure 3, the productivity increases with the increase of the dilution rate from 0,083 to 0,33.

3.2. Growth cycle of Tetraselmis tetrathele in outdoor mass culture.

The growth cycle of T. tetrathele starts from actively moving, quadriflagellate, young cells. These cells increase in volume and become rounded until they lost flagella giving rise to mature cells (cysts) ready to divide. Multiplication is by longitudinal or oblique division of mature cells and subsequent

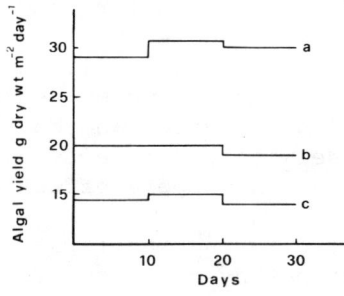

Figure 3 - Influence of the dilution rate on the productivity of Tetraselmis tetrathele culture in 80 sq metres ponds. Dilution rates: a= 0,33; b= 0,167; c= 0,083.

release of the two flagellate daughter cells.

In summer motile cells and cysts are present in roughly equivalent amount, although the relative proportion between the two types varies somewhat during the day. In the cold season the algal population is composed prevalently by non motile cysts probably because the low temperature inhibits cell division. This fact has several consequences on the behaviour of the culture during winter and early spring. Firstly it may be supposed that the great abundance of mature cells can limit the yield, since

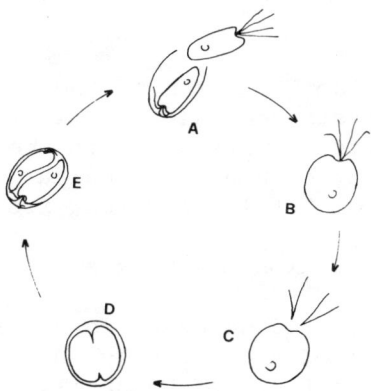

Figure 4 - Growth cycle of Tetraselmis tetrathele. A: release of the daughter cells (swarmers); B: mature cell; C: loss of flagella and cyst formation; D: protoplast division; E: flagellate daughter cells within the mother cell wall.

it is known that these cells have negligible photosynthetic activity. On the other hand, the cysts have an higher sedimentation rate compared with the young flagellate cells, so that the upper portion of the culture is available for the growth of floating algal species (see below).

3.3. Species control in board stirred ponds

In B ponds Tetraselmis cells are not uniformly distributed in the culture. In a given portion of the pond, after the passage of the mixing board, algal cells begin to sediment, until the subsequent passage of the board resuspends the cells homogenizing their vertical distribution. In such a system cells of algal species brought by the incoming sea water and having a sedimentation rate lower than Tetraselmis tend to accumulate in the upper portion of the culture and can compete effectively with Tetraselmis cells for light.

The two contamination events observed during the first year of operation with B ponds were carried out by small species. In September a massive invasion of Chlorella sp. occurred, while in March and April pennate diatoms (Nitsckia sp.) developed. In both cases the growth of the foreign species was temporarily controlled by removing a large portion of the culture suspension after the mixing board was stopped for the time required by Tetraselmis cells to settle at the bottom. The same diatoms had been observed also in ponds intensely mixed with the paddle wheel, but in this case the growth was restricted mostly on the pond walls at the air-liquid interface and on the paddle, while the growth in the culture suspension was more easily controlled.

These observations emphasize the need of some measures for controlling floating or slowly-sedimenting algal species.

4. Concluding remarks

The research carried out within the Project E of the EC Solar Energy R & D Programme has demonstrated the feasibility of the MCL process as a biofuel production system and, in the meantime, has lead to the development of a new mariculture based on microalgae and, therefore, completely different from the traditional mariculture.

The main points worthy of consideration may be summarized as follows:
(i) microalgae mass culture on fertilized seawater is a very efficient system for the conversion of solar energy into valuable biomass. In the climatic conditions of Calabria, by selecting the suited algal species it is possible the continuous operation of the outdoor culture;
(ii) a new mixing device, designed for minimizing the energy requirement of the culture, has been tested successfully on a significative scale. On the other hand, harvesting by sedimentation is practicable on a large scale if species with the proper settling rate are cultivated. In this way the energy requirement of the culture becomes a small fraction of the energy fixed in the algal biomass;
(iii) Tetraselmis tetrathele biomass is efficiently converted

into methane through anaerobic digestion. The efficiency can be improved by selection of algal strains having a lower protein and an higher carbohydrate content than T. tetrathele;
(iv) the biological control of outdoor mass culture on fertilized seawater seems relatively easy. Till now animal grazers or bacterial and fungal diseases did not interfered with the growth of the culture. Moreover the problem of epiphytism, common in macroscopic marine algae cultures, are avoided. Sometime microalgae with a low settling rate carried out with the incoming seawater can proliferate, but this phenomenon has been always controlled;
(v) nutrients from different sources can be utilized in the MCL process, including digested sludge and wastewater. The latter possibility is very actractive, since, besides energy, organic fertilizers and clean water are produced.

Intensive marine microalgae culture is a very versatile productive system that can find applications in other fields, including aquaculture, waste water reclamation, production of feed protein and chemicals. The work carried out till now allows the scaling up of the process to the full scale.

5. Acknowledgment

The AA. wish to thank the Regione Calabria and the Calabria's department of the Cassa per il Mezzogiorno for the support granted to this research.

6. References

(1) Florenzano G., Materassi R., Balloni W., Mozzoni M. (1979) Investigations on the productivity of outdoor mass cultures of marine algae in fertilized seawater for the project "Methane production by mariculture on land". 3rd Coordination Meeting of Contractors, EC Solar Energy Programme, Project E. Taormina 6-7 June.
(2) De Zarlo E., Tredici M., Balloni W., Materassi R. (1981) Investigations on the mass culture of marine algae in southern Italy. In: "Energy from biomass" (Chartier P. and Palz W.,Eds), Volume 1, pp. 70-75.
(3) Balloni W., Materassi R., De Zarlo S., Pelosi E., Sili C. (1982) Outdoor mass culture of algae in southern Italy utilizing seawater enriched with algal digested sludge. In: "Energy from biomass" (Grassi G. and Palz W., Eds), Volume 3, pp.107-113.
(4) Wagener K. (1981) Methane production by mariculture on land. In: "Energy from biomass" (Chartier P. and Palz W., Eds) Volume 1, pp.64-69.
(5) Murphy J., Riley J.P. (1958) A single-solution method for the determination of soluble phosphate in seawater. J. Mar.

Biol. Ass. UK, 37, 9-14.
(6) Nyns E.J., Naveau H.P. (1981) Methane production by anaerobic digestion of algae. In: "Energy"Solar Energy programme of the Commission of the European Communities. Project E: Energy from biomass. Summary of the work carried out during the first phase of the second programme. Extended abstracts. Commission of the European Communities Directorate General Scientific and Technical Information and Information Management, Bâtiment Jean Monet, Luxembourg, Ed.

EXPLOITATION OF LAGOON MACROALGAE FOR BIOGAS PRODUCTION

Authors   : Dr. A. VIGLIA
Contract number   : ESE-R-043-I
Duration   : 24 months   1.7.1981 - 30.6.1983
Total budget   : Lit. 288.000.000
CEC contribution   : Lit. 144.000.000
Head of project   : Vinicio Vianello - President C.S.A.R.E
Contractor   : C.S.A.R.E.- Centro per gli Studi e le
                     Applicazioni delle Risorse Energetiche
Address   : C.S.A.R.E. coop. a r.l.
                     Corso del popolo, 261
                     45100   ROVIGO (I)

PREPARATORY STUDY FOR A COMPREHENSIVE PROJECT TO EXPLOIT THE ALGAE IN THE LAGOON OF VENICE TO PRODUCE ENERGY AND IMPROVE THE ENVIRONMENT

Authors   : Dr. A. VIGLIA
Contract number   : ESE-R-076-I+D
Duration   : 10 months   1.3.1982 - 1.1.1983
Total budget   : Lit. 69.500.000
CEC contribution   : Lit. 69.500.000
Head of project   : Vinicio Vianello - President C.S.A.R.E.
Contractors   : C.S.A.R.E - Cooperative Dipl. Ing.
                     LINNEBORN - Prof. A. STREHLER
                     Landesanstalt für Landtechnik
                     (Universität München)
Address   : C.S.A.R.E coop. a r.l.
                     Corso del Popolo, 261
                     45100   ROVIGO (I)

Summary

In this report the main results and the stage of development of the research under the contracts ESE-R-043-I and ESE-R-076 I+D are presented.
The first contract refers to the project "Exploit of Lagoon macroalgae for biogas production", and the second is "Preparatory study for a comprehensive project to exploit the algae in the Lagoon of Venice to produce energy and improve the envoirnment". The latter project is being conducted in collaboration with Prof. Strehler from Munich University.
The results presented include:
- the study of the Lagoon of Venice
- mapping out, and how this can be used for the production of algae
- the basic features of the anaerobic digestion of the algae
- the biogas potentiality of the biomass collected
- the problems posed by the winter accumulation of biomass
- the use of the refuse waters coming from the methanation of the algae.

1. Introduction

The report means to refer on the situation regarding the technical results and the on at what stage the research refered to in the contracts: 1) ESE-R-043-I ; 2) ESE-R-076-I+D are.
The first refers to the project "Exploit of Lagoon macroalgae for biogas production", and the second refers to the project undertaken in collaboration with Prof. Linneborn and Strehler of the University of Munich, called "Preparatory study for a comprehensive project to exploit the algae in the Lagoon of Venice to produce energy and improve the envoirnment".
The reasons which led us to make one report of both studies can be found in the fact that the research done in collaboration with the above-mentioned Prof. Strehler of Munich Univeristy had as its main objective all the knowledge acquired from the first research starter by C.S.A.R.E. on the Lagoon of Venice in 1978.
The conclusions we came to with the research done in collaboration with the University of Munich, and with the possibilities of research we predict can still be done on the Lagoon of Venice, allow us to widen the horizons of enquiry we had, and propose furthur enquiry based on 2 different objectives:

one regarding energy and ecology, and the other regarding energy and agrozootechnology.

2. The Lagoon of Venice with reference to the other Italian Lagoons.

The Italian lagoons areas, including lagoons, coastal ponds and marshes, cover a surface of about 140.000.

The lagoons in the Veneto cover about 67.000 of these, of which the Venice Lagoon alone covers about 55.000. It is divided wite three main sub-basins. The average depth of this lagoon is 1 mt.

The algae mass (above all Ulva, Gracilaria and Valonia) present in the lagoon varies from 0,5 (winter months) to 2,5 (summer months) million tons, on the basis of wat weight.

Although the macroalgae production of the lagoon system is strongly influenced by various factors, the study showed that in the summer season the mass of algae collected without changing the waters in a certain area in made up in the space of 3-5 days, while in the winter months the rate of growth is much slower. These observations lead us to conclude that the potential amount of algae which can be produced annually by the lagoon system is somewhat larger than the amount resident, and so the projected taking of one million tons a year of algae biomasses (damp algae) need not necessarily alter the lagoon balance.

The lagoon areas which are mainly interested (accumulation areas) in an intensive collection of algae are probably the following: some areas of the North Lagoon, Palude di Cona, Palude Maggiore, Valle Cà Zane, the central area of the Lagoon (in particular the strip that goes from the Giudecca to the Canal of Poveglia and the Canal of S. Angelo) and finally the area included between the Canal of S. Secondo and the Canal Dese. In these latter areas, right from May, quantities ranging from 10 to 25 kg/m^2 were found. These observation have already led to the plotting of a map of the "accumulation areas", which is fundamental in programming the collection of algae biomasses to be used in the production of biogas. It is still worth underlining that this mapping out is to direct the collecting operations, since the population level found in the various areas is subject to fluctuations, determined by the water situation, which in turn is conditioned by the metereological conditions.

In fact the floating algae masses can be pushed by currents and winds in various directions until they reach a canal in which the current, because of the outgoing tide, takes them out of the lagoon it self.

3. Factors regarding methanation

3.1 Washing the algae.

For the production of biogas, the special algae taker in to consideration are Ulva, Gracilaria and Valonia.

The experimental phase of methanation was carried out both on biomasses washed in freshwater and on those taken direct from the Lagoon and drip dried.

The freshwater washing showed that it:
- took away mud and sand sticking to the algae
- eliminated the excess of salts present in the sea water taken in by the algae, and in particular lowered the level of sulphur (sulphates) with the consequent reduction of the level of H_2S in the gas produced.

However, a careful evaluation of the cost/benefit ratio of this washing operation showed us that for the amount in question (a million tons a year) it is preferable to use the algae well squeezed.

3.2 Breaking up of the algae

Tests of fragmentation and drying out of the algae biomasses were carried out to evaluate the usefulness of going on to the digestion of minute, damp, dry etc. material, so as to perfect the process of methanation. The results showed that an increase in the yield of the biogas is favoured by the breaking up of the fresh algae, as long as the granulometry is ≤ 1 mm and the cell walls are really broken down.

3.3. Insemination of the methanic fermentation

The possibility of insemination materials coming from other digesters already active for some time was taken into consideration to accelerate, in the stage of biological starting of the digesters, the processes of methane fermentation.

The following were tried:
- sludge coming from digesters used on urban refuse waters
- sludge coming from digestors which have already been used for some time on algae biomasses.

The results obtained confirmed on one hand that the insemination of sludge is useful in favouring a more rapid setting up of the conditions favouring methanation, but on the other hand there are no great differences between the various types of because they come from speed digesters.

3.4 Anaerobic fermentation of the algae biomasses.

The experiment referred to was conducted for about 12 months in laboratory digesters, with a volume of between 25 and 200 litres.

The parameters used are:

Temperature	35°C
Stirring	total mixing
HRT	15 days
Organic load	2 kg SV/m^3 a day
Total load	3,7 kg ST/m^3 a day
Type of algae used	fresh, squeezed Ulva
Type of material	fragmented and homogenous
N.ro of operativity weeks in digester	16

Under these conditions the following yields of conversion of algae biomasses into biogas were achieved:

Biogas produced	0,287 Nm^3/kg SV added
ST destroyed	44%
SV destroyed	63%
CH_4 in the biogas	65-67%
Calorific value of the biogas	5.600 kcal/Nm^3
Volatile fat acids in the digester	62,5 ÷ 236 mg/l as CH_3COOH

3.5 Checking the H_2S (in the biogas)

Since the amount of H_2S present in the biogas is basically tied to the level of S present in the substration treated, and that in the algae and in sea water there is a considerable amount of sulphates, experiments for checking the H_2S in the biogas produced were carried out.

The method which proved most economical and simple was adding $FeCl_2$ 3 ÷ 4 H_2O (in the quantity of 500 ppm a week) to the biomass in digestion.

With this concentration of iron, the amount of soluble sulphurs comes to around 50-100 mg/l; which do not seem to inihibit the process of methanation in digestors adapted to these levels of sulphurs.

3.6 Stoking the algae biomass

As we indicated previously, the average production ou the Lagoon of Venice in the colder winter months is not sufficient to guarantee in the planned areas of collection the 2700 tons a day coming from the hypothesis of final work.

The means collecting all the material necessary for the winter months (4 months) during the more favourable months.

The problems of stoking the algae biomass have therefore taken on a very important role as regards using them as economically and as well as possible.

Far from having found the final solution, the current idea is to use a horizontal ensilage on land and/or stoking in the lagoon in areas which are specially fenced off and equipped.

The final solution is tied to the results of an experiment still in process, which, excluding the use of vertical bins for the raw biomasses, would consider their use for the homogeneous biomasses or fot the liquid parts of these.

3.7. The digestion effluent

The values of BOD_5 and COD which are found in these effluents are in line with the characteristics which are normally found in the refuse waters coming from these types of treatment. Considering the effluent in its entirity (sludge and surface liquid) values of 3000 ppm for the BOD_5 and 9000 ppm for the COD are found.

After the separation of the two phases the above mentioned values are halved in the surface liquid. A study is being carried out on these latter refuse waters to sel if a secondary treatment by photobacteria to reduce the pollution forther and recycle the bacterial biomass in the digester is possible.

As regards the sludge, we are moving in the direction of using them in the formation of amendments to be used in agricultur, once we have estabished the heavy metals load, the pathogenous load and their chemical-physical balance on the basis of the specific agronomical needs.

3.8 The 5 cu.mt pilot plant

In the make of the operative and managerial parameters obtained from the laboratory digesters, there followed a study, project and realization of a 5 cu.mt pilot plant, which has been running for 4 weeks. On the pilot scale, once the running condition of all the data resulting from the laboratory scale will be made, and the various aspects concerning running on a larger scale will be defined.

The running of the plant will also permit the perfecting of the projecting and making digesters on a real scale.

4. New horizons in research on alternative used of the algae

We are aware that with research on producing biogas from algae, we wished to give at least one answer to the various questions which such a vast problem creates.

Ever while this research is still going on we are considering other answers to the question algae. One of the most concrete and interesting possibilities seems to us to be in the zootechnical and energy sector. This possibility as yet unexplored but potentially interesting is already being studied by C.S.A.R.E., particularly in the wake of the scientific interests evolung around the microalgae Spirulina as a food base.

5. Conclusion

This report wishes to illustrate the situation concerning the objects of the research which is being carried out under the 2 E.E.C. contracts mentioned in the introduction.

The current general situation of the Lagoon of Venice, under the ecological profile, and the growing awareness of the limitations on traditional energy resources, are by themselves sufficient to give reassurance to those who first tried to give a concrete answer to the "algal flowering" in the Lagoon, through anaerobic digestion.

The experimental results obtained so far, far from giving an exact answer to all the problems connected with such a vast and complex field of research; do, however, show how the road chosen, the production of biogas from algae, could be the right one.

Some points of no small importance remain to be cleared up particularly those concerning the use of the digested effluents. But isn't the spirit of research just this trying to resolve doubts?

PILOT DIGESTOR

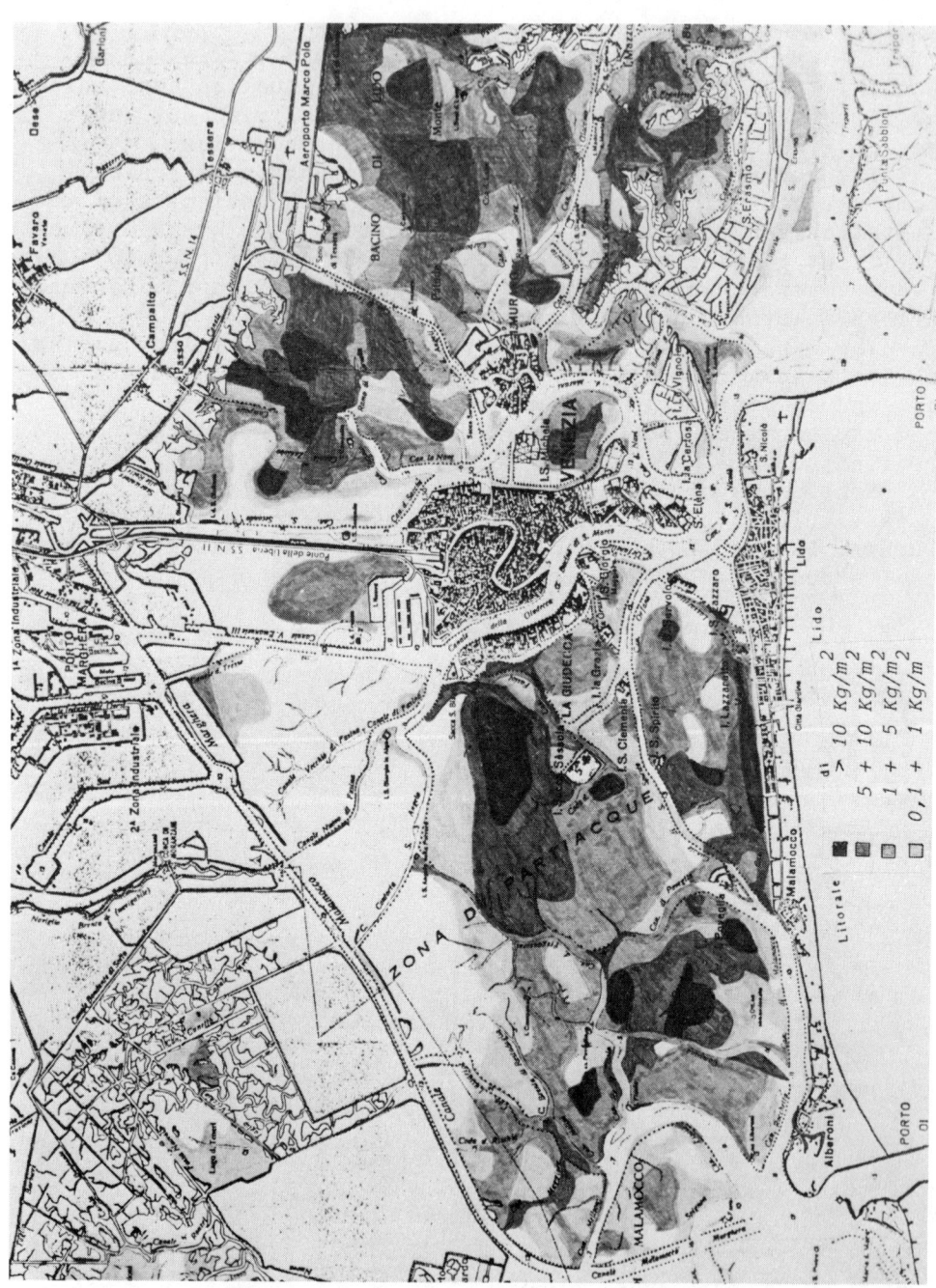

EXAMPLE OF DISTRIBUTION OF ALGAE IN VENICE LAGOON

ULVA RIGIDA (L.)

BOAT COLLECTOR (Projet n° 3)

BIOMASS FROM OFFSHORE SEA AREAS

Contract Number	:	ESE-R-021-UK*
Duration	:	33 months 1st Oct 1980 - 30 June 1983
Total Budget	:	£ 133,825 CEC Contribution : £ 60,000
Head of Project	:	Dr Joanna M. Kain (Jones)
Researcher	:	T.J. Holt
Contractor	:	University of Liverpool
Address	:	Department of Marine Biology University of Liverpool Port Erin Isle of Man

Summary

A successful method of artificially seeding horizontal ropes suspended below the sea surface with Saccorhiza polyschides, Laminaria saccharina and Alaria esculenta has been developed. The young stages of the algae are cultured on string in the laboratory for a minimum of 7 days, after which short lengths are inserted into the ropes.

Experiments to determine the effects of the time of seeding and the depth of the ropes have shown that plants seeded in spring can be severely inhibited by diatoms in the top 3 metres but not at depths of 5 or more metres.

The maximum biomass of Alaria seeded in this way occurred in late May to early June, while naturally seeded Saccorhiza and Laminaria peaked later in the summer. The effect of the time of seeding (November, December, February and April) upon the time and magnitude of maximum production is being observed in all three species using simple length measurements.

Groups of plants weighing 5.0 kg or more have been produced with all three species. Experiments to determine the effects of the proximity of the strings to each other and hence production/metre of rope are in progress.

Laminaria and Alaria seem more suitable for cultivation as they have a higher organic content and are more firmly attached to the string and therefore less prone to removal.

*See also report from Nottingham University (joint contract).

1. Introduction

The development of suitable techniques for the cultivation of the marine algae <u>Saccorhiza polyschides</u>, <u>Laminaria saccharina</u> and <u>Alaria esculenta</u> is continuing. The investigations have so far spanned two growing seasons, results from which are presented in this report. Experiments being performed in 1983 are described and preliminary results presented.

2. Techniques and Results

2.1 The Structures

The original design was generally satisfactory but became unstable once large quantities of algae were present. The vertical ropes were drawn inwards resulting in a downward bowing of the horizontal rope (Fig 1) which was noticeable from early June onwards. The depth of the bows was dependent upon the amount of algae, the height of the tide and the strength of the current which induced a horizontal bowing effect thereby reducing the amount of vertical bowing. Strong currents are common in the experimental area, and slack water never coincides with low tide, so that extreme vertical bowing was infrequent. Despite this, at slack water parts of a heavily colonised rope set at a depth of 2 m could occasionally reach as much as 7 m. Further problems were encountered during exceptionally strong currents at high tide since the drag on the algae was sometimes sufficient to completely submerge the structures. The collapsible nature of the buoys meant that once submerged they remained so. The addition of up to 10 extra buoys to each structure became necessary. In 1982-83 larger and more frequent, mainly rigid, buoys are being used to alleviate these problems (Fig 1c). Damage caused by pleasure boats should also be reduced.

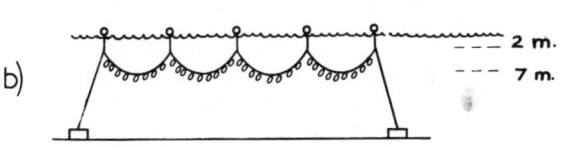

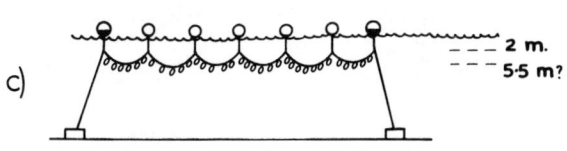

Many ropes were lost due to rapid rusting of the shackles. These are now used in pairs, and are wired, greased and checked regularly.

Fig 1 The effect of tidal height and current on heavily colonised ropes
a) high water, strong current
b&c) low water, no current
o 5 1. collapsible buoys
O 10 1. rigid buoys
● 27 1. collapsible buoys

2.2 Natural Colonisation

Colonisation of vertical ropes in exposed conditions was shown to produce a maximum biomass at a depth of 5-10 m in 1981 (1) and 4-7 m in 1982. The species composition was very variable and a maximum biomass of only 3.5 kg fresh weight/m of rope was attained, although a vertical rope in sheltered conditions bore over 14 kg/m of <u>Saccorhiza</u> near the surface in 1980.

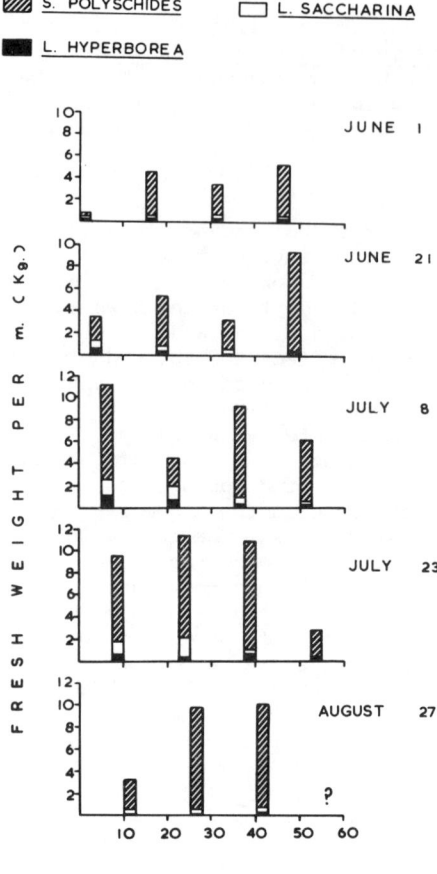

Horizontal ropes at 2 m depth in exposed conditions were more heavily colonised than the vertical ropes, with up to 11.0 kg fresh weight/m of rope being attained (Fig 2). There was much variation in the quantities produced along each rope and between different ropes (1).

2.3 Seeding Techniques

Attempts artificially to "seed" the horizontal ropes by the attachment of long lengths of string bearing the young stages of the plants (1) and by the attachment of sporing thalli to the ropes (2) were unsuccessful, but a third method involving laboratory culture of the young algae on strings, short lengths of which are then attached to the ropes, has been successfully developed.

Experiments were carried out to find the most suitable type of string (2) which proved to be a cheap synthetic fibre film. GeO_2 was added to the culture medium to inhibit diatom growth, and a controlled light regime was developed in order to prevent the cultures from being overrun by contaminating green algae. Continuous green light (wavelength 470-570 nm) at a fairly low photon flux density of about 15 $\mu E\ m^{-2} s^{-1}$ was found to be suitable.

The string was wrapped around a stiff frame which was immersed in the culture medium (Fig 3). Shortly before transfer to the sea, 25 cm lengths of string were produced by cutting along the edges of the frame. The bottom layer of strings were discarded since they were shaded by the top layer. The strings were inserted into the rope from a small

Fig 2. Fresh weight/m of the major colonisers on 2.5 m sections of a horizontal rope placed at 2 m depth in Nov. 1981, sampled during June, July and August 1982.

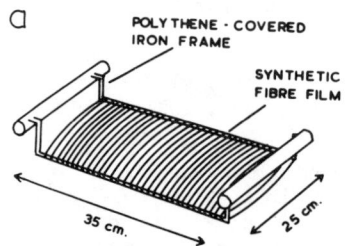
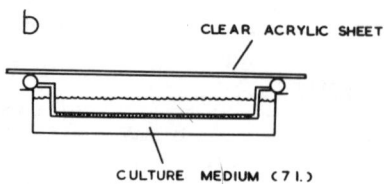

Fig 3 a) The frame around which the string was wrapped for culturing purposes. b) The frame in the culture medium.

			AERATED CULTURE			NON-AERATED CULTURE		
	DATE OF SEEDING	AGE OF CULTURE (DAYS)	NO. OF GROUPS GROWN	TOTAL NO. OF PLANTS	TOTAL WT. OF PLANTS (kg.)	NO. OF GROUPS GROWN	TOTAL NO. OF PLANTS	TOTAL WT. OF PLANTS (kg.)
a) L. saccharina	Mar. 1	5	0	0	0	0	0	0
	Mar. 19	23	8	438	36.82	9	53	6.36
	Mar. 26	30	10	?	?	-	-	-
	Apr. 22	57	0	0	0	0	0	0
	Mar. 26	9	0	0	0	-	-	-
	Apr. 22	36	0	0	0	-	-	-
b) A. esculenta	Feb. 17	7	10	671	4.29	10	970	7.46
	Mar. 1	19	-	-	-	10	431	7.08
	Mar. 19	37	-	-	-	10	?	?
	Apr. 22	71	-	-	-	10	677	3.67
	Mar. 1	5	0	0	0	0	0	0
	Mar. 19	23	10	?	?	10	?	?
	Mar. 26	30	10	1,256	34.92	10	1,521	44.14
	Apr. 22	57	10	1,373	14.35	10	882	7.89
	Mar. 26	9	0	?	?	-	-	-
	Apr. 22	36	10	485	3.57	-	-	-
c) S. polyschides	Feb. 17	7	10	?	?	10	?	?
	Mar. 1	19	-	-	-	0	0	0
	Mar. 19	37	-	-	-	6	20	10.98
	Apr. 22	71	-	-	-	10	120	8.19
	Mar. 1	5	0	0	0	0	0	0
	Mar. 19	23	2	5	4.84	5	?	?
	Mar. 26	30	6	30	13.95	0	0	0
	Apr. 22	57	9	97	4.40	-	-	-
	Mar. 26	9	0	0	0	-	-	-
	Apr. 22	36	6	120	6.17	-	-	-

Table 1. The number of groups of plants, total number and total weight of plants produced by batches of ten seeded strings. Each series of figures represents the same culture used at different ages. ? = groups which were left unmeasured.

DATE OF SEEDING	DEPTH (m)	LENGTH OF CULTURE (DAYS)	NO. OF STRINGS	NO. OF GROUPS (%)		
				ALARIA	LAMINARIA	SACCORHIZA
NOV. 25	2	37	50	100	100	100
DEC. 13	2	31	20	100	100	100
FEB. 16	2	34	20	100	100	0
APR. 5	1	46	20	65	0	0
	2	46	50	40	36	4
	3	46	20	100	100	45
	5	46	20	100	100	100
	10*	46	20	100	100	100

Table 2. The proportion of each batch of strings which produced groups of plants in the 1982-83 season.
* raised to 2 m after 43 days.

boat.

A trial of this method performed in November 1981 produced 45 discrete groups of Alaria attached to the rope (1, 2). The sisal string which was used had rotted and disappeared completely. All other experiments used the synthetic fibre film which did not rot or wear away, so that groups of plants were attached to both the rope and the string itself and hence consisted of larger numbers of plants.

Experiments begun in February 1982 showed that a culture period of 5 days is insufficient in all 3 species to produce adult plants, while 9 days is adequate (Table 1).

No clearly optimal length of culture was found although the largest quantity of each species was produced by cultures of 20-30 days.

There is no strong evidence for enhancement of production by aeration of the cultures.

Experiments in progress in 1983 use unaerated cultures. It was intended that a culture time of about 30 days be used in all cases, but postponements of seeding due to bad weather meant that the actual culture times were 31-47 days. The experiments were designed to determine the effects of the depth of the ropes (1, 2, 3, 5 and 10 m), the distance between the strings, (5, 10, 25, 50 and 100cm) and the time of year at which seeding takes place (November, December, February and April) upon production.

A larger proportion of strings have produced groups than in 1982 (Table 2) probably due in part to the age of the cultures used and to improved handling techniques when placing the strings on the ropes. However, Saccorhiza is poorly attached and several groups have been completely removed from the rope despite relatively calm weather. This was also observed during 1982.

The depth experiment performed in April has shown low numbers of groups being produced on those ropes nearer to the surface, particularly by Saccorhiza and Laminaria, (Table 2) which is in contrast to previous experiments performed at 2 m depth. The most likely reason is that the young plants were inhibited by the large quantities of filamentous diatoms which quickly covered the rope and strings during April. This possibility had been foreseen and a further rope had been placed at 10 m depth. This was relatively unaffected by diatoms and had produced groups of plants from every string after 43 days when it was raised to 2 m.

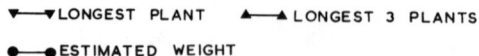
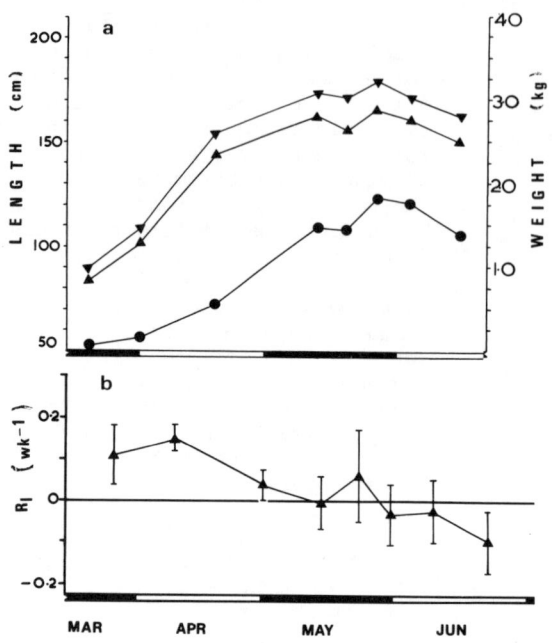

Fig 4 a) Mean estimated weights of 12 groups of Alaria with mean lengths of the longest 3 and the longest single plant in each group from March–June 1982. The groups were seeded in Nov. 1981 at 2 m depth.

b) Mean relative growth rate (R_1) of the longest single plants in each group, with 95% confidence limits.

2.4 Growth and Production

2.4 Growth

To produce reliable figures for the quantities of weed which can be produced it is necessary to know when the plants have reached a maximum biomass. Unfortunately the numbers of groups were so small that it was necessary to use non-destructive methods only, except in the case of the trial run which produced forty five groups of Alaria, allowing a certain amount of destructive sampling. This was used to make estimates of the fresh weight of 12 labelled groups from mid March until the end of June by a previously described method (2). There are probably considerable inaccuracies in estimates produced in this way, but despite this it seems likely that the groups reached a maximum biomass in the last week of May or the first week of June (Fig 4a). This is supported by the size frequency composition of the 3 sample groups removed periodically from the ropes (Fig 5). Fig 4a also shows that the lengths of the 3 largest groups measured in situ by divers are closely related to the weight estimates. Moreover the lengths of the single longest plant in each group, which can be measured more quickly and accurately, produce a similar graph.

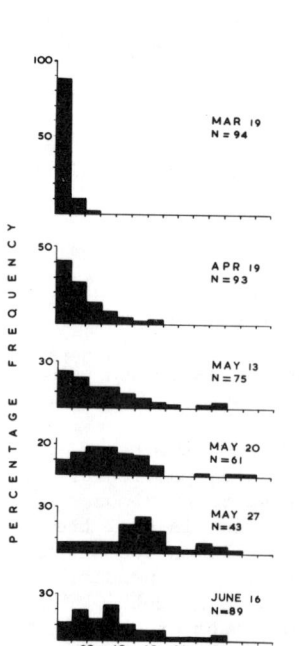

Fig 5 Fresh weight composition from Mar–June 1982 of samples of 3 groups of Alaria seeded in Nov. 1981 at 2 m depth.

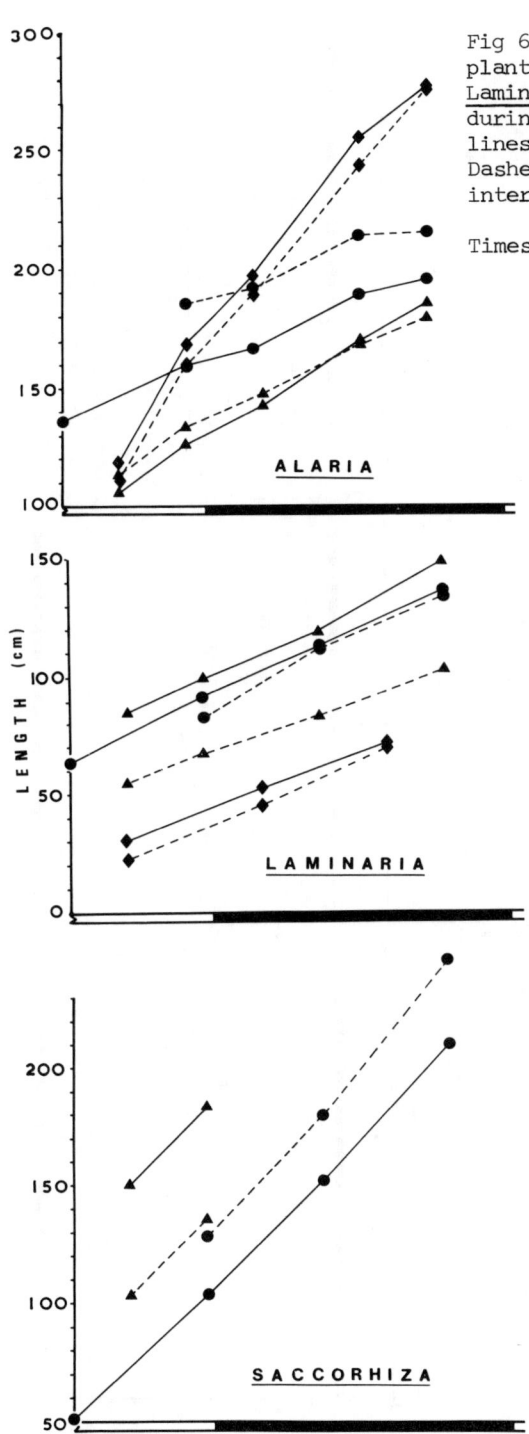

Fig 6 Mean length of the longest plant in each of 6-10 groups of <u>Alaria</u> <u>Laminaria</u> and <u>Saccorhiza</u> measured during April and May 1982. Solid lines - groups spaced at 1 m intervals. Dashed lines - groups seeded at 0.5 m intervals.

Times of seeding :
▲——▲ NOVEMBER
●——● DECEMBER
♦——♦ FEBRUARY

Consecutive length measurements were used to calculate the relative growth rate, R_l using the formula

$$R_l = \frac{\log_e l_t - \log_e l_o}{t}$$

where l_t = length at time t
l_o = length at time 0.

Fig 4b shows the mean R_l value for the 12 tagged groups from March-June, with 95% confidence limits, and gives an idea of the significance of the length changes in Fig 4a.

From April 1983 length measurements have been made of the longest plant in 15-20 groups of each species seeded in November, December and February and from late May April-seeded plants will also be measured. Figs 6 & 7 show the results obtained so far. Unfortunately the lengths of the plants (Fig 6) cannot be taken as an indication of the fresh weight of the groups because of the very variable number and shape of the plants. For example, although by mid-May the February seeded <u>Alaria</u> were very much longer than those seeded earlier, these groups consisted of numerous long, very narrow plants, while the November and December seeded plants consisted of fewer short but very wide plants. It is expected that <u>Alaria</u> will again peak in late May to early June, while length measurements on

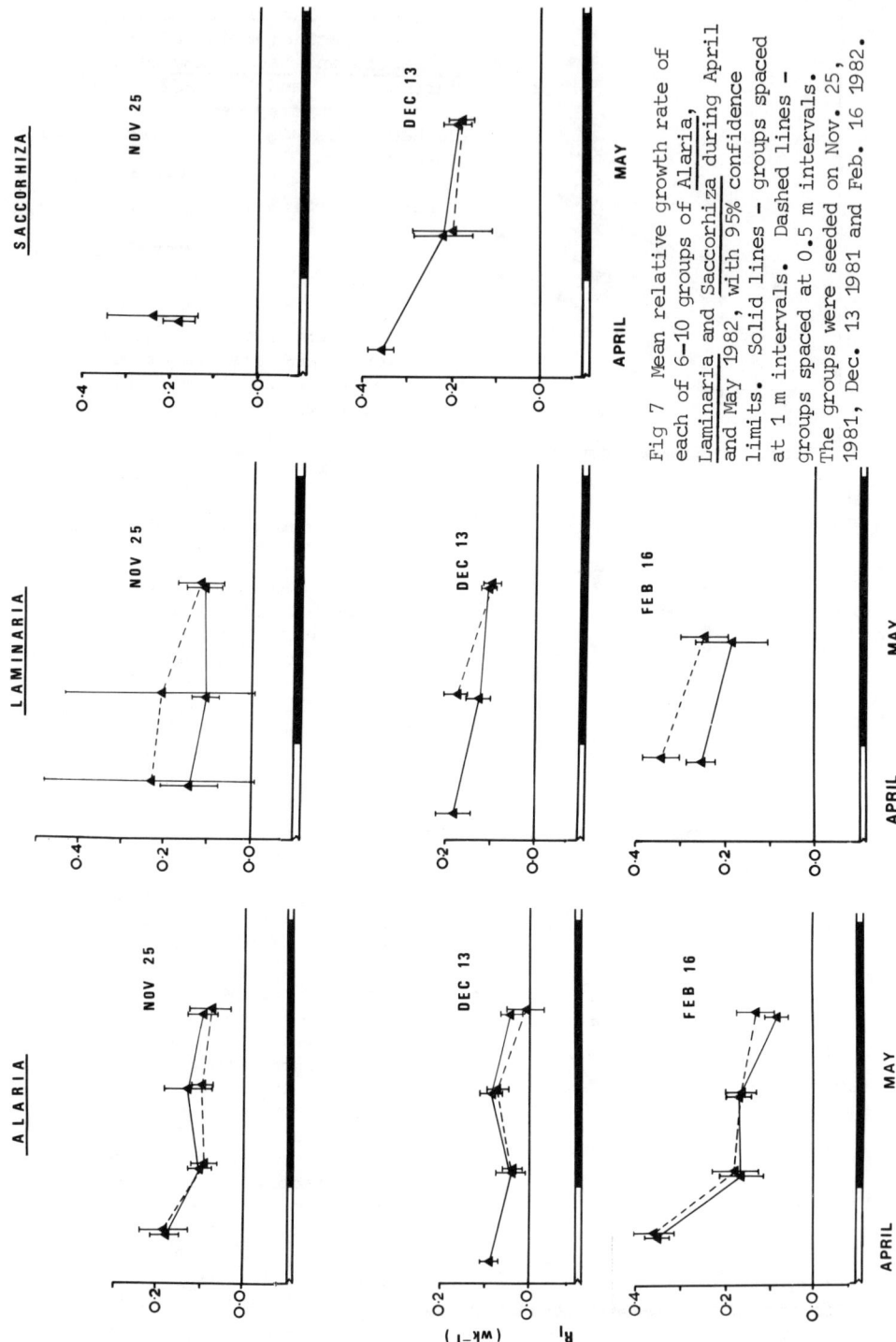

Fig 7 Mean relative growth rate of each of 6-10 groups of Alaria, Laminaria and Saccorhiza during April and May 1982, with 95% confidence limits. Solid lines - groups spaced at 1 m intervals. Dashed lines - groups spaced at 0.5 m intervals. The groups were seeded on Nov. 25, 1981, Dec. 13 1981 and Feb. 16 1982.

naturally colonised plants in 1982 suggest that large Saccorhiza plants will cease growing in June and Laminaria in August. It remains to be seen to what extent, if any, the time of seeding will influence the time at which the plants stop growing. The groups will be harvested and fresh and dry weights calculated immediately after growth ceases.

2.4.2 Production

Production in 1982 was very variable (Table 1). Individual groups weighing up to 6.6 kg (Laminaria), 5.5 kg (Alaria) and 5.0 kg (Saccorhiza) were produced. Harvesting took place in late July and early August, so that figures for Alaria are likely to be substantially underestimated as the measured groups peaked in late May to early June. It is expected that production in 1983 will be generally higher than in 1982 due to the improved handling techniques, longer time spent in the sea and harvesting at more appropriate times.

The effect of the proximity of the groups to each other upon production will be observed, and an estimate of possible production per metre of rope made. This is likely to be substantially higher than can be produced by natural colonisation.

2.5 Reseeding

At the end of the summer 1982, 10 groups of Laminaria saccharina were left on one end of one of the ropes to see if they could effectively recolonise the rope. The plants remained healthy and developed large patches of sporangia during the winter. Recolonisation of this rope will be observed during the summer of 1983.

Two further ropes bearing groups of Alaria and Saccorhiza were also left but these were lost during winter storms, due to insecure shackles.

2.6 Chemical

Analysis of the groups harvested in 1982 shows that the organic content of wet Saccorhiza is two thirds that of Laminaria and almost half that of Alaria (Table). Similar analysis will be carried out in 1983.

Table 3 - the dry and organic content of fresh algae harvested in 1982

	ALARIA	LAMINARIA	SACCORHIZA
DRY/FRESH WT. (%)	14.0	12.3	9.7
ORGANIC/DRY WT. (%)	65	61	52
ORGANIC/FRESH WT. (%)	9.1	7.5	5.0
	(1 :	0.82 :	0.55)

3. Conclusions

Groups of Alaria seeded in November 1981 reached a maximum biomass during late May and early June 1982. Naturally seeded Laminaria ceased growing in late August 1982, while large naturally seeded Saccorhiza ceased in early to mid June.

The string culture system was successfully used to produce groups of plants weighing 5.5 kg with Alaria and 6.6 kg with Laminaria. Improvements have been made in the proportion of strings successfully producing groups.

Groups of Saccorhiza up to 5.0 kg were also produced by this method, but this species produced a poorer proportion of successful groups than Laminaria or Alaria in 1983. It also has an organic content which is two thirds that of Laminaria and almost half that of Alaria and therefore seems to be the least suitable species for cultivation.

References

(1) Jones J.M. & Holt T.J. (1981). Biomass from offshore sea areas progress report in 'Energy From Biomass' 1981 Copenhagen 85-89.

(2) Jones J.M. & Holt T.J. (1982). Biomass from offshore sea areas progress report in 'Energy From Biomass' 1982 Brussels 120-125.

BIOMASS FROM OFFSHORE AREAS

Author.	J.G. MORLEY.
Contract Number.	ESE-R-021-UK*.
Duration.	33 months. 1st October 1980 - 30th June 1983.
Total Budget.	£133,825. CEC Contribution : £60,000.
Head of Project.	Professor J.G. Morley, University of Nottingham.
Contractor (Sub).	University of Nottingham.
Address.	Wolfson Institute of Interfacial Technology, University of Nottingham, University Park, Nottingham, NG7 2RD.

Summary.

A general description of a possible way in which marine biomass (seaweed) might be used as an energy crop is given. Growth on polypropylene rope is proposed with harvesting taking place at sea. Process requirements for the production of methane by anaerobic digestion are considered and the possibility of producing solid combustible fuel by simple drying processes is discussed. Elementary estimates of the likely commercial costs of various stages of the overall process are made where these can be based on existing technologies. It is suggested that major economies might be achieved through the design of combined offshore wind turbine and marine biomass energy farms.

* See also report from Liverpool University (joint contract).

1. Introduction.

This paper gives a preliminary general overview of the technical and commercial feasibility of utilising marine biomass (seaweed) as an energy crop.

2. Potential Scale of Resource.

Marine biomass offers a possible renewable energy resource which does not compete with the use of land for agricultural or other purposes. Energy production by this means could be as high as 1 kg coal equivalent per square metre per annum. Thus, in principle, an area of sea about 30 km x 30 km could produce the equivalent of 1 million tonnes of coal per annum. Various factors need to be taken into account in assessing how closely this theoretical figure can be approached in practical circumstances. Some of these are discussed below.

3. Proposed Arrangements for Seaweed Growth.

Large areas of sea surface would have to be utilised if a significant fraction of total energy requirements are to be generated by marine biomass. In addition it is desirable that the areas utilised should be located relatively near to population centres. These requirements make it necessary to consider the use of artificial substrates for the growth of seaweed, since natural sites are limited in extent and are not located in ideal positions. In addition it is considered that it may be more convenient to harvest from a regular artificial substrate than from the natural sea bed.

3.1. Type of Artificial Substrate Considered.

Polypropylene rope, moored and buoyed at appropriate intervals, is proposed for the substrate. An array of ropes probably arranged in parallel would form the energy farm (see Figure 2). It has been shown that seaweed can be grown successfully on polypropylene rope and growth rates on rope as high as 15 kg (fresh weight) per metre have been observed. (1). In energy terms this is approximately equivalent to 1 kg of coal.

3.2. Factors Affecting the Cost of the Artificial Substrate.

The cost of the substrate depends directly on the loads developed in the rope since this will govern both the cost of the rope and its moorings. At the present time it is felt that forces produced by the action of tidal currents on the seaweed overgrowth will be dominant. There is some uncertainty about the depth at which the horizontal ropes should be held for optimum seaweed growth and this will influence considerably the magnitude of the forces developed from wave action. These will act in different directions at different times and at different positions along a rope and the cumulative effect is not expected to be large. By contrast tidal currents will generate uniform drag forces over the whole length of a rope. These will act in a particular direction at any particular state of the tide.

3.3. Forces Developed by Tidal Currents.

Some simple experiments have been carried out, using material facilities and assistance provided by the University of Liverpool, in order to estimate the forces which would be developed in a weed covered rope. This is a function of the geometry of the rope and the mooring arrangements used as well as the direction and velocity of the current. Initial calculations indicated that a tidal current of 0.1 ms^{-1} would produce a tensile load of a few tonnes in a rope 1000 m long carrying an overgrowth of biomass of about 15 kg (fresh weight) per metre (2). Additional data has now been obtained for three species of plants (Figure 1). These data were obtained, as before, by attaching a number of plants to a wooden bar so that the assembly was almost completely immersed. The bar was then towed behind a boat at various speeds and the corresponding drag forces were measured. Different lengths of bar were used to establish whether any correction factor to account for the ends of the bar was necessary. The bar was also aligned transversely and parallel with the direction of water flow. The fresh weight of plants attached to the bar was close to the estimated maximum of 15 kg fresh weight per metre.

The drag force was observed to be approximately directly proportional to the mass of biological material present so where this differed from 15 kg/m the measured drag force was simply factored for this difference so that all the data points shown in Figure 1 are representative of a fresh plant weight of 15 kg per metre of bar. The drag forces developed by the wooden bars alone under similar circumstances were negligible and it is therefore considered that the data obtained is reasonably representative of weed covered polypropylene rope.

It was not practicable to measure drag forces at very low velocities because of errors caused by the irregular motion of the boat under these conditions. However the drag force values extrapolate to zero at zero velocity so that values for low current velocities can be estimated by interpolation without serious error.

Initial data showed no measurable difference between the drag forces developed when similar assemblies were aligned parallel to and transverse to the water flow (3). The more recent data (Figure 1) shows the drag forces to be substantially less when the bar is aligned parallel with the water flow, although the maximum drag forces observed for Sacchoriza polyschides are similar to the previously observed value of 22 kg at 1 ms^{-1} (3). The calculations previously made of the stresses developed in the ropes and anchor points for various current velocities and directions therefore probably over estimate these values.

3.4. Cost of Artificial Substrates.

The cost of the ropes and anchor points will be primarily controlled by the maximum forces developed. It was previously assumed that a maximum tensile load of about 2 tonnes would be developed in a rope anchored at 1 km intervals (2). This would be generated as a consequence of a drag force of 2 kg per metre of rope. The maximum tidal current velocity depends on the geographical area chosen and seems likely to be in the range of 0.1 ms^{-1} to 1.0 ms^{-1}. From Figure 1 the drag forces developed by current velocities near the lower end of this range correspond to those used in previous calculations. If conventional marine anchors of a suitable size are used then these will account for about 75% of the cost of the rope assembly which would then be in the region of £500. per kilometre of rope or the equivalent of £0.5 per square metre of sea surface utilised. This is less than, but comparable with, the cost of agricultural land so that attention is drawn to the need to investigate the feasibility of developing inexpensive anchors. The probable life of the polypropylene rope is unknown but is estimated as being about 10 years.

4. Harvesting.

4.1. General Arrangements.

It is envisaged that harvesting would be carried out at sea under suitable weather conditions using converted fishing boats. The ropes would be lifted out of the water allowing the weed overgrowth to hang down vertically from the rope. This is expected to facilitate the mechanical cutting of the weed from the rope. After the weed has been removed the rope would be returned to the sea in the expectation that the next crop would grow from self seeding.

Because of the large bulk and low value of the seaweed it seems essential that the harvesting boat should tow behind it a large capacity inexpensive barge for the storage of the biomass during harvesting. The seaweed could be transferred from the boat to a conventional steel barge using a suitable conveyor belt. When full it may be possible to convert the open barge into a digester tank by fitting a suitable cover and this would have the advantage of reducing capital costs. It is not known at the present time whether the digester tank would have to operate above ambient temperatures. If so the problem of thermal insulation and the supply of low grade heat will require investigation. Preliminary consideration has been given also to the use of large semi submerged flexible reinforced rubber containers, again to operate both as storage barges and anaerobic digester tanks. This is a known technology.

It seems probable that flexible containers of this type could be filled during harvesting by transporting the cut seaweed through a large diameter pipe to the container. Sea water would be used as the transport fluid, the excess water being vented from the container leaving the seaweed behind. Again it is envisaged that the flexible tank would be used for anaerobic digestion as well as for harvesting. An investigation of the problems of transporting the seaweed to the container and the extent to which the container can be filled uniformly with seaweed has been carried out using reduced scale models.

4.2. Model Studies of Harvesting Techniques.

Nylon ribbon 4 mm wide was cut into uniform short lengths of 7 cms for use as simulated seaweed. The ribbon was wetted by the water and had a slightly greater specific gravity than water. The feed head consisted of a cylindrical container 11 cm high and 22 cm in diameter having an outlet pipe in its base 22 mm in diameter and was mounted in a vertical position above the surface of the water in a flume in which the containers were immersed. Water was pumped from the flume so as to impinge tangentially at various rates along the inner surface of the feed head. Artificial seaweed was supplied to the feed head again at various rates. The vortex developed in the feed head tended to align the strands of simulated seaweed with the axis of the outlet pipe.

The pipe leading from the feed head was connected to the model container which was immersed in a stream of water in the flume. Two types of container were used. The first of these was approximately hemispherical in shape having a diameter of 23 cm and a height of 7.5 cm. This was mounted in the flume with the diameter plane on the hemisphere vertical and parallel with the direction of the water flow in the flume (compare with Figure 3). The glass sides of the flume made it possible to view the inside of the container directly. A circular piece of nylon gauze was attached to the base of the container to retain the seaweed and to allow excess water to escape without inhibiting direct visual observation of the interior of the container. The inlet pipe was positioned either at the top of the container or at its leading edge. It was envisaged that in a full scale system two containers of this type would be mounted base to base as indicated in Figure 3. Sufficient space would be left between the containers to allow excess water to be vented from the system.

Since only one half of such an assembly was studied in model form a baffle plate consisting of a transparent sheet of perspex was fixed to the container and separated from it to allow venting of excess water. This plate simulated the median plane between two containers, and in this way the behaviour of a single container simulated with tolerable accuracy the behaviour of two coupled containers. The baffle plate could be set at various angles to change the distribution of local water pressure across the nylon gauze which was developed as the water in the flume flowed past the container.

As was expected, the large area of gauze helped to prevent a local accumulation of seaweed in the container. The effects of varying the inlet port position and the separation and angular position of the baffle plate were studied. The maximum packing densities were achieved with the entry pipe positioned at the top of the container when the baffle plate set parallel with the sheet of gauze 1 cm from the base of the container. For these conditions the efficiency of transport of the artificial seaweed into the container was independent of the flow rate of water into the feed head over a range of about 0.08 to 0.4 litres per second, and for seaweed feed rates ranging from 5 gm per minute to 30 gm per minute. The capacity of the container was 2.3 litres and it was filled uniformly with simulated seaweed under these conditions. It was encouraging to discover that the distribution of artificial seaweed in containers of this type can be controlled in a systematic manner by changing the geometry of the system. The packing efficiency of the second type of submerged container has also been studied. This type is illustrated in Figure 4. The same type of feed head was used to introduce the simulated seaweed into the container. Again a relatively large area of gauze was mounted at the rear of the container and premature blocking of the gauze by the artifical seaweed did not occur. Similar packing densities to those observed for the first type of container were observed.

4.3. Estimated Costs of Harvesting from Ropes.

It is estimated that the cost of operating a harvesting boat will be approximately £10. per kilometre of rope harvested. The value of the biomass from 1 kilometre of rope would be worth three or four times this figure in energy terms.

5. Energy Production.

5.1. Anaerobic Digestion.

The favoured route for the conversion of marine biomass to a useful energy source is to convert a portion of it to methane by anaerobic digestion. This has been demonstrated to be feasible on a laboratory scale (4).

5.2. Capital Cost of Harvesting Barges/Digester Tanks.

The cost of semisubmerged flexible containers appears to be of the order of £10 per cubic metre which is likely to be several times the value of the seaweed contained with the same volume. Hence it will be necessary for digester/barges to have a high utilisation rate if the effect of their high capital costs is to be minimised. Data on the effect of temperature on the anaerobic digestion of seaweed is unknown to the author and no estimates have been made of costs which might be incurred in heating the seaweed during digestion.

5.3. Air Drying.

It has been observed that the water content of marine biomass can be removed in air at modest temperatures quite quickly. The drying speed is

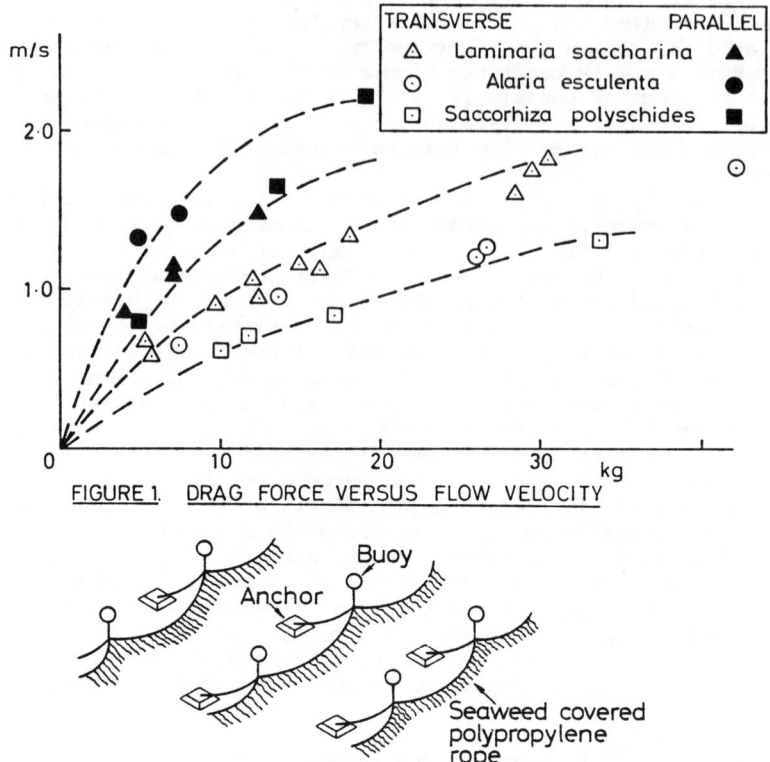

FIGURE 1. DRAG FORCE VERSUS FLOW VELOCITY

FIGURE 2. SCHEMATIC ARRANGEMENT FOR PROPOSED ARTIFICIAL SUBSTRATE

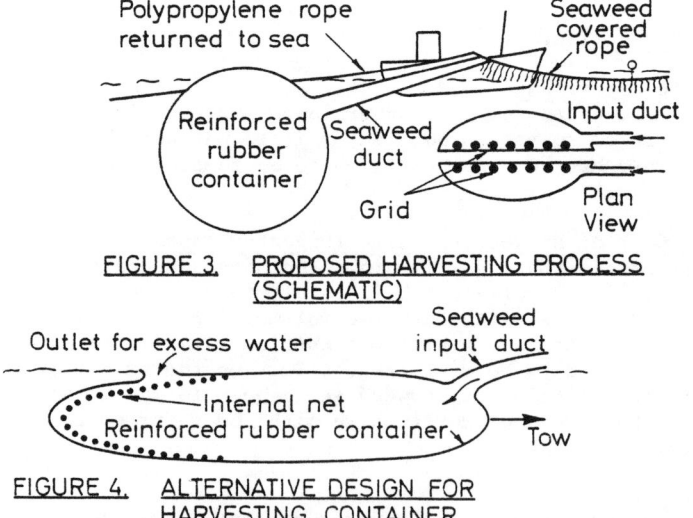

FIGURE 3. PROPOSED HARVESTING PROCESS (SCHEMATIC)

FIGURE 4. ALTERNATIVE DESIGN FOR HARVESTING CONTAINER

increased by increasing the temperature and the air flow speed and by decreasing the relative humidity of the air flowing over the seaweed (5). It is observed that the drying rates are not strongly dependent on these parameters except at high relative humidities. Typically at an air temperature of $20^{\circ}C$, relative humidity of 50% and an air flow rate of 1.7 ms^{-1} drying is almost complete after 2 hours. A sample of Laminaria saccharina studied had a calorific value, when dry, almost 50% that of the same weight of coal.

6. Further Considerations.

Consideration is being given to the use of offshore sea areas as energy farms using aero generators (6). It seems reasonable to suppose that substantial economies could be made if the same area of sea surface were also used for the production of marine biomass. In Section 3.3 it was pointed out the major cost of artificial substrates using rope seemed likely to be set by the cost of anchoring. Should it prove possible to utilise the aerogenerator masts as anchor points the cost of the artificial substrate might be substantially reduced. At the same time the production of significant amounts of additional renewable energy from the site would improve the overall commercial feasibility of the aerogenerator proposal.

7. Conclusions.

The technical feasibility of using moored and buoyed ropes as artificial substrates for the growth of marine biomass seems to be well established. Also harvesting procedures based on existing technology exist. Anaerobic digestion of seaweed to produce methane has been demonstrated in the laboratory and it has been shown that seaweed can be dried quickly with little difficulty to produce a combustible fuel. The commercial feasibility of the approach is more difficult to establish at the present time but preliminary estimates indicate that further enquiries would be justified particularly where benefits might be gained through the development of combined systems.

References.
1. Jones J.M. and Holt T.J. (1982). Biomass from offshore sea areas. Progress Report on ESE-R-201-UK. May 1982.
2. Morley J.G. (1982). Biomass from offshore sea areas. Progress Report on ESE-R-201-UK. May (1982).
3. Morley J.G. (1982). Biomass from offshore sea areas. Progress Report on ESE-R-201-UK. January (1982).
4. R.A. Troiano, D.L. Wise, D.C. Augestein, R.G. Kispert and C.L. Clooney. Fuel gas production by anaerobic digestion of kelp. Resource recovery and Conservation 2, (1976), p. 171-176.
5. S.I. Sallis. Assessment of seaweed as a fuel (1982), B.Sc. Thesis. Department of Mechanical Engineering, University of Nottingham.
6. D.J. Millbarrow, D.J. Moore, N.B.K. Richardson and S.C. Roberts. The U.K. offshore wind power resource. Paper D4, Fourth International Symposium on wind energy systems (1982). BHRA Fluid Engineering, Cranfield, Bedford, England.

SOLAR BIOTECHNOLOGY STUDY AND DEVELOPMENT OF TUBULAR SOLAR RECEPTORS FOR
CONTROLLED PRODUCTION OF PHOTOSYNTHETIC CELLULAR BIOMASS FOR METHANE
PRODUCTION AND SPECIFIC EXOCELLULAR BIOMASS

Authors   : C. GUDIN, D. CHAUMONT

Contact number : ESE-R-073-F-(SD)

Duration   : from 1st July 1982 to 1st July 1983

Total budget   : 3 921 260 FF   CEE contribution : 1 561 750 FF

Head of project : C. GUDIN

Contractor   : Association pour la Recherche en Bioénergie Solaire

Address   : A.R.B.S. (C.E.A.)
                  C.E.N. Cadarache
                  Laboratoire de Biotechnologie Solaire
                  B.P. n° 1
                  13115 SAINT PAUL LEZ DURANCE
                  FRANCE

Summary

From the concept developed in 1978 of a double layer cultivator and the experimental results obtained in 1979 at a 1 m^2 scale a microalgae growing unit of 6 m^2 has been set up in 1982 and a pre-industrial plastic solar receptor prototype was built in 1983.

The economic analysis done in 1980 showed the necessity to focus on processes for production of exocellular chemicals representing at least 30 % of the total biomass with a selling price ⩾ 10 FF/kg and a small to medium size market. The best target seems to be *carragheenan* type polysaccharides used in food industry as gelling agents.

After having successfully tested the continuous culture of *Porphyridium cruentum* on various scales up to 6 m^2, the computer control system with its full instrumentation was connected and operated on line.

For the first time, *Botryococcus braunii* was cultured in natural solar conditions on 6 m^2. Experiments showed the intense sensitivity of this microalgae to contaminants.

Partition of the solid phase from the liquid phase of *Botryococcus braunii* and *Porphyridium cruentum* cultures using ultrafiltration and rotary filter (diatoms bed) was investigated.

INTRODUCTION

From the concept developed in 1978 (1) of a tubular double layer cultivator in order to associate a culture of microalgae and a culture of photosynthetic bacteria for a better use of solar energy, and from experimental results obtained in 1979 (2), with a microalgae unit of 1 m^2 (fig. 1), a tubular glass cultivator of 6 m^2 (fig. 2) and a 10 m^2 outdoor plastic one (fig. 3) have been built respectively in 1982 and in 1983.

In 1980 a technico-economic analysis on microalgae production in a tubular system (3) leads us to only consider the use of microalgae species able to produce chemicals in a proportion equal or greater than 30 % of their dry weight. With such a growing unit, a yearly average photoconversion yield of 4 % of total solar energy was obtained, which represents, in our mediterranen conditions, a productivity in total biomass of 76 t/ha/year (2) which leads to a possible production of at least 25 t/ha/year of chemicals.

Two types of products can be considered (4) :

- Exocellular products (with an easy recovery step) such as :

. Polysaccharides : several microalgae species are able to produce exocellular polysaccharides which can represent up to 50 % of the total biomass : *Porphyridium cruentum, Porphyridium aeruginum, Rhodosorus marinus, Rhodella maculata, Chlamydomonas mexicana*. Such polysaccharides are interesting in food industry (as gelling agents) and in oil recovery (to increase viscosity of solutions) (5).

. Pure hydrocarbons. The production by green microalgae *Botryococcus braunii* can reach 30 to 40 % of the total biomass (6).

- Endocellular products (with a difficult recovery step) :

. Triglyceric lipids (*Chlorococcum* or *Neochloris*).

. Lipids (*Phaeodactylum*).

. Osmoregulators in hypersalinity conditions : glycerol (*Dunaliella*), Sorbitol (*Stichococcus*), mannitol (*Platymonas*), Cyclohexanetetrol (*Monochrysis*).

. Acrylic acid (*Phaeocistis*).

MATCHING BIOLOGICAL OFFER WITH INDUSTRIAL NEEDS

Whatever the type of production, after recovery of the interesting chemical, 70 % of the total biomass produced consists of residual cellular biomass (50 t/ha/year) which can be either converted by fermentation into alcohol or methane, or directly used as fish food.

To adapt the biological process to the industrial constraints, three technologies are available.

1 - The open pond culture technique such as the Mariculture in Land project (7). If this system of culture seems to be suited in very specific cases, to energy production, effluent treatment or food production, it does not appear to be well adapted to the well controlled production of a specialty chemical.

2 - The tubular culture system can lead to a steady production if all the culture conditions are monitored (8) and particularly if the thermal problem can be solved. Production costs in such a system are estimated between 3 and 7 FF/kg of dry total biomass (9). Consequently it appears that this technology can be used for achieving our present economic goal of a high value product with a market price above 10 FF/kg.

If we consider the possible biological processes, presently, only polysaccharides production (as visquous or gelling agents) with a selling price from 30 to 80 FF/kg could be exploited with this technology. The glycerol is a possible production within a 10 years term. Starting from propylene as raw material, the petrochemical production cost is 6,2 FF/kg. A bioproduction with microalgae cultures is estimated at 5,2 FF/kg, but everything else being equal, capital investments would be 4 times as high as the one necessitated by the petrochemical process (based on the same market volume). Sorbitol can be compared to glycerol for production and sell prices. Hydrocarbons (produced by a culture of *Botryococcus braunii* in a tubular cultivator) with a production cost estimated at 10 FF/kg cannot be considered, at the present time, as an interesting energy vector and within the next 10 years could become a rather pure raw material for the cosmetic industry.

With regard to industrial needs, two criteria have to be taken into account :

- Size of the market : this new sophisticated technology will be developed and improved only if it is applied to small markets requiering small size production units. Taking the example of carragheenan type polysaccharides the world market is 10000 t/year and the French production 2000 t/year (10). 80 ha of microalgae cultures would cover the French needs and production units of only a few hectares would be economically viable. In the case of sorbitol the production of USA ans Europe together is 60000 t/year. This market would require 2400 ha of microalgae culture, which at the present state of the technology cannot be realistic.

Thus, in a first step, in order to test and improve the production technology and therefore reduce the production costs, only high added value compounds and small markets will have to be considered.

- Energetic balance. Even in the case of the production of a high added value compound, the residual cellular biomass can be considered as a by-product of the process and converted to a high energy containing product such as ethanol or methane. The energy balance, that is to say the ratio of energy harvested (as exocellular and cellular biomass) over the energy spent to produce it (plastic materials, nutrients...) must be greater than 1. The energy analysis made in 1980 indicate that this balance is close to 1 (9). This analysis points out the bottlenecks and the eventual improvements in the technology.

3 - The controlled continuous production in a biophotoreactor using immobilized cells of microalgae

The utilisation of immobilized photosynthetic cells producing the desired chemical in the culture medium allows one to avoid the separation step of the liquid from the solid phase and appears as an interesting solution for the future. More work must be devoted to the subject in order to improve the technique.

Production of exocellular sulfated polysaccharides by immobilized cells of *Porphyridium cruentum* was already obtained (11) (12) (13).

Production of exocellular hydrocarbons by immobilized cells of *Botryococcus braunii* entraped in polyurethane foams was demonstrated by electron microscope studies.

The technology making polyurethane foams was improved. We are now able to fabricate polyurethane films with a porous face (open porosity) containing immobilized cells and a water proof translucent face (non open porosity).

RESULTS

1 - 6 m^2 microalgae growing unit

This unit was already described in a former report (January 1983) (15). It is composed of a tubular solar receptor made of glass tubes with a modulable surface area from 1 to 6 m^2. All the measurement instruments and analyzers are connected to a data logging system. Results are computed, printed and plotted an a real time bases.

Productivities in total biomass are calculated from experimental ponctual measurements (optical density and dry weight) and from computed data : carbon balance (from gaseous CO_2) and dissolved oxygen balance. More work will be devoted to improve the accuracy of dissolved CO_2 and gaseous oxygen balances. However, it can be emphasized that mass balances already allow us to obtain a rather good estimate of the actual productivity and therefore will be used in the future in order to control and hopefully optimize the overall biomass production (fig. 4).

- Culture of *Porphyridium cruentum* in the microalgae growing unit

    Continuous cultures (with residence time of 4 days) of *Porphyridium cruentum* were performed.

    . at the 1 m^2 scale from January to April 1982. During this period the average productivity was 8 g/m^2/day,

    . at the 2 m^2 scale (Mai - June 1982) with an average productivity of 15 g/m^2/day,

    . at the 3 and then 6 m^2 during few days (mechanical tests).

    The main purpose of this set of experiments was to test the 6 m^2 unit, to set up all the analyzers (instrumentation hardware part of the computer system) to check that all the equipment could be used whatever the surface of solar receptor.

- Culture of *Botryococcus braunii*

For the first time *Botryococcus braunii* was cultured on a continuous basis in natural sunlight conditions. (December 1982, January, February, April 1983) at the 6 m^2 scale (in cooperation with ENSCP, Mme CASADEVAL, contrat n° ESE-R-022-F).

The serie of test runs shows the intense sensitivity of this microalgae to contaminants : the proportion of *Selanestrum, Chlamydomonas, Scenedesmus, chlorella,* reached up to 50 % of the total biomass. Several improvements were done to dominate this problem : new system for water purification (used for culture medium) UV sterilisation of medium, filtration of the gas mixture at the inlet of the cultivator. The idea of recycling of the overfloating part of a settling column in which harvested culture is stored (to enrich the culture with young cells of *Botryococcus braunii* poors in hydrocarbons) was discarded (this technique resulted in a simultaneous enrichment with contaminants).

These cultures allowed us :

. to improve the growing unit on a mechanical point of view (continuous culture system),

. to test the software structure (data logging system),

. to set up a calibration process for all the measuring electrodes (dissolved O_2 and CO_2, pH, temperature),

. to develop an apparatus for continuous measurement of the optical density of the culture.

The purpose is to follow growth oscillations during day and night alternations and to convert the cultivator from a chemostat system into a turbidostat system. Alternation of cell division periods and increase in size periods observed with a continuous culture of *Scenedesmus acutus* in the 1 m^2 growing unit was confirmed : growth oscillation of cultures of *Porphyridium cruentum, Rhodosorus marinus, Chlamydomonas mexina* have been studied with coulter countersize analysis, hematocryte technique and optical density mesaurement (16).

A set of 5 turbidostat growing units (400 ml of culture each) was installed beside the 6 m^2 cultivator. They will be used in the same culture conditions (in particular solar energy) to determine the best dilution rate for an optimal exocellular production.

. To begin preliminary investigations on a CO_2 feed-rate control system. The idea, here, is to feed CO_2 to the culture accordingly and proportionally to the intensity of the solar energy.

. To compare different methods used to estimate biomass production. If there is a good correlation between gaseous CO_2 and dissolved oxygen balances, CO_2 balance (calculated from CO_2 concentration in the gas mixture at the inlet and the outlet of the cultivator) seems actually to be the most accurate.

2 - <u>Techniques for partition of solid from liquid phases have been investigated</u>

- Ultrafiltration (on porous membranes from 50 to 600 Å). Good results were obtained with *Botryococcus braunii* cultures : no clog up of membrane (200 and 600 Å) was observed after 36 hours of experimentation and a concentration coefficient of 22 to 24 was obtained. With cultures of *Porphyridium cruentum* (visquous solutions) clog up of membrane appeared after only 2 hours of experiment.

- Rotary filter (with a diatoms bed) is under experimentation and seems to be suitable for these two organisms.

THE 10 m^2 PROTOTYPE CULTIVATOR

An outside stand for testing pre-industrial prototypes of solar receptor was set up. It can be connected to most of the hardware of the 6 m^2 growing unit (data logging system, mixing column, measuring apparatus...).

A solar receptor made of plastic tubes of 200 μ in thickness was constructed (by CEA PIERRELATTE). It is composed of 8 tubes of 19 m length and 6,5 cm in diameter, which represent a 10 m^2 surface exposed to solar irradiation and a 500 litres culture volume.

Modifications are in progress to improve the hydrodynamics of the system. The total pressure loss of the 10 m^2 cultivator is about 180 mbars (for a liquid flow of 600 l/h) with 60 mbars for the plastic solar receptor.

Thermal analysis of the process will start soon. Temperature regulation will be performed using a 2 m^3 water bath, which itself is cooled by addition of cold water (maximum flow 700 l/h).

A prototype made of plastic tubes of 400 μ in thickness is under study.

CONCLUSIONS

The final target is the energy production, but the suitable technology will not be developed before the next 10 years. At present, microalgae culture in a tubular unit is already a possible way of bioproduction of high added value specialty chemicals.

Surface area of cultivation was progressively extended to 10 m^2 with steps of 3 and 6 m^2.

The system is highly instrumented and computerized so that we can make accurate mass balances. This will allow us to continue in depth investigations for the optimization of biomass production.

Different means of controlling growth and metabolism are now available to us :

- variable dilution rate (chemostat),
- variable biomass concentration (turbidostat),
- variable CO_2 feed rate.

Both test cultures (*Botryococcus braunii* and *Porphyridium cruentum*) were performed at the 6 m^2 scale. Possible methods for recovering the interesting chemical (hydrocarbons or polysaccharides) as well the cellular biomass itself have been studied.

REFERENCES

1 - GUDIN C., CHAUMONT D.
   Bioconversion de l'énergie solaire dans un système double couche. "Héliosynthèse et Aquaculture". MARTIGUES (FRANCE) 1978 - Edition CNRS.

2 - GUDIN C., CHAUMONT D., DESANTI O., PIOLINE D.
   Pour une biothechnologie solaire basée sur les microalgues. Revue Internationale d'Héliotechnique (FRANCE) 2ème semestre 1980.

3 - GUDIN C., CHAUMONT D., DESANTI O., PIOLINE D.
   Culture continue d'organismes cellulaires chlorophylliens pour des productions spécifiques. Revue du Palais de la Découverte n° Spécial 21. Juillet 81.

4 - GUDIN C.
   Microalgae for biomass, chemicals and high energy products (Present state and future developments). Energy from biomass - 2nd E.C. Conference Berlin 1982. Applied Science Publishers.

5 - BRAUD J.P., CHAUMONT D., GUDIN C.
   Microalgues productrices de polysaccharides exocellulaires. Colloque sur la Valorisation des Végétaux Aquatiques Bonbannes FRANCE (16-19 Novembre 8 Sous Presse.

6 - CASADEVAL E.
   Renewable hydrocarbon production by cultivation of the green algae *Botryococcus braunii*. Investigation of the factors affecting hydrocarbon production. Energy from biomass. Serie E. Proceedings of the EC contractors' meeting held in Brussels. May 1982. p 142-149. D. REIDEL Publishing Company.

7 - WAGENER K.
   Methane production by mariculture on land. Energy from Biomass. Serie E. Proceedings of the EC contractors' meeting. Brussels. May 1982, p. 102-106. D. REIDEL Publishing Company.

8 - GUDIN C., CHAUMONT D.
   A biotechnology of photosynthetic cells based on the use of solar energy. Biochemical Society Transactions 8, (4), August 1980.

9 - GUDIN C., CHAUMONT D., BERRA E.
   New concepts in solar biotechnology. Energy from biomass. 1st E.C. Conference - Brighton 1980 - Appl. Sci. Publish.

10 - GUDIN C.
   La biomasse algale. Cahier de l'AFEDES (FRANCE) n° 6. 1981.

11 - GUDIN C., THEPENIER C., THOMAS D., CHAUMONT D.
   Solar and CO_2 biotechnology:production of exocellular biomass by a biophotoreactor of immobilized cells. Solar World Forum. Brighton. August 1981.

12 - GUDIN C., THOMAS D.
   Production de polysaccharides sulfatés par un biophotoreacteur à cellules immobilisées de *Porphyridium cruentum*. Comptes rendus de l'Académie des Sciences. Série III, vol. 293, 1981 (p. 36-38).

13 - GUDIN C., CHAUMONT D.
   For a biotechnology based on microalgae. Energy from biomass. Serie E. Proceedings of the EC contractors' meeting.- held in Copenhagen - June 1981 - p. 81-84.

14 - THEPENIER C., GUDIN C., THOMAS D., BROUERS M., BARBOTIN J.N.
   Immobilisation de microalgues et productions exocellulaires. Colloque sur la valorisation des Végétaux Aquatiques Bombannes (FRANCE).
   16-19 Novembre 1982. Sous presse.

15 - GUDIN C., CHAUMONT D.
   Solar biotechnology study and development of tubular solar receptor for controled production of photosynthetic cellular biomass for methane production and specific exocellular biomass. EC contract n° ESE-R-073-F (SD). Half term report - January 1983.

16 - CHAUMONT D., GUDIN C., CHASSIN P., LEMAIRE C.
   Production de métabolites et croissance rythmique des microalgues en photoréacteur. Colloque annuel de la Société Française de Microbiolobie "Potentialités biotechnologiques de la photosynthèse - 10 et 11 Mars 83. Université de Technologie de Compiègne".

Fig.1 - Microalgae growing unit of 1m² (1979)

Fig.2 - Microalgae growing unit of 6m² (1982)

Fig.3 - The 10m² prototype of solar plastic receptor (1983).

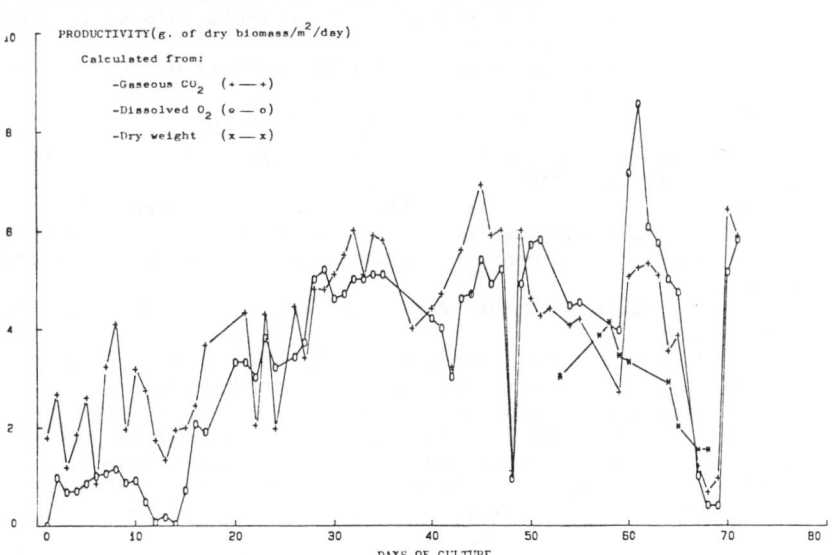

Fig.4 - Continuous culture of *Botryococcus braunii* in natural sunlight conditions. Comparison of different methods for the estimation of biomass produced.

CULTURE OF A HYDROCARBON PRODUCING ALGA, BOTRYOCOCCUS BRAUNII, AT PILOT LEVEL

AUTHORS : DESTORDEUR M, ROSSI M.E.
CONTRACT N° : ESE-R-074-B
DURATION : 18 months (1.1.82 → 30.6.83)
TOTAL BUDGET : 4.920.425 BF, CEC contribution : 25 %
SPONSORS BESIDE CEC : INIEX
　　　　　　　　　　　　University of Liège
HEADS OF THE PROJECT : NOEL R. : INIEX
　　　　　　　　　　　　　SIRONVAL C. : University of Liège
CONTRACTORS: INIEX
　　　　　　　　Rue du Chéra 200
　　　　　　　　B-4000 LIEGE

　　　　　　　　UNIVERSITE DE LIEGE
　　　　　　　　Lab. de Photobiologie B22
　　　　　　　　B-4000 LIEGE (Sart-Tilman)

ABSTRACT

　　The research consisted in cultivating the alga B. braunii in an open air pilot installation in order to see if it is possible to make it produce important quantities of hydrocarbons.
　　A culture installation was developed for the culture of this alga at pilot level (500 ℓ tanks).
　　The mass culture experiments mainly demonstrated that :

1. the alga is capable of growing in open air from April to October (temperature between ± 8°C and 32°C).
2. the B. braunii culture is rapidly contaminated (2 to 3 weeks after seeding) by other species such as Scenedesmus and Chlorella. No correlation was established between the moment the contaminated appeared, and temperature or rainfall.
3. the culture's hydrocarbon outputs don't exceed 10 % as compared to the dry weight of the algae.

　　The essence of the research consisted in the description of the evolution of the culture (number of cells, rainfall, temperature, culture medium analysis, etc...), as well as in finding means to avoid the poliferation of algae contaminating the culture need to increase the hydrocarbon production of B. braunii. These means as well as experiments aiming at the valorization of algae of which the hydrocarbon have been extracted and their qualitative analysis are described in the present report.

INTRODUCTION

　　The aim of the research was to cultivate the hydrocarbon producing algae, Botryococcus braunii, in an open air pilot installation in order to use the hydrocarbons synthetized by the alga.

In the past the numerous research projects undertaken on this alga were mainly concerned with appropriate culture mediums for the alga, as well as with the chemical study of the hydrocarbons produced by the alga cultivated in the laboratory or taken from its natural medium. The highest poliferation of the alga and its highest hydrocarbon content having been observed in the case of algae taken from nature, we wanted to cultivate this alga in mass and in open air in order to attempt to obtain results which would be comparable to those sporadically observed in the natural medium.

The first objective to be reached was the design of a culture installation adapted to the alga. This installation, its functioning should be simple and at low cost because the expected products are hydrocarbons, products with low added value.

The second objective to be realized was the acquiring of methods allowing to ensure the maintaining of the culture (evaluation of the growth of the algae, their hydrocarbon content, determination of the main nutritional elements of the medium, measuring of the main atmospheric parameters on the culture level).

1.2. Description of the culture installation

The alga suspension is cultivated in stainless steel 500 ℓ tanks (dim; : 1 x 1 x 0,75 m). A valve in the bottom of each tank makes the algae harvest easier. A tube with holes of increasing diameter, running along the shells of the tank, at 20 cm distance, ensured the agitation and the aeration of the culture thanks to the compressed air circulating inside it. The culture is aerated 24 hours on 24 by means of air not enriched with CO_2.

This installation type allows the maintaining of a colonial behaviour of the alga.

1.3. Results

1.3.1. Cultures

The first open air culture test was carried out by inoculating, in the summer, a tank by means of semi-sterile B. braunii cultures prepared in the laboratory. At the beginning of the experiment, the number of B. braunii cells was $8.10^4.mℓ^{-1}$. During this experiment, the medium conditions were voluntarily kept unchanged in order to observe the culture's natural evolution. This experiment has shown that :

a) Botryococcus can grow in the open air; the strongest growth was observed during the first month following inoculation (1) (evolution from 8.10^4 cells.$mℓ^{-1}$ to 22.10^4 cells.$mℓ^{-1}$ in one month).
b) Only 2 species of algae compete Botryococcus in its medium : Scenedesmus and Chlorella. In wintertime no apparition of different species is observed in the medium. The temperatures registrated at the culture level varied from - 15°C to + 25°C.
c) After 6 weeks of culture, a fall in the number of algae is registrated during 2 months. This fall can be explained by the fall of the average temperature as well as by the exhaustion of the main nutrition elements of the medium. Indeed, after 7 days of culture, the phosphatic concentration of the medium diminishes from 22 mg/ℓ to 1 mg/ℓ, and after two months the nitrates concentration of the medium falls from 150 mg/ℓ to 20 mg/ℓ (1).

Following their progressive decrease, the number of algae practically stabilises for 3 months.

After the thaw of the culture, the 3 species (Botryococcus, Scenedesmus, Chlorella) are found in identical proportions as before the frost. No greater <u>resistance to low temperature</u> and <u>lack of nutrition elements</u> has been observed in Botryococcus or its contaminants.

d) For <u>the collecting of the algae</u> (harvesting), the agitation and airing are brought to an end for 24 hours. The precipitation of most of the algae is then observed. A hollow cylinder (height : 10 cm) has been adapted to the draining valve of the culture device, thus allowing for the elimination of the elements floating on the surface. When the cylinder has been removed the algae still on the bottom of the tank flow out via the valve and are stocked in 10 ℓ carboys. After 24 h of decanting, elements come to float on the surface, which are eliminated through pumping. The algae which precipitated, were then lyophilized. The dry weight of the harvest contained 6.10^4 cells of Botryococcus/mℓ; 65.10^4 cells of Scenedesmus/mℓ and 35.10^4 cells of Chlorella/mℓ, is 0,18 g/ℓ when the weight of the insoluble atmospheric fallout elements having reached the culture has been taken into account, i.e. 198 mg/m^2.day or 5,94 g/m^2.month.

Rainfall varied from 10 to 120 ℓ of water/m^2.month.

e) The <u>hydrocarbon content</u> of the culture appeared at every moment of the culture (sampling and analysis every fortnight) to be very low (1,5 % of the dry weight of the collected algae).

The existence of a causal relation between the Scenedesmus contamination on the one hand and the low hydrocarbon output of Botryococcus on the other, is highly probable. It is well known, indeed, that Scenedesmus secretes antibiotics in the medium (5), and that a Botryococcus stock which has practically no bacteria, does not produce hydrocarbons (case of the Göttingen stock) (6).

Laboratory experiments were conducted to try to eliminate the contaminants or to prevent their development, by the introduction into the Botryococcus culture of a green filamentous algae : Hydrodictyon reticulatum. It has, indeed, been observed that in a mass open air culture of these algae, the culture is contaminated by algae which differ strongly morphologically from Botryococcus (and are thus easily separable by a simple physical means), and no proliferation of Scenedesmus or Chlorella is observed.

Experiments showed that :
1) in the case of cultures already contaminated by Scenedesmus and Chlorella at the moment of introducing Hydrodictyon reticulatum in the medium, the presence of this alga does not restrain the proliferation of Scenedesmus and Chlorella;
2) in the case of uncontaminated Botryococcus cultures at the moment of introducing Hydrodictyon reticulatum in the medium, the presence of Hydrodictyon reticulatum prevents the apparition of Scenedesmus and Chlorella in the culture, whereas in all the check-up cultures (identical cultures without Hydrodictyon reticulatum), Scenedesmus and Chlorella appear and proliferate.

Laboratory experiments have been conducted in order to test the influence of parameters (2.3.4.) which are supposed to increase the algae's synthesis of lipids, precursors of hydrocarbon on the production of these by Botryococcus.

The following parameters were tested : lack of phosphates and/or sulphates in the medium.

The results show that, compared to the check-up (complete media), the hydrocarbon content is higher in the impoverished cultures. The hydrocarbon output decreases as follows; cultures without PO_4 and without SO_4, cultures without PO_4, cultures without SO_4 and check-up cultures.

On the other hand, in order to check if a different Botryococcus stock from the one which was cultivated up to now (LB 807/1 Cambridge) could not produce more hydrocarbons, a stock from the University of Göttingen was cultivated (Cambridge stock without the bacteria which colonize it, except for one species). This stock appeared to be very poor as regards hydrocarbons (hydrocarbon output of more or less 1 % of the dry algae weight), even when cultivated in a medium with poor PO_4 and/or SO_4.

The preventive introduction of filamentous algae, Hydrodictyon reticulatum into the B. braunii culture medium was applied at pilot level.

Concerning the preventive effect of the Hydrodictyon introduction with regard to the proliferation of Scenedesmus and Chlorella, the experiment has shown (with regard to the reference culture without Hydrodictyon) that the introduction of Hydrodictyon :
1. slows down the B. braunii growth (Hydrodictyon takes nourishing elements away, darkening of the culture medium);
2. delays the appearance of Scenedesmus and Chlorella, but does not prevent the appearance and the proliferation, in opposition to what happens in laboratory cultures. In the reference tank, the Scenedesmus and Chlorella appear after about 3 weeks culture, while in the other tank they appear after 4-5 weeks culture;
3. after about 1 week culture causes the appearance of zooids in the medium (proceeding from Hydrodictyon) as big as Chlorella.

After 2 weeks culture, part of the algae are sampled in the "reference" tank (without Hydrodictyon) and in the "test" tank (with Hydrodictyon) (10 ℓ in both cases) to compare the hydrocarbon yields.

In the case of the "test" tank, the zooids are found in the filtrate with B. braunii, since they are not held back by the filter during the collecting.

In this case the evaluation of hydrocarbon yields produced by B. braunii, realized by extraction and the ratio of their weight over the dry weight of the collected algae is tainted with an error by default because of the presence of the zooids.

The hydrocarbon yields for both cultures are roughly identical (7,3 % for the "reference" tank, and 7,1 % for the "test" tank). These yields indicate that the hydrocarbon production by B. braunii is comparatively more important in the case of the "test" culture than in the case of the "reference" tank. This result can be explained by the fact that the Hydrodictyon deprives B. braunii from food material (nitrates, phosphates, etc...).

Indeed, it was shown in the previous report that the absence of PO_4 or SO_4 in the culture medium causes an increase of the hydrocarbon yield of B. braunii.

The effect of PO_4 and SO_4 starvation was studied at the level of open air culture on the growth of B. braunii and of the contamination as well as on the hydrocarbon production of B. braunii.

In the previously described open air culture installation, two tanks were seeded with a pure B. braunii culture; the "pilot tank" containing the complete culture medium (Chu medium), and the "test tank" containing the same culture medium without SO_4 and PO_4.

The results are the following :
1. As compared to the pilot tank, the growth of B. braunii is slightly lower than in the test tank,
2. Scenedesmus and Chlorella appear simultaneously in both tanks,
3. The hydrocarbon content is largely identical in both tanks, i.e. 6,5 % for the pilot tank and 7,8 % for the test tank. The hydrocarbon output in-

crease in the cultures without SO$_4$ and PO$_4$ is lower in the case of the open air culture than in that of the laboratory cultures. This may be explained by the PO$_4$ and SO$_4$ input of the rain. Indeed, the analysis of rain samples at the culture level indicates a PO$_4$ content of 0,09 mg.ℓ^{-1} up to 0,33 mg.ℓ^{-1} following the harvesting period. The non-stavred Chu medium contains 13 mg of PO$_4$ per ℓ^{-1}.

The SO$_4$ content of the rain is 1 to 14 mg.ℓ^{-1} following the harvesting period.

The SO$_4$ content of the non-starved SO$_4$ medium is 20 mg.ℓ^{-1}.

The rainfall varies from 0 to 60 ℓ/m^2 and per week, whereas the initial culture volume is 200 ℓ. The quantity of nutritional elements introduced by the rain is not negligeable, and the experiment is not quite comparable to the one carried out in laboratory culture since in the case of the open air culture there is no real starvation.

1.3.2. Valorisation of the algae after hydrocarbon extraction

The highest heat capacity at constant volume of dry Botryococcus of which the hydrocarbons have been extracted (pure semi-sterile laboratory cultures) were determined following the Belgian norm NBN M02-010.

The result was 3,850 kcal/kg (16,17.10^6 J/kg). As a comparison : dry wood has, in the same conditions, an average maximum heat capacity of 4,700 kcal/kg.

The total protein content of B. braunii after hydrocarbon extraction, determined by the Lowry method, is evaluated at about 28 % average with regard to the dry algae weight.

1.3.3. Qualitative analysis of the hydrocarbons

The absence of absorption peaks characteristic of groups which are foreign to the C-C or C-H bonds, at the registration of the IR-Spectrum (external and internal) proves that pure hydrocarbonated extracts are obtained. The presence of absorption peaks characteristic of the vinyl group (= CH$_2$) indicates the existence of olefines in our extracts (fig.).

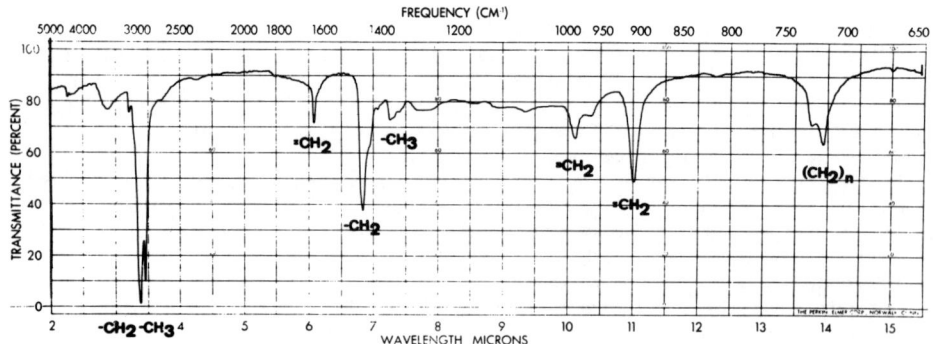

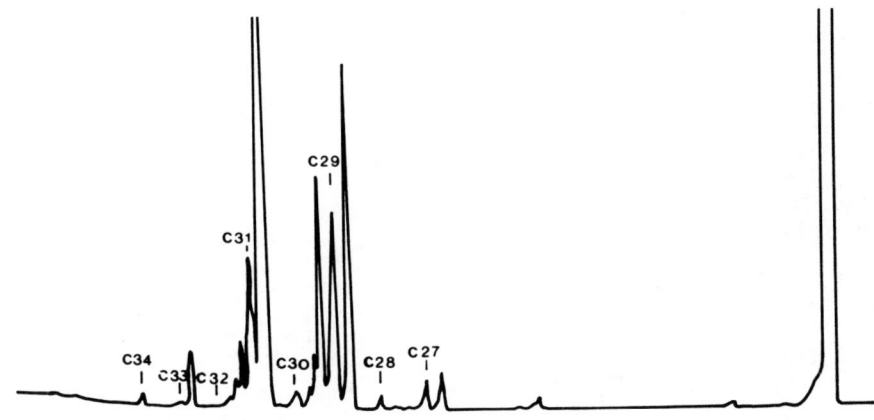

The GLC of the extracted hydrocarbons (external and internal) shows that it is a complex mixture of hydrocarbons (fig.). The most important among them have a number of carbon atoms equal to 27-29-31. The internal hydrocarbons sometimes have, besides the hydrocarbons which are commun to the external hydrocarbons, shorter hydrocarbons (C_{15} - C_{17}). No correlation was observed between one or other culture condition and the presence of one or other hydrocarbon.

CONCLUSION

The mass culture experiments of the alga B. braunii show :
1. that it is possible to make this alga grow in open air in the Belgian climate conditions, from April to October.
2. The difficulty to obtain a pure monoculture associated to high hydrocarbon outputs (> 10 % of the dry weight of the algae).

The solutions applied to the problems mentioned under 2) above were successful in the case of laboratory cultures (closed cultures), but did not work in the case of the open air culture because the culture is in an open system.

If it seems, then, that the described open culture system is of a simple design and economic functioning, it has nevertheless not allowed to produce an important biomass associated to high hydrocarbon outputs.

On the basis of the results we obtained, it would be interesting to cultivate the alga in mass in 2 culture systems, i.e. a system allowing the production of an important biomass (complete medium), and a culture system allowing part of the biomass produced in the first system to produce important quantities of hydrocarbons (starved medium).

REFERENCES

1. DESTORDEUR - ROSSI - SIRONVAL (1981)
   Energy form biomass
   Solar Energy R et D in the European Community
   Serie E/Vol. 2, p. 153 - 165

2. HEALEY, HENOZEL (1975)
   J. Phycol. 11, 303 - 309

3. SPOEHR, MILNER (1949)
   Plant Physiol. 24 : 120 - 149

4. OTSUKA (1961)
   J. Gen. Appl. Microbiol. 7 : 72 - 77

5. JØRGENSEN (1962)
   Physiol. Plant, 15

6. MURRAY, THOMSON (1977)
   Phytochem., 16 : 465 - 468

7. LOWRY - ROSEBROUGH - FARR - RANDALL
   J. Biol. Chem. 193, 265 - 275 (1951
   Protein measurement with the Folin phenol reagent

RENEWABLE HYDROCARBON PRODUCTION BY CULTIVATION OF THE GREEN ALGA
BOTRYOCOCCUS BRAUNII

INVESTIGATION OF THE FACTORS AFFECTING HYDROCARBON PRODUCTION

Contract Number : ESE-R-022-F(S.D.)

Duration   : 36 months   1 January 1981 - 31 December 1983.

Total Budget   ; F 778 720   contribution C.E. 30 %.

Head of Project : Dr. E. CASADEVALL (Directeur de recherche CNRS).

Contractor   : Ecole Nationale Supérieure de Chimie de Paris.

Address   : Laboratoire de Chimie Bioorganique et Organique
                  Physique
                  E.N.S.C.P.
                  11, rue P. et M. Curie - 75231 PARIS CEDEX 05

Summary.

The green unicellular alga Botryococcus braunii with a hydrocarbon content as large as 75 % of the dry weight holds the first place among energy-rich biomass. It can be assumed that cultivation of this alga would afford a direct and renewable source of "solar fuel". With the object of checking this hypothesis it was planned first, to test the feasability of an industrial production of hydrocarbons by large scale cultivation of Botryococcus and then, to determine the culture conditions allowing to reach economic rentability.

Owing to our poor knowledge about this alga, basic studies on its morphology and metabolism along with researchs on parameters of its culture have been carried out. Our mains results showing :
- exocellular localisation of hydrocarbons,
- highest hydrocarbon productivity during active stage of growth and substantial improvements of productivity by variation of culture conditions, prove, indeed, that large scale cultivation of Botryococcus for industrial production of hydrocarbons can be realized.

Recently, culture at a prepilot scale (6 m^2) has been experimentated in collaboration with GUDIN team in the tubular photoreactor located at Cadarache (CEA). It can be considered that positive results have been obtained.

At this time all the objectives of our C.E. contrat are on the point of being reached.

INTRODUCTION.

L'algue verte unicellulaire Botryococcus braunii (B.b.) présente la particularité jusqu'içi unique parmi les organismes photosynthétiques examinés, d'avoir un contenu en hydrocarbures pouvant atteindre jusqu'à 75 % du poids de la biomasse sèche. Elle accomplit ainsi directement la bioconversion de l'énergie solaire en un produit qui ayant la même nature chimique que les hydrocarbures fossiles, pourrait en constituer le milieu succédané.

En raison des productivités élevées en biomasse que présentent les cultures d'algues unicellulaires, on pouvait à priori faire l'hypothèse qu'il serait possible d'obtenir, par culture de masse de Botryococcus, une source renouvelable "d'hydrocarbures solaires" susceptibles de se substituer au moins en partie, aux hydrocarbures fossiles.

Le but de nos recherches est de vérifier le bien fondé de cette hypothèse. Pour cela nous avons entrepris :

- de tester la faisabilité d'une production industrielle d'hydrocarbures par culture à grande échelle de Botryococcus.

- de rechercher les conditions pour lesquelles cette production pourrait être économiquement rentable.

Cependant, en raison du peu de connaissances relatives à cette espèce, il était nécessaire, pour atteindre le but fixé, d'effectuer en premier lieu une recherche de base. C'est pourquoi le projet qui a fait l'objet de notre contrat C.E. comportait tout d'abord l'étude des caractères morphologiques et métaboliques de Botryococcus, ainsi que celle des paramètres biologiques et physico-chimiques des cultures susceptibles d'influer sur la production des hydrocarbures. Il impliquait aussi l'expérimentation en laboratoire de différents systèmes de cultures discontinus (batch) et continus (chemostat). Enfin il prévoyait d'effectuer, en collaboration avec l'équipe du CEA (Cadarache), des essais de culture en éclairement naturel à l'échelle du m^2 dans le cultivateur tubulaire mis au point par Claude GUDIN pour la culture industrielle des algues unicellulaires. Les résultats que nous allons rapporter montrent que tous ces objectifs ont dans leur ensemble été atteints.

RESULTATS.

1 - Caractères morphologiques et métaboliques.

A partir d'échantillons de Botryococcus prélevés dans la nature, deux types d'hydrocarbures avaient été identifiés (1).

- Des hydrocarbures linéaires normaux à longue chaîne : alcadiènes impairs de C_{25} à C_{31} et un triène en C_{29}.

- Des hydrocarbures ramifiés polyinsaturés de structure triterpéniques de C_{30} à C_{37} (formule générale C_nH_{2n-10}) : les Botryococcènes.

L'étude ultra structurale que nous avons effectuée sur des Botryococcus en provenance de la collection de Cambridge (souche LB 807/1), couplée à un examen in vivo des cellules par spectroscopie Raman à microsonde nous a permis de localiser la majorité (95 %) des hydrocarbures linéaires, seuls produits par cette souche, dans la paroi externe de l'algue. Des expériences d'incubation de ces mêmes algues avec des marqueurs radioactifs ont montré que ces hydrocarbures linéaires ont l'acide oléïque comme précurseur direct et que la paroi externe est, non seulement leur site d'accumulation, mais aussi celui où sont localisés les systèmes enzymatiques responsables de la transformation (élongation puis décarboxylation) de l'acide oléïque en hydrocarbures linéaires.

Une étude ultra structurale effectuée sur les souches en provenance des collections de Göttingen, Thonon et Austin, qui produisent aussi des hydrocarbures linéaires, et sur des souches sauvages, qui produisent des Botryococcènes, a mis dans tous les cas en évidence la même localisation exocellulaire des hydrocarbures. Des expériences d'incubation des algues produisant des Botryococcènes (souches sauvages), avec de l'acide mévalonique radioactif, a apporté par ailleurs la preuve que la biosynthèse de ces hydrocarbures s'effectue par la voie isoprénique (condensations successives d'unités C_5), ainsi que leur structure terpénique le suggérait.

Le point important mis en évidence par cette étude est la localisation exocellulaire des hydrocarbures, qui permet d'envisager l'utilisation non seulement de moyens de récolte simplifiés, mais encore de procédés de production des hydrocarbures à partir de cellules immobilisées. L'immobilisation de cellules entières de Botryococcus dans des supports poreux transparents à la lumière a fait l'objet d'une étude qui a été précédemment rapportée (2).

2 - Paramètres des cultures.

2.1 - Paramètres biologiques. Influence de la souche sur la nature des hydrocarbures.

Nous avons indiqué plus haut que Botryococcus produit deux types d'hydrocarbures. Des publications antérieures (3) suggéraient que la structure (linéaire ou ramifiée) des hydrocarbures serait fonction de l'état physiologique de l'algue : hydrocarbures linéaires produits pendant la phase verte de croissance active ; botryococcènes pendant la phase rouge de repos. Nous avons montré par la culture en laboratoire de différentes souches d'algues (souches de collections et souches sauvages), que le type des hydrocarbures produit ne peut être relié ni à l'état physiologique, ni à la pigmentation des algues, mais à la souche mise en culture : une souche donnée ne produit tout au long de la culture, qu'un seul type d'hydrocarbures. Sa pigmentation peut passer, en particulier pour les souches à Botryococcènes, du vert en début de culture au rouge en fin de culture, sans que cela traduise un changement de la nature des hydrocarbures produits. Il y aurait donc deux variétés de Botryococcus se différenciant par la nature des hydrocarbures produits. Le choix de la souche permettra donc d'opter pour la production de l'un ou l'autre de ces deux types d'hydrocarbures.

Influence des bactéries contaminantes sur la productivité.

Des données de la littérature (4) suggéraient que la présence de bactéries associées aux cultures serait indispensable pour une forte production d'hydrocarbures par Botryococcus. L'étude la productivité en hydrocarbures de souches axéniques (tableau 1) nous a permis d'apporter la preuve que la présence des bactéries n'est pas nécessaire à une production importante d'hydrocarbures par l'algue. L'étude d'associations volontaires Botryococcus - bactérie (cinq bactéries différentes, choisies en fonction de leur caractère spécifique, ont été testées) a mis nettement en évidence l'influence, soit favorable, soit défavorable, d'une espèce bactérienne donnée sur la production de biomasse (et) ou d'hydrocarbures (figure 1) pour des cultures non carencées en CO_2. Par contre, pour des cultures où le CO_2 est limitant, on observe un effet généralement favorable des bactéries qui interviennent vraisemblablement en enrichissant le milieu en cet élément (5).

Influence de la souche sur la productivité.

La comparaison des productivités rapportées sur le tableau I pour trois souches axéniques fait nettement apparaître des aptitudes très différentes de ces souches à produire des hydrocarbures. La sélection de souches très productives constitue donc un problème important à résoudre avant d'entreprendre une culture à grande échelle.

TABLEAU 1 - Productivité en hydrocarbures de souches axéniques déterminée pour une culture en batch (durée de culture identique pour les trois souches).

	% Hydrocarbures Biomasse	Productivité en mg/l/j
$T_{ax.}$	4	3
$G_{ax.}$	19	46
$A_{ax.}$	26	35

<u>Influence de l'état physiologique de l'algue sur la productivité en hydrocarbures et en polysaccharides</u> :

La productivité en hydrocarbures a été étudiée en fonction de l'âge de la culture (culture en batch) et reliée aux différentes phases du développement : phases de latence, exponentielle, linéaire, de ralentissement et de repos. Parallèlement un examen en microscopie électronique des cellules pour ces différentes phases a permis de visualiser les modifications morphologiques qui traduisent les états physiologiques successifs. Cette étude a montré que la productivité est maximale, quelle que soit la souche examinée, pendant la phase de croissance active (phase exponentielle et début de la phase linéaire). Il apparaît donc qu'une production élevée en hydrocarbures résulte chez Botryococcus de son métabolisme normal et n'est pas liée à un arrêt des divisions cellulaires comme on l'observe le plus souvent chez les espèces qui accumulent des lipides sous forme d'acides gras. Cette étude nous a également permis de mettre en évidence la production, par Botryococcus, de polysaccharides exocellulaires. Cette production faible pendant la phase de croissance active, semble augmenter lorsqu'on atteint la phase de repos, comme on peut l'observer sur la figure II.

Il y a donc la possibilité d'obtenir successivement à partir d'une même culture, deux types de composés exocellulaires intéressants.

2.2 - Paramètres physico-chimiques.

Botryococcus était décrite comme une algue à croissance lente. Des temps de doublement de 8 jours avaient été rapportés (6). Une étude préliminaire de l'influence des paramètres physico-chimiques du milieu (éclairement, concentration en nutriments majeurs NO_3^-, PO_3^-, CO_2, agitation, aération) qui a fait l'objet, d'un précédent rapport (7) nous avait conduits à définir des conditions de culture en batch pour lesquelles le temps de doublement de la biomasse était abaissé à 2,5 j. Ce temps de doublement a d'ailleurs été confirmé à partir de cultures continues. Dans le dernier rapport nous avons ensuite présenté (8) une étude plus approfondie de l'influence des concentrations initiales du milieu en phosphates et en nitrates sur la production de biomasse et d'hydrocarbures. Les résultats de cette étude soulignent que pour les types de milieu utilisés la teneur initiale en phosphate n'a qu'une faible influence sur la productivité des cultures, alors que celle des nitrates joue un rôle prépondérant.

Nous avons poursuivi cette étude de l'influence des nitrates et montré que pour une souche axénique, la production globale d'hydrocarbures comme d'ailleurs celle de biomasse est, pour une même durée de culture (culture en batch), d'autant plus forte que la concentration initiale (0,2 à 3 g/l) en nitrate est élevée. Le tracé des courbes de croissance (figure III) pour des concentrations initiales de 0,2 - 1 et 3 g par litre en NO_3K met en évidence que la teneur initiale la plus faible (0,2 g) devient assez rapidement limitante, alors que les teneurs plus élevées prolongent très nettement la phase de croissance active. Pendant cette phase le contenu en hydrocarbures des cellules se situe à un taux supérieur à 10 %. Il croit progressivement en cours de culture car la productivité en hydrocarbures diminue plus lentement que la productivité en biomasse (tableau 2), pour atteindre de 20 à 30 % en phase de repos.

TABLEAU 2 - Influence des teneurs initiales en nitrate du milieu sur les productivités (en/mg/j rapportés à 1 g de biomasse) de biomasse (B) et d'hydrocarbures (H) (culture en batch - souche Austin axénique).

temps de culture \ (NO_3K) g/l	0,0 *		0,2		1,0		3,0	
	B	H	B	H	B	H	B	H
7 j.	434	58	514	75	580	80	680	74
19 j.	36	11	278	45	325	48	350	38

* Dans ce cas en début de culture la croissance s'effectue aux dépens de l'azote contenu dans l'inoculum.

Pour les mêmes teneurs initiales en nitrate (0,2 - 1 et 3 g/l) nous avons suivi l'évolution, en fonction de l'âge de la culture, de la concentration en nitrate du milieu et de la teneur en azote interne des cellules (tableaux 3 et 4). On observe que cette dernière valeur, voisine de 6 % au début de la phase exponentielle, s'abaisse à 3 % au moment où la courbe de croissance s'infléchit. A 3 % d'azote interne, les algues commencent à jaunir. A ce point de la culture, la teneur en nitrate du milieu est inférieure à 10 ppm (limite de détection). La teneur en nitrate du milieu apparaît donc bien dans ces conditions comme le facteur limitant.

TABLEAU 3 - Variation des teneurs (%) en azote interne des cellules en fonction de l'âge de la culture pour différentes teneurs initiales en nitrate du milieu (culture en batch - souche Austin axénique).

temps de culture \ (NO_3K) g/l	0,0	0,2	1,0	3,0
7 j.	3,4	6,2	5,5	5,9
13 j.	2	4,1	6,2	6,2
19 j.	1,7	1,9	5,3	5,5
27 j.	1,8	1,8	3,4	4,6

TABLEAU 4 - Variation des teneurs en azote du milieu (mg/1 NO_3H) en fonction de l'âge de la culture pour différentes concentrations initiales (culture en batch - souche Austin axénique).

temps de culture \ (NO_3K) g/1	0,2	1,0	3,0
7 j.	79	665	2025
13 j.	16	465	1870
19 j.	traces	40	1650
27 j.	"	traces	1040

Nous avons confirmé pour des cultures continues l'influence déterminante des concentrations en nitrate du milieu sur les productivité en hydrocarbures. Cette étude des paramètres en mettant en évidence
 - que Botryococcus peut avoir en culture une croissance rapide (temps de doublement de la biomasse < 2,5 j).
 - qu'une production élevée d'hydrocarbures découle de son métabolisme normal.
 - qu'une productivité maximale en hydrocarbures se manifeste pendant la phase de division active des cellules, apporte la preuve qu'il sera possible d'obtenir en culture continue des états stables présentant une productivité élevée. Cette donnée très importante montre qu'une production à grande échelle d'hydrocarbures par culture de cette algue est tout à fait envisageable.

3 - Essais de culture en éclairement naturel dans un réacteur tubulaire (en collaboration avec l'équipe de C. GUDIN - C.E.A. Cadarache).
Ce réacteur tubulaire a déjà été décrit et expérimenté pour l'algue rouge Porphyridium cruentum (9). On dispose actuellement de 2 unités de culture de 3 m^2 de récepteur solaire. Chaque unité peut être modulée par tranche de 1 m^2 correspondant à 30 1. de culture. Les essais que nous avons effectués dans le cultivateur étaient destinés à déterminer dans quelles conditions on pouvait l'adapter à la culture de Botryococcus. La souche mise en culture est Göttingen LB 807/1. Le milieu utilisé est celui précédemment décrit pour les cultures en laboratoire (10). L'évaluation de la biomasse produite peut être faite, soit à partir du bilan carbone (analyse par IR des concentrations en CO_2 gazeux à l'entrée et à la sortie du cultivateur) ou du bilan en O_2 (évaluation par électrode à oxygène de l'oxygène dissous à l'entrée et à la sortie du cultivateur, soit par mesure de la densité optique de la culture à 560 ou 670 nm, ou par détermination du poids sec de biomasse continu dans un volume donné de culture, enfin par comptage des cellules. La culture est effectuée dans des conditions non axéniques, en éclairement naturel, sans aucun apport de lumière artificielle.
Run 1 sur cultivateur de 3 m^2. La température était régulée à 24° ± 1° pendant le jour, et à 19° ± 1° pendant la nuit pour éviter de fortes pertes de biomasse par respiration. La durée de la photo période était de 8,6 H (énergie totale 476 Kcal/m^2/j, dont visible 194 Kcal/m^2/j). Le pH non régulé s'établissait à 7,3 pendant le jour, à 8,3 pendant la nuit. L'aération était effectuée par de l'air enrichi en CO_2 (1 %). Le système fonctionne en circuit fermé (batch). L'évolution de la biomasse est rapportée ci-après. L'essai a été limité à 10 jours, car un incident mécanique (rupture d'une électro-vanne) a entraîné l'arrêt de l'expérience.

Temps	Biomasse en mg/l	Hydrocarbures en mg/l	Hydrocarbures / Biomasse %
0	740	190	26
6 j	700	56	11
10 j	730	63	9

On observe une baisse des poids de biomasse et d'hydrocarbures en début de culture, suivie d'une faible remontée. Cette disparition d'une partie de la biomasse peut être imputée, d'une part à la faible intensité d'éclairement (activité respiratoire supérieure à l'activité photosynthétique), d'autre part à la présence importante de bactéries susceptibles de se développer aux dépens de la biomasse et des hydrocarbures.

Run 2 sur cultivateur de 3 m². La température était régulée à 23° ± 1° pendant le jour et 16° ± 1° pendant la nuit. La durée de la photo période était de 9,5 h (énergie totale 679 Kcal/m²/j, dont visible 277 Kcal/m²/j). L'aération était effectuée avec de l'air enrichi en CO_2 (1,3 %). Le pH évoluait pendant le jour de 7,3 à 7,7, pendant la nuit de 8,3 à 8,4. La culture a été maintenue en batch pendant 7 jours, puis on est passé en culture continue avec un temps de renouvellement du milieu de 7 jours. Dans ces conditions on observe une augmentation régulière de la biomasse dans le cultivateur, qui laisse penser qu'un temps de renouvellement plus court pourrait permettre d'atteindre un état stable. Après 36 jours de culture on observe une brusque contamination par une micro chlorophycée du genre Selenastrum. Bien que l'on ait supprimé rapidement la cause de cette contamination (elle était introduite avec le milieu neuf), et tenté d'arrêter son extension, l'algue contaminante dominait après 44 jours (80 % de la biomasse). Avant contamination, un rendement de conversion énergétique (rapporté à l'énergie solaire totale) égal à 3 % avait été mesuré.

Run 3, début de culture sur 3 m², extension à 6 m². La température était de 25° ± 1° pendant le jour et de 21° ± 1° pendant la nuit. Le pH variait de 7,3 à 7,5 pendant le jour et de 8,0 à 8,2 pendant la nuit. La photopériode était de 13 h (énergie totale 1348 Kcal/m²/j, dont visible 558 Kcal/m²/j). L'aération était effectuée avec de l'air enrichi en CO_2 (1,7 %). Après une culture en batch de 5 j sur le cultivateur de 3 m², on passe en culture continue avec un temps de génération de 7 j. Après 13 j. de culture à 3 m², on étend la culture à 6 m². 2 j. après cette opération, on observe une contamination massive par des Chlamydomonas, Scenedesmus et Chlorella. La contamination est totale après 5 j. Il semble que cette contamination se soit introduite par les circuits d'aération.

Ces premiers essais, bien qu'ils n'aient pas été entièrement couronnés de succès ont montré que la culture de Botryococcus en cultivateur tubulaire était possible. Le fonctionnement en batch du cultivateur a conduit à des résultats satisfaisants. En culture continue, l'obtention pour le deuxième run, avant contamination par Selenastrum, d'un rendement de conversion énergétique de 3 %, est tout à fait encourageant.

Parmi les problèmes rencontrés au cours de ces essais, celui de la contamination par d'autres espèces d'algues unicellulaires, semble pouvoir être assez facilement résolu par l'utilisation de filtres efficaces à l'entrée de milieu et sur les circuits d'aération.

Le deuxième problème lié d'ailleurs au premier, résulte du fait que nos essais ont été principalement réalisés dans une période de faible éclairement, ce qui n'a pas permis un développement rapide des cultures. En effet, nous avons pu constater précédemment au laboratoire que de

Fig.I
Influence de la souche bacterienne associée sur la production de biomasse et d'hydrocarbures par des cultures en batch de Botryococcus.

A: Botryococcus axénique
Ps: Pseudomonas oléovorans
Pi: Pityrosporum ovale
Co: Corynebacterium aquatile
Az: Azotobacter chroococum
Fl: Flavobacterium aquatile

☐ Biomasse
■ Hydrocarbures

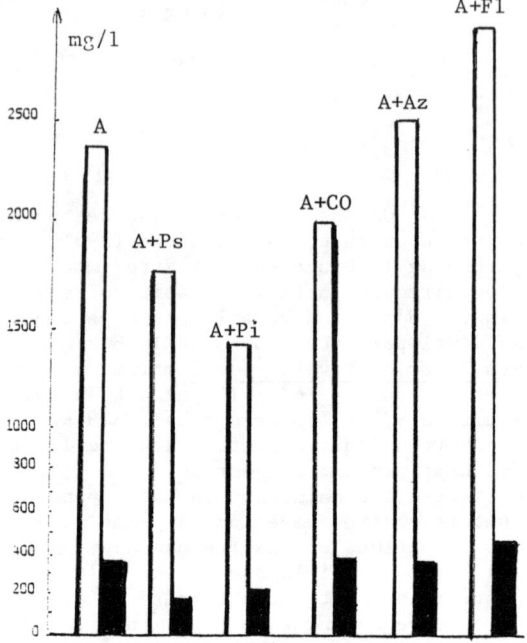

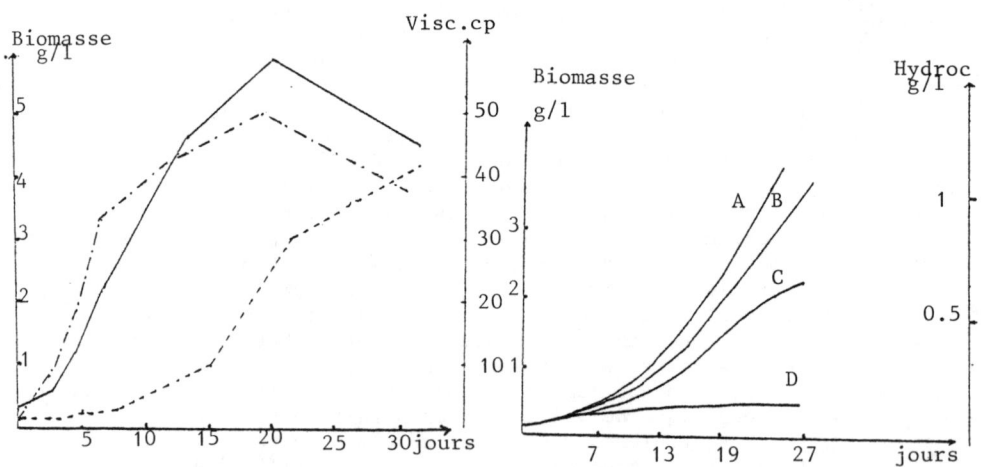

Variations en fonction du temps de culture des poids de biomasse(—) d'hydrocarbures (—·—·) et de la viscosité du milieu (·····)(polysaccharides)
Fig.II

Courbe de croissance pour des concentrations différentes en KNO_3 (A=3g/l; B=1g/l; C=0,2g/l; D=0,0g/l)
Fig.III

– 208 –

forts éclairements avec des photo périodes longues sont favorables à une croissance active de Botryococcus et à une forte production d'hydrocarbures. En conséquence, dans ces conditions de lumière très limitantes, la culture s'est trouvée être particulièrement sensible à la contamination par des espèces moins exigentes sur le plan de l'énergie lumineuse.

Parallèlement aux essais en cultivateur, qui nous l'espérons pourront se poursuivre, il sera maintenant nécessaire d'approfondir en laboratoire l'étude des paramètres pour des cultures en continu avec la lumière comme seul facteur limitant, ce qui n'a pas été fait jusqu'ici. Par ailleurs il conviendra d'essayer de mettre au point un milieu sélectif (addition de certains sels, etc...) et de définir des conditons de culture (pH, température etc...) qui pourraient assurer une protection "naturelle" de la culture contre les contaminations par d'autres algues. Il conviendra également d'expérimenter des agents habituellement utilisés pour la protection des cultures, tels que fongicides, bactérides, etc...).

CONCLUSION.

Les recherches et les expérimentations qui ont été effectuées, nous ont permis d'atteindre les objectifs fixés dans notre contrat. Ainsi nous avons non seulement apporté de nombreux résultats montrant qu'une production d'hydrocarbures par culture de masse de Botryococcus était réalisable, mais nous avons également accompli les premières étapes sur la voie de cette réalisation.

BIBLIOGRAPHIE

1) MAXWELL J.R., DOUGLAS A.G., EGLINTON G. et Mc CORMICK A. 1968 Phytochemistry, 7, 2157. COX R.E., BIERLINGAME A.L., WILSON D.H., EGLINTON G. et MAXWELL J.R. 1973 J. Chem. Soc. Chem. Comm. 284.

2) BAILLIEZ C., LARGEAU C., CASADEVALL E. et BERKALOFF C. 1983 C.R. Acad. Sc. t296, série III, p. 199 et référence 8 ci-dessous.

3) BROWN A.C., KNIGHTS B.A. et CONWAY E. 1969 Phytochemistry, 8, 543.

4) MURRAY J. et THOMSON A. 1977 Phytochemistry, 16, 465.

5) CHIRAC C., CASADEVALL E., LARGEAU C. et METZGER P. 1982 C.R. Acad. Sc. t295, série III, p. 671.

6) BELCHER J.H. 1968 Arkiv. fur Mikrobiol. 61, 355.

7) CHARTIER P. et PALZ W. Ed. Energy from Biomass - D. Reidel publishing company, 1981, vol. 1, p. 96.

8) GRASSI G. et PALZ W. Ed. Energy from Biomass - D. Reidel publishing company, 1982, vol. 3, p. 141.

9) PALZ W. et GRASSI G. Ed. Energy from Biomass - D. Reidel publishing company, 1981, vol. 2, p. 128.

10) LARGEAU C., CASADEVALL E., BERKALOFF C. et DHAMELINCOURT P. 1980 Phytochemistry, 19, p. 1049.

METHANE PRODUCTION BY ANAEROBIC DIGESTION OF ALGAE

I. PILOT PLANT BIOMETHANATION OF CULTIVATED MARINE ALGAE TETRASELMIS FOR ENERGY PRODUCTION IN SOUTHERN ITALY

Authors	: A. Legros, M.R. Tredici, C.-M. Asinari, F. Collard, E. Dujardin, C. Sironval, G. Florenzano, E.-J. Nyns and H. Naveau
Contract number	: ESE/R/025/B (RS)
Duration	: 2 years 1st July 1981 - 30th June 1983
Total budget	: F.B. 26.500.000 C.E.C. contribution : 50 %
Head of project	: Prof. E.-J. Nyns, Unit of Bioengineering
Contractor	: Catholic University of Louvain
Sub-Contractor	: Prof. C. Sironval, University of Liège
Address	: Place Croix du Sud, 1 bte 9 B-1348 Louvain-la-Neuve, Belgium

Summary

It has been shown previously that marine algae Tetraselmis can be transformed into methane by a one step completely-mixed biomethanation process, in a reliable way and with good yields and good methane production rates. This process can be adapted to work equally well in sea water.

The final goal of this research being to install at the pilot scale level an integrated plant for energy production by biomethanation of cultivated marine algae Tetraselmis in Southern Italy, it was first necessary to reach this goal to determine the final possible running conditions for the biomethanation of Tetraselmis algae. A biogas production rate of 1.33 m^3 gas per m^3 digester and per day (66 % methane) has been obtained with a volumetric loading rate of 4 kg volatile solids per m^3 digester and per day and a mean retention time of 14 days. A yield of 0.25 m^3 methane per kg volatile solids added, has been obtained at a salt concentration of more than 10 g sodium per liter with a volumetric loading rate of 3 kg volatile solids per m^3 digester and per day and 14 days of mean retention time.

Based on those and previous laboratory results, a pilot-scale 1 m^3 digester has been installed by the authors at Lamezia-Terme (Calabria, Italy). The digester is actually fed since more than one year with Tetraselmis algae produced by 400 m^2 of culture ponds built and operated by Professors K. Wagener (Aachen, Germany), Florenzano and R. Materassi (Firenze, Italy). It has been run on a moderate volumetric loading rate (0.5 kg volatile solids per m^3 digester and per day). The same methane yield and same methane production rate have been reached with the 1 m^3 pilot digester than with the laboratory scale installations. The one year work with this pilot plant has also given the possibility to integrate all the steps of this energy production system and to show its technical feasability.

1. INTRODUCTION

1.0. Past results

It has been shown previously (1) that methane production from biomethanation of marine algae Tetraselmis can be obtained with good rate and yield and good reliability in a one step continuous and completely-mixed methane digester maintained at 35 °C. E.g., at a low volumetric loading rate of 1.9 kg VS_o x m^{-3}ML x d^{-1}, a yield of 0.26 $m^3 CH_4$ x kg^{-1} VS_o* (volatil solids introduced) can be obtained with a mean retention time of 14 d. Moreover, it has also been shown (2) that this digestion system can be adapted to work equally well in sea water, a point which is essential for the economy of the system in southern Italy.
E.g., the same yield of 0.26 $m^3 CH_4$ x kg^{-1} VS_o can be obtained with a volumetric loading rate of 2 kg VS_o x m^{-3}ML x d^{-1}, a mean retention time of 14 d and a salt concentration of 20 g NaCl x l^{-1} ML.
During the biomethanation of marine algae Tetraselmis, significant differences have been observed between fresh and air-dried algae (2). Higher yields are obtained with fresh algae (0.33 lCH_4 x kg VS_o^{-1}) than with air-dried algae (0.26 lCH_4 x kg VS_o^{-1}).

1.1. Objectives of the research

Four objectives have been assigned to the present contract : (1.) biomethanation of the marine algae Tetraselmis at the 1 m^3 pilot scale at Lamezia (Calabria); (2), general improvement of the methane digestion systems devoted to algae, (3) production of fresh water algae, and (4), evaluation of the potential for methane production of other algal species.
This first part of the report will deal only with results obtained at the laboratory scale showing the final improvement of the biomethanation of Tetraselmis algae and results obtained with the pilot scale installation. Results on production of fresh water algae (objective 3) are presented in the second part of the report (see next paper by Collard and al.).

2. LABORATORY SCALE RESEARCH

2.1. Materials and Methods

2.1.1. Algae : Unicellular marine algae Tetraselmis are grown in ponds at Lamezia-Terme, (Calabria, Italy) (Prof. K. Wagener and R. Materassi, contact ESE/R/O2O/D). Once harvested they are sun-dried or kept fresh after centrifugation. An analytical characterisation of these algae is given in Table I.
Methane digesters : One step continuous and completely-mixed methane digestions with no active biomass recycle were conducted in 2 l all glass vessels described previously (2).
Analytical methods : All analytical methods used in this work have been previously described (2, 3).

---

* for abbreviations and symbols see Table 3.

Table I : Characterisation of *Tetraselmis* algae

VS	Ashes	CO_2 as $CaCO_3$	N_{TK}	N_{NH4^+}	COD
		(g . kg^{-1} TS)			(gO$_2$. kg^{-1} TS)
659	225	162	62	0.72	881

Abbreviations, symbols and units : see Table 3.

2.2. Results

Final improvement of the performances in biomethanation of *Tetraselmis* algae has been done at laboratory scale with and without salt addition. The results of biomethanation of dried algae are reported in Table II.

Table II : Results of the biomethanation of dried *Tetraselmis* algae at high loading rates in one step completely mixed digester

	B_V		$r_{V.gas}$ (% CH_4)	Y_{ec}	Y_{conv}	VA_2	VA_3	pH
	$\dfrac{kg\ VS_o}{m^3 ML \times d}$	(d)	$\dfrac{m^3 gas}{m^3 ML \times d}$ (%)	$\dfrac{m^3 CH_4}{kg\ VS_o}$	$\dfrac{kg\ VS_e}{kg\ VS_o}$	$\dfrac{kg}{m^3}$		
(1)	4	14	1.33 (66)	0.22	0.37	2.62	0.26	7.8
(1)	3	14	0.91 (66)	0.20	0.40	± 3.00	0.20	7.8
(2)	4	14	1.03 (66)	0.17	0.43	2.65	1.92	7.7
(3)	3	14	1.15 (66)	0.25	0.41	0.20	± 2.00	7.7

(1) : without salt addition
(2) : [Na$^+$] = 13.9 g x l^{-1} ML
(3) : [Na$^+$] = 10.6 g x l^{-1} ML and [Mg^{++}] = 1.27 g x l^{-1} ML

2.3. Conclusions

During the biomethanation of *Tetraselmis* algae, a maximum biogas production of 1.33 m^{3}gas x m^{-3}ML x d^{-1} can be obtained with a volumetric loading rate of 4 kg VS$_o$ x m^{-3}ML x d^{-1} and a mean retention time of 14 d. A yield of 0.25 m^{3}CH$_4$ x m^{-3}ML x d^{-1} can be obtained with a concentration of sodium in the mixed liquor of 10.6 g x l^{-1}ML, a mean retention time of 14 d and a volumetric loading rate of 3 kg VS$_o$ x m^{-3}ML x d^{-1}. In the same conditions of biomethanation but without sodium in the mixed liquor a yield of 0.20 m^{3}CH$_4$ x kg VS$_o^{-1}$ has been reached. The lower value is due to the use of a different sample of algae.

3. RESEARCH CONDUCTED AT THE PILOT SCALE

3.1. Description of the installation

This installation has been set up at Lamezia-Terme (Calabria-Italy) at the beginning of April 1982. It has been built on a surface of non-arable land close to the sea ($\pm$ 800 m). A global scheme of the installation is proposed in Fig. 1. The installation has been set up on a platform of 1000 m^2 (20 m x 50 m) made of concrete. On this surface six ponds of 40 m^2 (2 m x 20 m) and four ponds of 80 m^2 (4 m x 20 m) have been built by the team of Professors K. Wagener (University of Aachen, Germany) and R. Materassi (University of Florence, Italy) and equipped with a mechanical mixing device. Near the ponds, a digester of 1 m^3 working volume, loaned by the Region Wallonne of Belgium has been set up by the authors. This digester is described in Fig. 2.

Marine algae Tetraselmis are unicellular organisms. A harvesting system based on a two step sedimentation process has been set up. A preliminary sedimentation is done in a pond in which a given volume of algal suspension is introduced each day, five days a week. After this preliminary sedimentation, the supernatant is taken off or recycled into the culture ponds and the somewhat concentrated algal suspension is introduced in a sedimentation tank. This tank is the second step of the harvesting system and it is described in Fig. 3. The final concentrated algal mixture is taken out ou the sedimentation tank and used to load the digester.

3.2. Operation of the installation

The operation description will be restricted in this paper to the digester's operations. Complete description and operation of the ponds will be presented in this conference by Wagener and al. in an other paper.

3.2.1. Starting up of the digester

The digester was filled on day 0 with 1 m^3 of sludge from a stabilisation tank of piggery wastes situated in a farm at Lamezia-Terme. From day 0 to day 11 the digester mixed liquor was let to ferment at 35 °C without loading until the biogas production had stopped. On day 11 the pH of the mixed liquor was 8.1 and the percentage of methane in the biogas was 94 %. On day 12 the digester was loaded as described in (3.2.2).

3.2.2. Running conditions of the digester :

The digester was run at a temperature of 35 °C.
After a three months starting and trials period and from day 85 to day 220, an hydraulic retention time of 25 days was used. The amount of concentrated algae to be added to freshwater to reach a volume of 40 liters was estimated so as to reach a volumetric loading rate of 0.5 kg VS_o x m^{-3}ML x d^{-1}. The real concentration was measured on a sample (see § 3.3.) and used for all further calculations of parameters. From day 280 to day 310, the same procedure was used with the exception of the mean retention time which was 50 days. From day 310, the volumetric loading rate has been maintained between 0,4 and 0,6 kg VS_o x m^{-3}ML x d^{-1} (average $\pm$ 0.5) while the volume added and hence the mean volumetric retention time could be changed to accomodate to the concentration in COD of the algal sediment as measured daily.
A concentration of NaCl of 5 g x l^{-1} was kept by using freshwater for dilution of the concentrated algae.

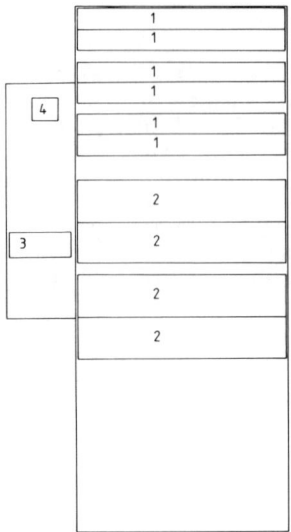

Fig. 1. : Global scheme of the installation
1. ponds of 40 m^2; 2. ponds of 80 m^2; 3. 1 m^3 digester; 4. sedimentation tank.

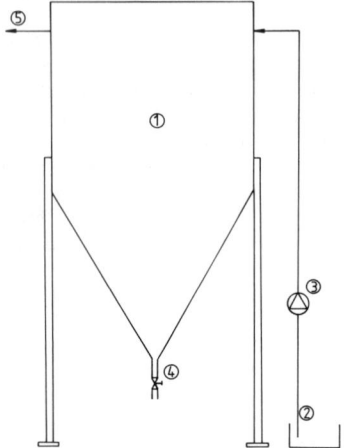

Fig. 3. : Sedimentation tank
1. sedimentation tank; 2 and 3. Algal suspension inlet; 4. Algal sedimentation outlet; 5. Supernatant outlet.

Fig. 2. : Description of the digester
1. Digester
2. Load inlet
3. 4. and 5. Heating system
6. Oulet of mixed liquor
7. and 8. Sampling
9. Overflow
10. Pump for mixing
11. Storage of load
12. Pumping of load
13. Sediment from sedimentation tank
14. Effluent storage tank
15. Control of mixing

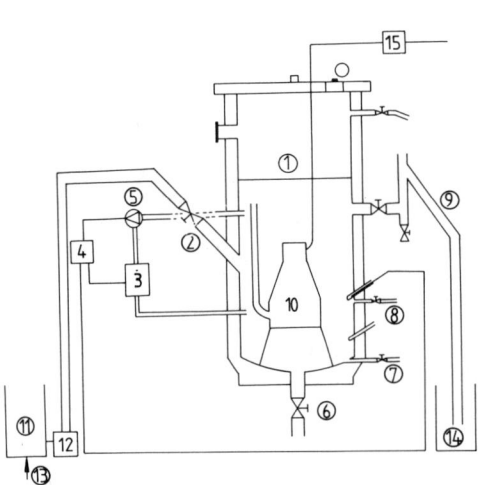

3.2.3. Daily operations of the digester
All the installation is run on a semi-continuous basis with loading once a day, five days a week. Daily operations can been summarised as follow : each morning the concentrated algae are taken out of the sedimentation tank. While mixing the content of the digester, the loading is introduced into the digester and effluent mixed liquor goes out through the overflow. The mixed liquor is further homogenised for 10 sec each hour.

3.3. Analytical control of the digestion

The following determinations were made each day : biogas production, percentage of methane, pH and temperature of the mixed liquor. Gravimetric analyses giving the concentration of the organic matter in the load and the effluent of the digester were made once a week from day 0 to day 310 on the total volume of the samples collected each day, 5 days a week and placed at -20 °C until analysis (method previously described (2)). In addition, from day 310 to now, a daily COD analysis of the samples was made to give a better control of the volumetric loading rate of the digester (see § 3.2.2.).

3.4. Results of digestion at the pilot scale

The results obtained during the first year of experiment are summarised in Fig. 4.

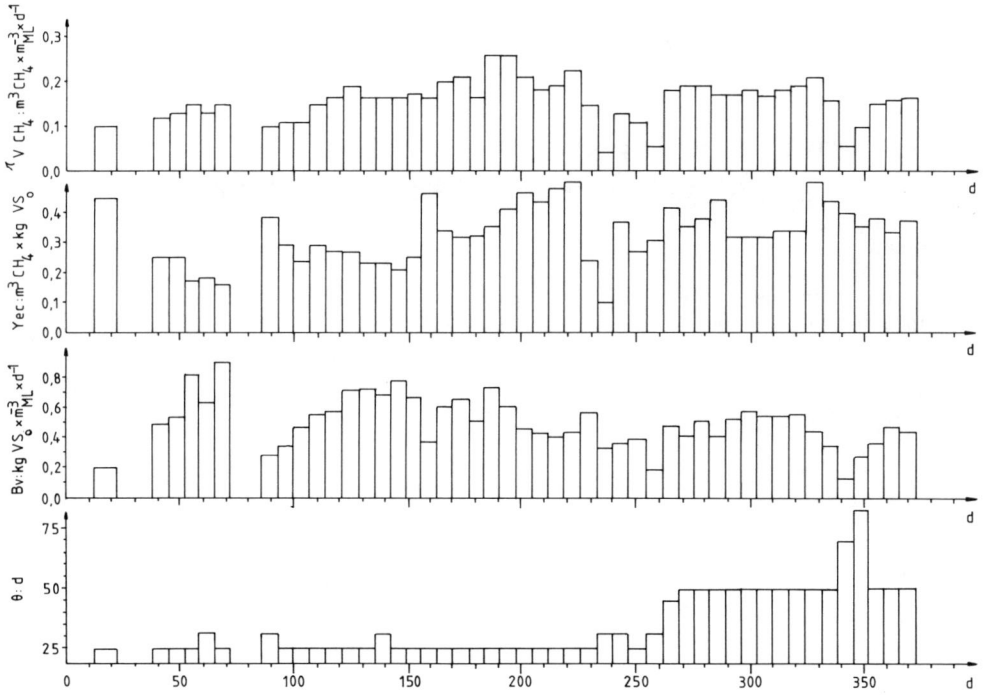

Fig. 4. Results of biomethanation obtained during the first year work.

- 215 -

From those results it can be seen that this digester is working since more than one year without problems and with better yields than those obtained at the laboratory scale. A mean methane yield of 0.35 $m^3 CH_4 \times kg^{-1}$ VS_o has been obtained.

This first year of pilot-scale work has also given us the possibility to integrate all the steps of this system and to show its technical feasability : culture of algae, harvesting system, semi-continuous digestion, sun-drying of the algae and characterisation of the process at all levels.

Table III : Abbreviations, symbols and units

- Concentrations (kg/m^3)

TS : total solids
VS : volatile solids
COD : chemical oxygen demand
VA_2 : acetic acid
VA_3 : propionic acid

- Rates

B_V : volumetric loading rate (kg VS or kg COD $\times m^{-3}$ ML $\times d^{-1}$)
$r_{V.gas}$: biogas production rate (m^3 gas $\times m^{-3}$ ML $\times d^{-1}$)
$r_{V.CH4}$: methane production rate (m^3 $CH_4 \times m^{-3}$ ML $\times d^{-1}$)

- θ : mean retention time (d)

- Yields

Y_{ec} : methane yield (m^3 $CH_4 \times kg^{-1}$ VS_o or $m^3 \times kg^{-1}$ COD_o)

Y_{conv} : conversion (kg $VS_e \times kg^{-1}$ VS_o or kg $COD_e \times kg^{-1}$ COD_o)

- Indices

o : in inlet (in feed)
e : in effluent

- Various

ML : Mixed liquor (useful volume)

References

(1) Methane production by anaerobic digestion of algae, (1981)

C.-M. ASINARI, A. LEGROS, C. PIRON, C. SIRONVAL, E.-J. NYNS and H.P. NAVEAU.

In "Energy from Biomass", série E, vol. 1, Chartier P. and Palz W. eds, Reidel D. Publishing Company, Dordrecht, Holland, 113-120.

(2) Methane production by anaerobic digestion of algae, (1981)

E.-J. NYNS and H.P. NAVEAU.

In "Energy" Solar Energy Programme of the commission of the Summary of the work carried out during the 2nd phase of the second Programme - Extented abstracts.

Commission of the European Communities, Directorate General Scientific and Technical Information and Information Management, Bâtiment Jean Monet, Luxembourg, ed.

In the press.

(3) Fermentation profiles in bioconversions, (1982)

A. LEGROS, C.-M. ASINARI di SAN MARZANO, H.P. NAVEAU and E.-J. NYNS.

Biotechnology Letters, $\underline{5}$ (1), 7-12.

METHANE PRODUCTION BY ANAEROBIC DIGESTION OF ALGAE.

II. PRODUCTION OF ALGAE.

AUTHORS : E. DUJARDIN, F. COLLARD, A. LEGROS, C.M. ASINARI DI SAN MARZANO, E.J. NYNS, H. NAVEAU and C. SIRONVAL.
CONTRACT N^r : ESE-R-25-B (G)
DURATION : 36 months   1 July 1980 - 30 June 1983
TOTAL BUDGET : 34.958.000 FB   CEC contribution : 50%
HEAD OF PROJECT : Prof. E.J. NYNS
CONTRACTOR : Catholic University of Louvain
ADDRESS : Unité de Génie Biologique, Place Croix du Sud, 1, bte 9 B-1348 Louvain-La-Neuve, Belgium.
SUBCONTRACTOR: Prof. C. SIRONVAL - University of Liège.

SUMMARY.

- Algae have been grown in waste luke-warm waters at temperatures between 20° and 35°C in order to recuperate calories by increasing the yield of biomass and to produce methane at the expense of the calories.

- The algae have been cultivated in lagoons (2200 m²) irrigated by the water of the river Meuse warmed up by circulating through the 3rd cooling circuit of Tihange nuclear power-plant. The culture yielded of the order of 10 T dry biomass. hectare^{-1}. year^{-1}.

- A rate of biogas production of 1.7 m³ gas.m$^{-3}_{ML}$. d^{-1}, with a methane concentration of 65% can be obtained.

- The process achieves a chemical and a thermal depollution of the water, -a circumstance which may represent a spare of expenses for the process.

- The different operations which are involved in the biomass harvest have been fully assessed.

- The chemical and biological compositions of the harvested biomass have been established.
Yearly changes have been described. Biomass utilizations other than methanization have been experimented, among which animal feeding is a promising one.

INTRODUCTION.

We have proved that it is possible to recover as biomethane a part of the waste energy rejected as heated water by an industry, by growing algae in this water and by subjecting the harvest to biomethanization (anaerobic fermentation).

It is well known that the algae grow at a particularly high rate in favourable conditions, of which the temperature is one of the most critical. The water of the river Meuse, warmed by circulating through the 3rd cooling circuit of the nuclear power plant of Tihange, provides favourable temperatures for the growth of the algae. The temperature of this water is 12°C above that of the river.

It was found that part of the pollutants (organic matters and minerals) rejected above Tihange in the river Meuse are used as fertilizers by the algae. In addition, by circulating the water slowly through the lagoons where the algae are cultivated, one decreases its temperature to a level compatible with the ecological equilibrium of the river. As a consequence, the harvested biomass (10-12 T. ha^{-1}. year^{-1}) becomes cheaper, as its production allows to depollute the water chemically and thermally.

The biomass collected in Tihange allows to produce methane gas with a good yield (see this contract report, part.I). The fraction of this biomass which is not subjected to methanization has several other applications. It may be used for extraction of interesting compounds, for feeding animals, especially fishes, etc... The residue of the methanization may be used as a fertilizer; we are able to demonstrate an integrated industrial process as summarized in the following scheme:

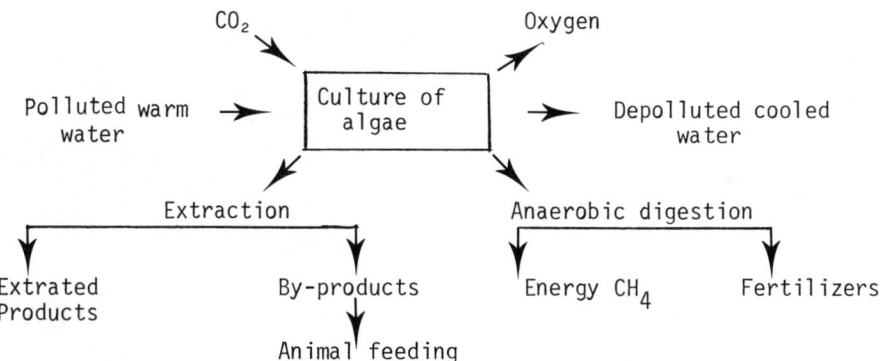

1.- THE LAGOONS.

In 1981, the area of the lagoons has been increased from 400 m² to 2205 m². The lagoons are different in length one from another, but all are 5 m wide and 0.5 m deep (see fig.1). The water enters the lagoons at a temperature varying between 20°C in the early spring and the autumn, and 30°C in the summer (Table I).

Months	Mean daily temperature of the water in °C.
May	20 - 20
June	26 - 30
July	25 - 27

TABLE I.-

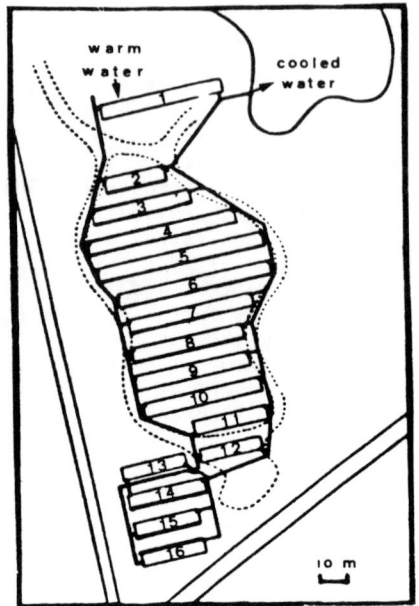

Figure 1. The lagoon system in Tihange (Belgium).

It flows very slowly (50 l.min^{-1}). At the exit of each lagoon, the water is collected and rejected in a 2000 m² pond.
In 1984, we intend to add 7500 m² to the culture area. This will allow us to consider the culture problems on a pilot-plant scale and to assess properly the actual yield of the biomass collected in 1 hectare.

2.- THE HARVEST.

During the sunny period, the algae accumulate quickly on the surface of the lagoons. They are scraped from the water surface using a rake provided with floats (fig.2).
The collected algae are centrifuged in order to eliminate the bulk of the water from the fresh biomass. They are then dried in the open air on a tray system allowing a uniform ventilation. In high production periods, the frequency of the harvests is about four or five times a month. We are mechanizing presently the harvest system in order to follow up more closely the harvest of the biomass.

Figure 2. The harvest of the biomass showing the rake which scrapes the water surface.

During the years 1981 and 1982, the biomass was harvested from the end of May until the end of October. The cumulated biomass corresponded to a yield of the order of 10 T. hectare^{-1}. year^{-1}(fig.3).

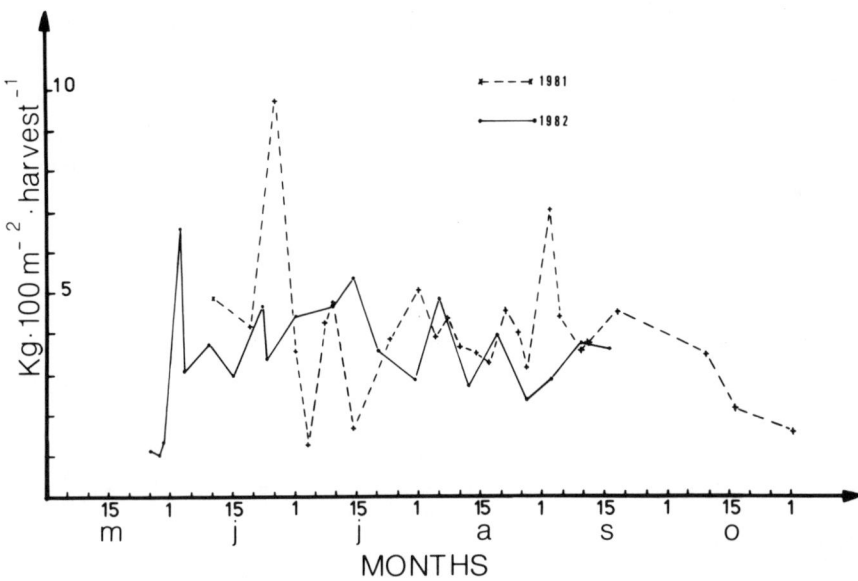

Figure 3. Evolution of the dry weight of the biomass harvested from May to October (kg per 100 m² and per harvest). In September 1982 the power-plant did not warm up the water.

The first harvest occurs ordinarily at the end of May. Three periods may be distinguished:
1.- there is first a rapid increase of the biomass production which brings the weight collected per harvest to a high level at the beginning of June.
2.- Then the production remains more or less steady during three months.
3.- The production decreases in October in relationship with the decrease of the water temperature.

3.- THE BIOLOGICAL AND CHEMICAL COMPOSITION OF THE BIOMASS.

3.1. Biological composition.

The method which we have chosen for the analysis of the population composition of the biomass is the following:
- every week fresh samples (about 100 g) are collected at random. They are washed several times with tap water. This water carries away the sediments; it is filtered on paper; the filtration residue is made essentially of the sediments. The washed biomass is then examined. The following species are found most frequently:
  - the green alga : Hydrodictyon reticulatum
  - the aquatic angiosperm : Lemna
  - the Gastropods.

The algae are separated by hand from the Lemna, from the Gastropods and from other organisms. The fractions are dried at 85°C until constant weight. The contribution in weight of insects and larvae is negligible. They are not taken into account for the calculations.

Table II shows the analysis of the biomass harvested in 1982. It gives, for each harvest, the proportion of three main biomass fractions, expressed in percent of the total dry weight, sediments excluded. The main constituent was the green alga Hydrodictyon which makes up an average of 75% of the dry biomass. Lemna follows: 15% of dry biomass on an average. The Gastropods represent some 10% of dry biomass on an average.

The sediments amounted to about 31% of the total (biomass + sediments).

Number of the sample	date of harvest	Hydrodictyon	Lemna	Gastropods
6	9.6.82	99	1	0
7	15.6.82	85.2	5.6	9.1
8	22.6.82	67.9	7.2	24.9
10	30.6.82	84.2	9.2	6.6
11	9.7.82	85	11	4
12	12.7.82	79	12.8	8.2
13	22.7.82	67.2	5.4	27.4
14	29.7.82	78	20.8	1.2
15	5.8.82	68.7	20	11.3
16	12.8.82	71.6	18.3	10.1
17	19.8.82	55.5	36.5	8
18	26.8.82	53.5	40	6.5
19	2.9.82	83	8.8	8.2
20	9.9.82	66.6	23.4	10

TABLE II. *Biological composition of the biomass harvested in 1982 (in % of the dry weight, sediments excluded).*

The proportion of the algae in the biomass tended to decrease a bit during the year as already observed in 1981. The proportion of Lemna was rather high in some harvests, especially in August.

3.2. Chemical analysis.

Table III gives the proportions of the main organic constituents in % of dry weight of the total (biomass + sediments). The high ash content (35-43% of dry weight) is due partly to contaminations from the bottom of the lagoons.

	% dry w.
Proteins	14-20
Lipids	5-7
Total carbohydrates	15-18
Soluble carboh.	5
Pigments	1
Water	5-8
Ashes	35-43
Others	10

TABLE III. *Chemical composition of the crude biomass.*

Table IV shows that the calcium level is particularly high, reflecting the high calcium content of the river Meuse. Minerals accumulate in the biomass demonstrating that the culture is an excellent tool for recuperating them from the river.

The aminoacid pattern of the total proteins is given in Table V. The percentage of lysine (6.52%) reaches the same value as in Soybean proteins, or proteins of the alga Spirulina used for centuries in human nutrition. The methionine content (2.17%) is even higher than in Soybean (1.6%).

% dry weight

Major elements		Minor elements	
P	0.17	Cu	0.008
S	1.5	Zn	0.03
I	0.10	Hg	0.028
K	3.14	Pb	0.003
Na	0.14	Cd	not detectable.
Mg	0.02		
Ca	5.1-17.3		

Aminoacid	%
Lys	6.52
Met	2.17
His	2.17
Thre	5.43
Arg	5.43
Asp ac.	11.95
Ser	5.43
Glu ac	13.04
Pro	6.52
Gly	5.43
Ala	7.60
Cys (1/2)	0.91
Val	4.34
Ileu	4.34
Leu	8.69
Tyr	4.34
Phe	6.52

Table IV. *Mineral composition of the crude biomass (Data determined in collaboration with the "Institut National des Industries Extractives, Liège").*

Table V. *Aminogram of the total protein fraction of the crude biomass (in % of each AA in the total proteins).*

4.- UTILIZATIONS.
a) Methanization.

In parallel with the biological analysis, samples of crude biomass were sent to the Catholic University of Louvain-La-Neuve in order to estimate some parameters of methane production. Some of the characteristic features of the content of these samples, as measured in Louvain-La-Neuve, are given in Table VI.

In a two stages digestion system (with two digesters completely mixed: liquefactor + digester), the following performances can be obtained at 35°C:
Mean retention time (θ) ; 16 d
Volumic load (B_v): 4 kg $VS_o \cdot m^{-3} \cdot d^{-1}$.   Conversion yield ($Y_{con}$): 0.65 kg $VS_e \times kg^{-1} VS_o$
Economic yield (Y_{ec}): 0.28 m^3 CH_4 x $kg^{-1} VS_o$
A rate of biogas production of 1.7 m^3 gas$\cdot m^{-3}_{ML} \cdot d^{-1}$, with a methane concentration of 65% can be obtained.

Year 1982	TS	MS	VS	Ash	CO_2	$CaCO_3$	N.T.K. mg Kg^{-1}	COD Mg Kg^{-1}
Lot 6. 9.06	949.3	520.1	429.2	360.0	160.2	364.2	30855	628134
Lot 10. 30.06	938.9	494.5	444.3	337.2	157.3	357.5	29364	621291
Lot 13. 22.07	945.2	582.5	362.6	397.8	184.7	419.9	22980	501486
Lot 17. 19.08	947.0	600.6	346.4	432.9	168.2	382.3	22732	481000
Lot 18. 26.08	943.2	527.9	415.9	385.6	147.8	322.2	28395	583080
Lot 19. 29.09	938.8	600.8	338.0	453.8	147.0	334.0	23035	421519
Lot 21. 10.09	943.0	573.1	369.9	397.8	173.4	399.6	19964	508479

TABLE VI. *Characteristics of the biomass subjected to methanization.*

TS = Total solid
MS = Mineral solid
VS = Volatile solid
Ash, CO_2, $CaCO_3$ in g.kg^{-1} of biomass.

NTK = Total Kjeldahl Nitrogen
COD = Chemical Oxygen demand   in mg.kg^{-1} of biomass.

The fermentation residue is a good fertilizer for traditional agriculture as well as for the culture of algae themselves.

b) Other uses.

Part of the harvest has been extracted by various solvents to yield pigments and other high added-value products. The extraction by-products have been processed into food for animals in conjunction with the total biomass.

1.- Laying hens were fed with a diet containing extraction residues. The results prove that this diet was perfectly suitable.
Compared with the control, the weight of the hens increased faster and reached higher values in the essays than in the control. On the other hand, after an adaptation period, the number of eggs in the essays was higher, the egg shell was smoother, thicker and stronger, and the yolk weight was higher, compared to the control (fig.4).

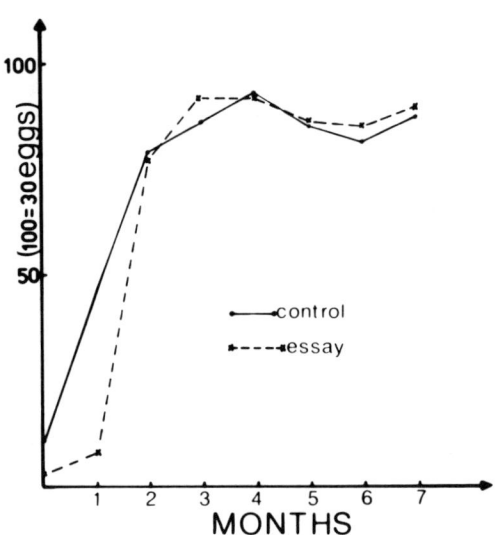

Figure 4.

Effect of introducing algae extraction residues in the diet on the production of eggs by laying hens.

Wheat (20% of the weight of the control diet)and calcium carbonate (4% of that diet) have been replaced by algae in the Essay(24% of the Essay diet)

(100% production is taken as being 30 eggs- hen^{-1}. month^{-1}).

2.- Fishes (aquarium fishes like Gouppy and Tilapias) were fed with a diet in which part of the proteins were algae proteins. The results showed a stimulation of the growth due to this diet. No abnormal diseases or deaths were observed.

CONCLUSIONS.

1.- Warming up by waste calories of industrial origin benefits the growth of filamentous algae. Yields of the order of 10 Tons dry weight.hectare^{-1}. year^{-1} have been achieved using calories rejected by the nuclear plant of Tihange (Belgium). The biomass produced in this way is convenient to achieve methane production. A rate of biogas production of 1.7 m³ gas.m^{-3}$_{ML}$. d^{-1}, with a methane concentration of 65% can be obtained.The whole process is a recuperation of rejected calories, which appear ultimately as stored in methane by the intermediary of

algal growth.
2.- Thermal, as well as chemical depollution of the water is achieved by the process. This represents a negative production cost.
3.- The different operations involved, from algae production to methane are well known. They have been tested by a three year practice. Improvements are however to be worked out in order to scale the production up; especially a complete mechanization of the harvest has to be achieved.
4.- Chemical analysis as well as some particular experiments have pointed out several possible utilizations of the algae biomass apart from biomethanization. Animal nutrition (fishes and poultry) is one of the most promising of these other utilizations.
5.- The yearly changes of the biological composition of the harvested biomass have been established. In Belgium, the harvest starts in May. The proportion of the algae in the crude biomass is high at the beginning of June and then decreases slowly during four months. Biomass production ceases in October.
6.- We are now increasing the lagoon surfaces up to 1 hectare, in order to reach the size required for demonstrating the whole processes on an industrial scale following scheme 1.

ACKNOWLEDGEMENTS

We thank P. Delcambe, C.Lacrosse and C.Piron - Fraipont for scientific contribution and J.Allard, F.Coulibaly, J.Schmitz and S.Sougné for technical assistance

REFERENCES.

Asinari di San Marzano C.M., A.Legros, C.Piron, C.Sironval, E.J. Nyns and H.P. Naveau (1981)
In "Energy from Biomass", series E .Vol.1. P.Chartier and W.Palz eds, Reidel D. Publishing Company, Dordrecht Holland. pp.113-120.

E.Dujardin, P.Delcambe, C.Lacrosse and C.Sironval (1982).
in "Energy from biomass" series E vol.3. G.Grassi and W.Palz eds. Reidel D. Publishing Company.Dordrecht, Holland. pp.135-141.

E.Dujardin (1981)
"Actes des Journées Européennes de Bioénergie" (R.Buvet et J.P. Massué(Eds) pp.63-66.

Piron-Fraipont C., E. Dujardin and C. Sironval (1981).
In "Energu from Biomass" 1st E.C. Confer. (W.Palz, P.Chartier and D.O.Hall, Eds) Appl. Sc. Publ. LTD London .pp.703-708.

E.C. Dujardin and R.A.G. Devreux.
Belgian Pat. n^r 890, 359, sept 15th 1981;
Patents pending for other countries.

BIOMASS CONVERSION (THERMOCHEMICAL ROUTES)

The development of furnace/heat exchanger systems in which chopped cereal straw is the fuel

The use of gas scrubbers for heat extraction from straw furnaces

Heat energy from animal waste by combined drying, combustion and heat recovery

High temperature straw granulation

Mobile pyrolysis plant

Modelling of a fluidized bed wood gasifier

Biomass gasification - programme status

Hydrogen absorption in $LaNi_5$ dispersed in liquid

Catalytic liquefaction of wood material

Study on the pyrolysis of agricultural wastes

THE DEVELOPMENT OF FURNACE/HEAT EXCHANGER SYSTEMS IN WHICH CHOPPED CEREAL STRAW IS THE FUEL

Contract Number	: ESE - R - 026 UK
Duration	: 1 July 1980 - 30 June 1983
Total Budget	: £43,000 CEC Contribution : £21,500
Head of Project	: B. Wilton, Faculty of Agricultural Science, Sutton Bonington, Loughborough, Leics., England.
Contractor	: University of Nottingham
Address	: University Park, Nottingham

Summary

The sales and use of straw-burning boilers in the United Kingdom remain at a low level. Apart from the fact that conventional fuels are still readily available the main reasons for this apparent lack of interest are high equipment costs and the regular chore of handling bales and ash. It is suggested that a mechanised system of handling chopped straw, a cheap means of storing it and an automatically controlled method of feeding it into a furnace (rather than a boiler) may well revive interest.

An enclosed trailer with a controllable positive unloading system has been contructed and linked to a sloping grate furnace; the latter supplies hot flue gas to a free-standing heat exchanger. The trailer, furnace and heat exchanger have all been operated satisfactorily. Control systems for both the trailer and furnace have been installed and have also proved to be satisfactory.

1.1 Introduction

Despite the fact that several importers of straw-burning equipment have been marketing in the United Kingdom for almost ten years, they have met with only modest success. They have been joined, in recent years, by a few UK manufacturers to compete for what seems to be a somewhat limited market. It appears that the total number of units regularly fuelled on straw, or a mixture of straw and wood, is perhaps 5,000. From average straw consumption figures it therefore appears that the total amount of straw used as fuel is only some 5% of that burned in the field.

Although new domestic units are being installed, it is known that some of the early ones are not now used. Smoke, low efficiency, inadequate capacity, the need to stoke batch fed units regularly, the chore of moving straw and ash, and the difficulty of handling these two materials cleanly have combined to give small units a bad name. There have also been accidental fires.

A few larger commercial units have been supplied for heating glasshouses and drying hay and grain: some homemade models have also been constructed by enthusiastic farmers and growers. Although some reports of satisfactory performance have been made, growers in particular have encountered problems - insufficient capacity, failure to last through the night without re-stoking, combustion difficulties, and in some cases completely unsatisfactory procurement, storage and straw handling facilities.

Initially it was regarded as an environmentally respectable move to have a straw-fired unit; now, with other more convenient sources of energy still freely available and not particularly expensive, this attitude is less in evidence.

In the long-term it is reasonable to assume that more straw will be burned usefully : legislation to prevent field burning would help and rises in the true costs of alternative fuels will also encourage a change of attitude.

One major problem with simple domestic-scale units is that most of them are batch fed and that combustion usually takes place in a water-jacketed chamber. Automatic feeding of chopped straw derived from bales is possible and improves boiler perfomance, but this more than doubles the cost. Pelleted straw also improves the situation, but this is not readily available and in any case it is considerably more expensive than unprocessed straw on the farm. The pelleting process itself absorbs energy.

If straw is to be burned on the premises of the farm where it is produced - and this seems to be the most logical place to start - then steps must be taken to encourage farmers to burn it. One way to do this would be to subsidise equipment; another would be to make the handling a less unattractive process than it is at the moment. It is with the latter task in mind that the present work was undertaken.

It can be both economic and convenient to transport straw by road over distances of say 50-100 kilometers in the medium density bales commonly produced on farms. As transport distances increase so does the need for densification, either by a move to very high density bales or to pellets. On a farm, however, it can be argued that even conventional bales are not necessary if a satisfactory method of handling loose chopped straw can be found.

On cereal-growing farms space for low-grade storage (which may be uncovered) is no problem, and if efficient combustion is promoted by the use of chopped straw it would seem sensible to chop at the time of harvesting and then to store, handle and feed in this form if possible. A primary requirement is then to provide a clean, mechanised handling system which will feed chopped straw into a combustion chamber, on demand, with a high level of reliability.

Heating equipment is usually sited relatively near to the place to be heated: for reasons of both safety (from accidental fire) and cleanliness near the farm house, the separation of the straw store and heater is advisable. With these factors in mind it was decided to build an enclosed, hopper bottomed, self-unloading trailer for chopped straw: this would usually be large enough to sustain a typical heater for 24 hours, capable of being filled and towed by a tractor, and equipped with electrically driven and controlled unloading mechanisms.

Straw storage during the work at Nottingham has been on a concrete pad surrounded by low walls. The self-unloading trailer has been made and fitted with an unblockable discharge mechanism, controlled by thermostats in a heating system driven by a stepped grate furnace.

2.1 Straw storage and handling

In the autumn of 1981 a forage harvester was used to gather and chop wheat straw from combine swaths, and to load it into tipping trailers. Two 5-tonne heaps were made on a concrete base surrounded on three sides by low wooden walls. One heap was covered with a weighted polythene sheet, the other was left uncovered.

At the time of collection the mean straw moisture content was 20-22%. After five months some 95% of the covered stack had a moisture content below 20% - the range of individual determinations was from 17.5% to 41.3%. The latter sample was taken from a limited zone where rainwater had been shed by the sheet onto an uncovered edge of the heap.

There was surprisingly little penetration of water into the uncovered heap, and although the depth of wet material was variable it was seldom more that 20cm. The reason for this was that the heap settled fairly rapidly, the material being well chopped, and the surface was rearranged by wind to form something approaching a thatched cover. In fine weather the outer surface dried out; while in the body of the heap (some 85% of the total) moisture contents stayed below 20%, on the surface they ranged from 30% to 70%. In a larger and higher heap the ratio of 'wet' outside to 'dry' centre would obviously decrease. It was not considered necessary to repeat this exercise in 1982/3 in view of the fact that the results were in broad agreement with others produced in Denmark (1).

The trailer (Figure 1) was fitted with a 0.2m diameter auger which ran along the whole length of the bottom of the hopper (2.5m) and then extended 0.5m to deliver into the inclined feeder auger on the furnace(Figure 2). It was driven by a 0.7 kW motor at 16 rpm. A number of narrow plates were fitted across the top of the auger trough to hold back straw at this point by encouaging it to bridge. This was done to minimise the risk of overloading the auger motor and to cut down the rate at which the trailer could discharge.

A track was mounted along the centre of the trailer and some 1.0m above it: a slider located in the track originally carried a chain to which was attached a 10 kg weight, positioned so that it just cleared the plates above the auger. The slider was made to travel back and forth along the track by a 0.3kW reversible geared motor and an adjustable delay at the end of each pass of the slider was provided by means of an electrical timer.

Whereas a single weight was adequate when the trailer was only partially filled it was found necessary to use three further small weights, each on a separate chain, to avoid bridging when the trailer was filled to capacity.

Two minor problems were encountered with the trailer, otherwise experience during prolonged operation has been satisfactory. One problem is that occasionally when a weight is pulled from the top of the heap of straw, it can swing and make contact with the hopper side: this makes a noise that could be annoying. A covering of rubber can reduce this problem. The second problem was encountered once when the top slider tipped unduly under load, fouled the frame and caused the reversing motor to burn out. A redesigned slider or a better electrical overload protection device would have prevented this from happening.

2.2 Furnace/heat exchanger unit

The furnace, designed to be capable of construction from readily available standard materials, has an adjustable sloping grate 1.05m long and 0.45m wide, housed in a firebrick structure having overall dimensions of 1.75 x 1.50 x 0.75m (see Figure 2). The top of the structure was originally made from a steel plate to the underside of which firebricks were attached. An arched firebrick top was subsequently installed because heat distorton made it difficult to seal the joint between the furnace walls and top. The grate surface is of trimmed firebricks spaced on a steel frame to give a horizontal gap of some 10mm between each course of bricks.

The grate and the inclined auger feeder (0.2m diameter, driven at 26 rpm by a 0.3kW motor) are both mounted on skids so that the whole unit can be withdrawn for maintenance.

A 2-pass heat exchanger of more-than-adequate capacity has been mounted above the furnace: flue gases pass through twelve 0.05m diameter tubes in one direction, up through a transfer cover and then back through twelve similar tubes to a manifold which leads to a 0.25m centrifugal fan driven by a 0.6 kW motor at 600 rpm. The size of the water jacket around the tubes is 2.5 x 1.2 x 0.8 m and it is supported on legs which straddle the furnace. A short flue passes through the roof of the building housing the unit. The heat exchanger is linked to a fanned water-to-air heat exchanger fitted with finned tubes, water being circulted by a centrifugal pump around a circuit which includes a flowmeter.

The second heat exchanger was originally used to provide a load for a much larger furnace and had a capacity approximately six times that required. Since it was mounted in an open building it was decided to set its variable speed fan at a low rate and to use a high/low thermostat on the water return to control the furnace feed system. The heat output was determined by considering the circulation rate and the flow and return temperatures.

In operation the furnace was first lit by hand and the uninterrupted straw feed rate was such that the system's maximum working temperature ($65°C$) was reached within aproximately one hour from cold: after this point was reached the feeder would run between 20% and 40% of the time, the actual proportion depending on the ambient temperature (ie the temperature of the air entering the second heat exchanger). By switching the system off manually it was found that the furnace would re-light up to about two hours after stoking ceased, indicating that stoking for a period of say five minutes every hour would make the fitting of a gas-fed pilot light unnecessary. Daily removal of ash by hand was required and weekly inspection and cleaning of the first heat exchanger is advisable.

The starting sequence was to switch on the flue gas fan, the inclined auger feeder, the trailer auger and the trailer agitator motors in order at approximately two second intervals. The closing down sequence was in

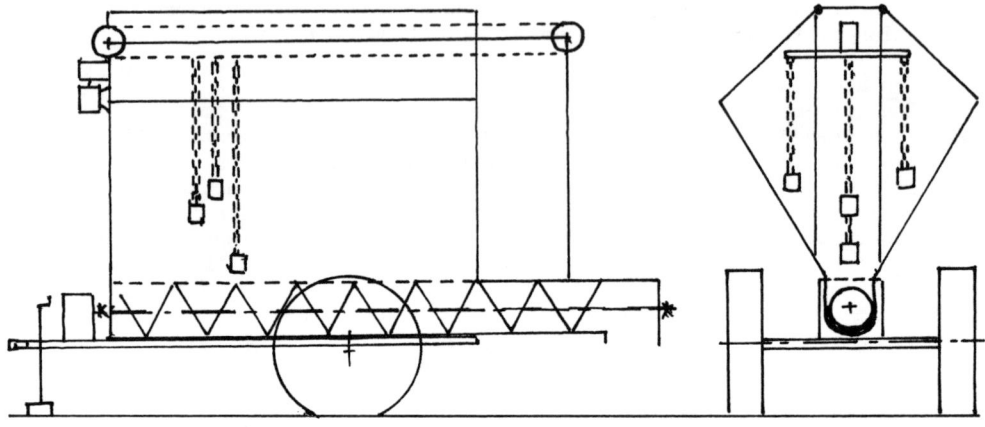

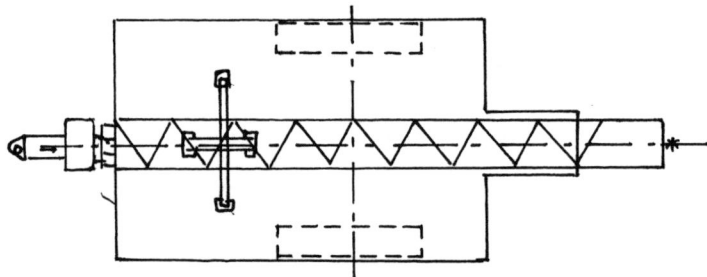

Figure 1. The enclosed self-unloading straw trailer (not to scale)

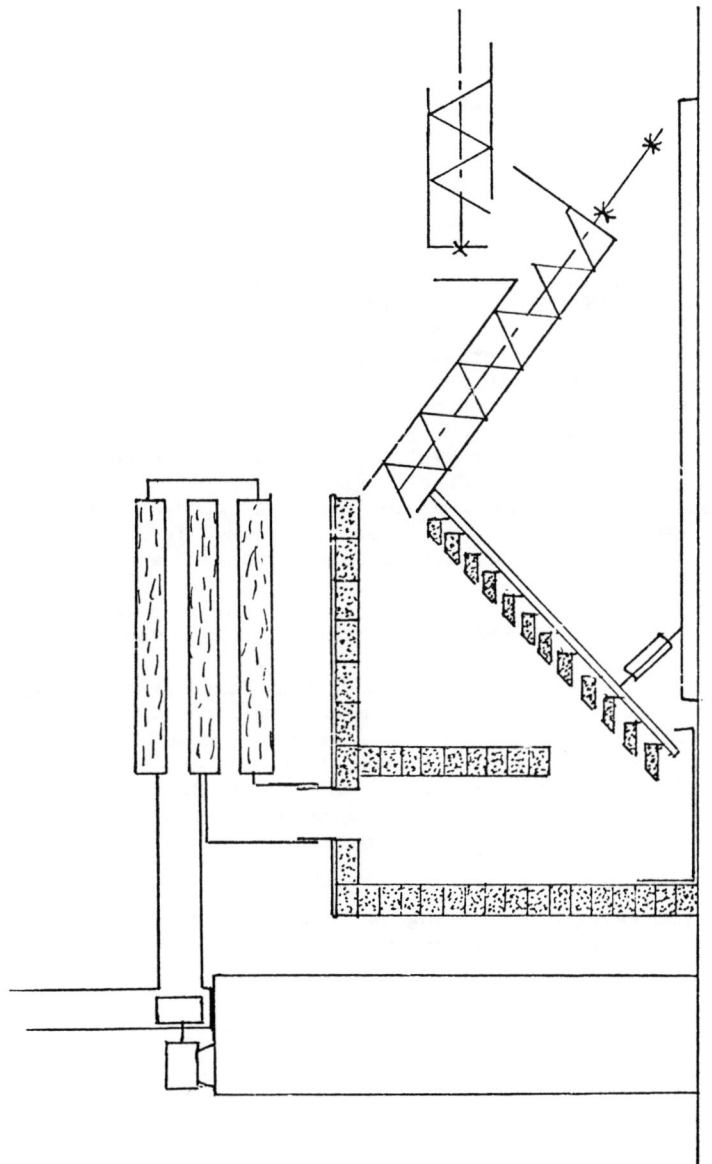

Figure 2. The sloping grate furnace and flue gas/water heat exchanger (not to scale)

the reverse order, but a ten second delay was introduced between each stage to minimise the risk of either motor overload on restarting or burning back. The circulating pump and heat exchanger fan ran continuously.

Overall efficiency of the furnace and system was not particularly high although flue gas analysis indicated that combustion was efficient. The main source of heat loss was through the uninsulated furnace walls and top - consisting of either one or two layers of firebrick which became too hot to touch after about half an hour - and from the surface of the first heat exchanger which was also not insulated. In a commercial unit these components would of course be insulated.

3. Conclusions

(i) Outdoor storage of chopped straw in large uncovered heaps is adequate in UK conditions. A concrete base with a sightly domed profile is the ideal: this will shed rainwater and avoid the possibility of stones which could damage augers by becoming entrained in the straw. At least one side of the storage area should have a wall to facilitate the trailer loading operation. It would be advisable to keep a few tonnes of straw under cover for use at times when snow has fallen because of the difficulty of loading snow-free straw from an outdoor heap at such times.

(ii) Spillage of straw from an auger discharge trailer is nil and the tendency for straw to bridge in such a design can be used to prevent the unloading mechanism's motor from overloading. It is essential to have a controllable mechanical agitator system to overcome the effect of bridging when straw is required by the furnace.

(iii) Combustion on a sloping grate furnace is good and it is both possible and desirable to make the grate of firebrick. It is significant that most of the manufacturers of straw-burning equipment are now moving away from combustion in a water-jacketed zone.

(iv) It may be cheaper to construct separate furnaces and heat exchangers: it certainly gives more flexibility, however more insulation is required.

(v) Control systems for small domestic straw-fired systems can be relatively simple, but it is essential to match the capacity of any flue gas fan to the stoking rate. In most situations it should not be necessary to provide either a variable speed flue gas fan or separate fans for running and idling conditions; however sufficient clearance should be allowed around the fan to prevent it forming an obstruction to natural draught in the flue when the unit is idling.

Reference

(1) Persson, K. and Have, H. Handling and storage of chopped straw for heating purposes. CEC project no. 324-78-ES-DK, (1980).

"THE USE OF GAS SCRUBBERS FOR HEAT EXTRACTION FROM STRAW FURNACES"

Authors   : H. Kofoed Nielsen and H. Nielsen

Contract Number: ESE-R-027-DK (G)

Duration   : 36 months   1 July 1980 - 30 June 1983

Total Budget   : Dkr. 1,571,710   CEC contribution 50%

Head of Project: Prof. T. Tougaard Pedersen

Contractor   : Jordbrugsteknisk Institut
                  Royal Veterinary and Agricultural University

Address   : Rolighedsvej 23
                  DK-1958   Copenhagen V

Summary

The purpose of the project is to improve the efficiency of straw furnaces by cooling down the flue gas below dew point and thus make the upper calorific value of straw a realistic target for energy utilization. The heat exchanger to be used is a gas scrubber.
The course of the project has been:
a) Basic experiments with gas scrubbers as heat exchangers for very hot air.
b) Design, construction and test of a prototype flue gas scrubber for a commercial straw furnace. The scrubber improved the efficiency from 70% to 90% of the lower calorific value of the straw supplied. No major problems were found.
c) Design, construction and test of a special straw combustor with the flue gas scrubber as sole heat exchanger. This system is cheap and has, compared to other furnaces a high efficiency (approximately 79%). Apart from being an efficient heat exchanger the scrubber reduces the particle emission by 80-90%.

1. Introduction.

At present more than 12,000 straw furnaces are installed in Denmark (1). About 10% are of the automaticly stoked type, while 90% are of simple construction with efficiency below 50 percent. These furnaces are inefficient because of incomplete combustion and a high flue gas temperature. The purpose of the project has been to investigate a method of improving the performance of such furnaces.

In the first two years we have developed and tested a flue gas scrubber for a conventional water jacket straw furnace. The scrubber was connected to the flue of the furnace, fig. I. The scrubber worked very efficiently as heat exchanger and in addition, it reduced the particle emission considerably (2), (3) and (4).

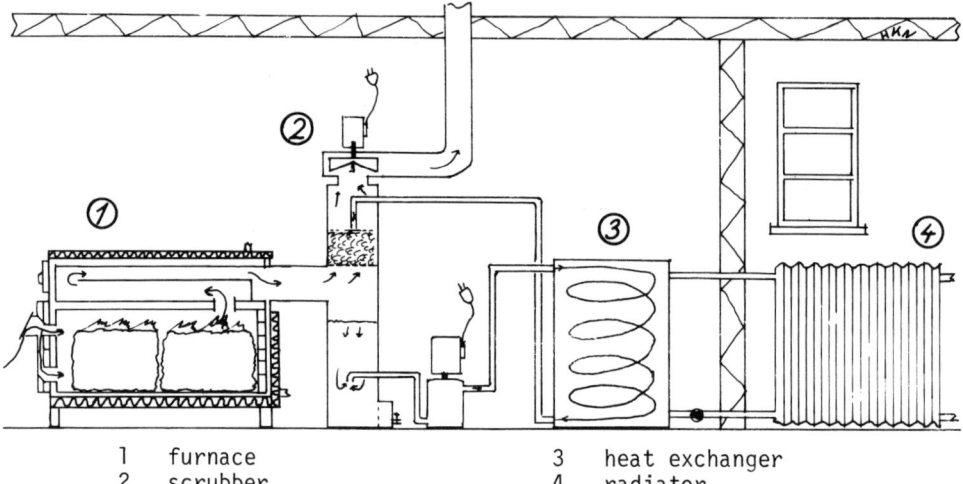

1 furnace
2 scrubber
3 heat exchanger
4 radiator

Fig. I. Scrubber fitted to ordinary furnace.
A fan is mounted to pull through the flue gas. Water is circulated to the packing elements by a centrifugal pump. Heat is transmitted from the wash water to cold water through a copper spiral.

The scrubber can accept flue gas temperatures up to 1000 $^{\circ}$C or more. Therefore, it can be used as the only heat exchanger in a combustion system, and the combustion chamber can cheaply be made from refractory material. This again means that the combustion unit may be used for straw with higher water content than usual, since the heat losses from the combustion chamber are small. On the basis of this the last year of the project has been used to design a special straw combustor for the flue gas scrubber.

2. Description and function of the cyclone furnace.

Some experiments made in USA (5) indicate that a burner of the vortex or cyclone type could be very efficient for our purpose. Therefore we have studied this principle and constructed the cyclone furnace shown in fig.II. The straw is supplied continuously by a stoker auger to the top or bottom of the cyclone. The air inlet is tangential to the combustion chamber and normally in the top section. The inside of the combustor is made from refractory material. During testing the furnace was connected directly

– 236 –

to the flue gas scrubber as shown in the figures.
   The combustion has three overlapping phases:
a) Evaporation of the moisture.
b) Distillation of the volatile gasses from the straw and combustion of these gasses.
c) Combustion of the fixed carbon.

   With bottom-feeding of the straw all phases except the gas combustion take place in the middle of the cyclone as a turbulent tight flame spiral. Thus, any unburned material or fly ash is centrifugally separated from the flame and thrown into the outer downward vortex air stream. This material will be reignited at the top of the fuel bed in the bottom.

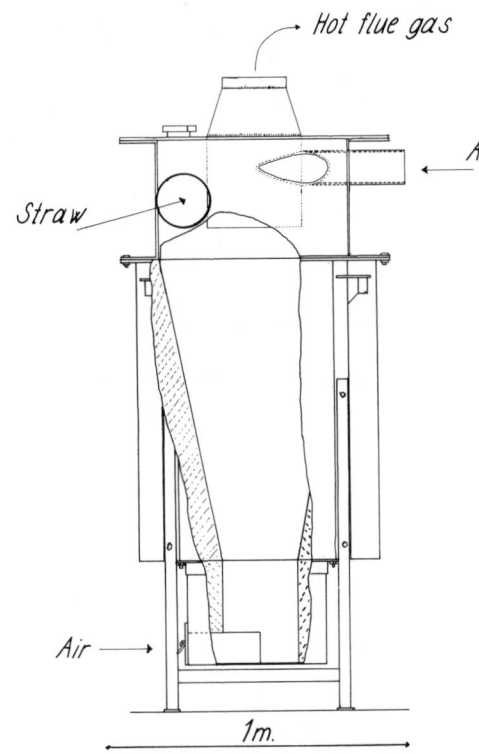

Fig. II Cyclone furnace. Volume of the combustion chamber 0.31 m³.

   With top-feeding some of the evaporation (a) and the distillation (b) takes place in the downward vortex air stream by convective and radiant heat transfer from the flame spiral.

3. Results.

During the winter 1982-83 the cyclone furnace was operated a total of 180 hours at our subcontractor Statens jordbrugstekniske Forsøg. Under these tests straw was fed to the top of the furnace only as initial tests showed that the bottom of the combustion chamber was too narrow.
   The furnace was operated with two dimensions on the cyclone outlet tube. The diameters were 215 mm and 308 mm . These pipes caused different diameters of the flame spiral in the cyclone and thus different retension time in the combustion chamber. Also two diameters of the air inlet pipe

were tested in order to see the effect of different rotational speed in the cyclone. The diameters were 63.5 and 100 mm respectively.

During the test a total of 1500 kg of straw was burned and the following results were obtained:

1. The maximum thermal power output was about 45 kW

2. The overall thermal efficiency based on the lower calorific calue was 79% in the best power range from 20-45 kW, fig. III. About 10% was lost through the flue gas, when it was cooled down to 32 °C.

3. The highest average flue gas temperature for a 2 hour operation period was 700 °C. While the highest value measured was 955 °C.

4. The particle emission was measured simultaneously by two instruments before and after the scrubber. The emission from the combustor was 900-1500 mg/normal m^3 by 12% CO_2. The scrubber retained 80-90% of this emission. In absolute figures, the lowest average of 5 repeatings was 31 mg/normal m^3 (12% CO_2).

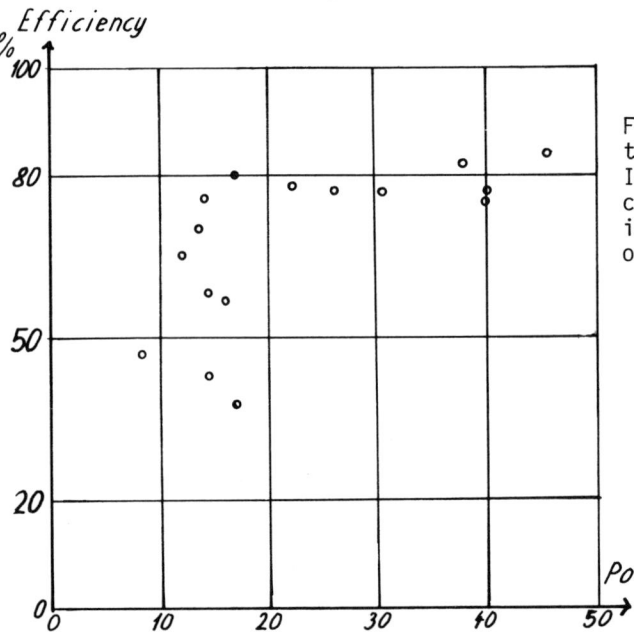

Fig. III Efficiency and thermal power output. It can be seen that a constant efficiency only is achieved with power output above 20 kW.

5. The retained emission products were removed by the wash water and partly settled in the condensate container. Analysis of wash water, which had been recirculated for a longer period, showed the following results:

   a) A pH-value of 5.2. This value could be increased one unit by aeration. Thos shows that it depends on CO_2 from combustion.

   b) The amount of nitrogen is shown in table I. Of the total 5.5 and 11.0% was nitrate and nitrite by use of the big and the small cyclone outlet tube respectively.

   c) Also the amount of tar depends on the outlet tube dimensions. With a 308 mm tube in the cyclone 28.2 mg tar/kg straw was retained.

With the small 210 mm tube 620 mg/kg or 22 times more was retained in the wash water.

d) Plant nutrients of the straw are accumulated in the ashes and condensate. The values were subject to large variations, but table I shows the average figures.

	Straw (6)		Ashes (6)	Accumulation of plant nutrients in the wash water after combustion of:		
				50.5 kg	301.6 kg	1466 kg
	mg/kg	%	%	%	%	%
N	6799	100	0.5	5.7	0.9	1.2
P	700	100	87	3.0	0.7	0.5
K	7700	100	89	42	15	2.7

Table I.

It appears that almost the total amount of potassium and phosphorus are regained. Of nitrogen very little is retained. It can be seen that the washing effect decreases as more and more straw is burned. Thus clean water washes best.

In general the results are promising, but further development is desirable.

4. Present stage of commercial utilization.

The flue gas scrubber is already available from two manufacturers. The price is about twice a chimney, which can be omitted because of the low flue gas temperature. Several scrubbers are already operating in connection with ordinary straw furnaces in Denmark. In a short time also in other countries.

As far as the cyclone furnace is concerned one private manufacturer is interested in production of a system consisting of the cyclone furnace with the flue gas scrubber as sole heat exchanger.

5. References.

(1) Maskintjenesten (1983): Meddelelse No. 794. - Landskontoret for bygninger og maskiner, p. 1.

(2) Commission of European Communities (1981): Energy from Biomass - Proceedings of the EC Contractors' meeting held in Copenhagen, 23-24th June, 1981. - Serie E, Volume 1. - D. Reidel Publishing Company, Dordrecht: p. 163.

(3) Commission of European Communities (1982): Energy from Biomass - Proceedings of the EC Contractors' meeting held in Brussels, 5-7 May 1982. - Serie E, Vol. 3. D. Reidel Publishing Company, Dordrecht: pp 165-169.

(4) Commission of European Communities (1983): Energy from Biomass - 2nd EC Conference. Applied Science Publishers Ltd., London: pp 479-483.

(5) Claar II, Paul W. et al. (1981): Crop-Residue-Fired Furnace for Drying Grain. - Paper No. 81-1032. For presentation of the 1981 Summer Meeting American Society of Agricultural Engineers.

(6) Kjellerup, Viggo (1983): Halmfyringsaskens indhold af plantenæringsstoffer; Statens planteavlsforsøg. - Meddelelse No. 1710.

"HEAT ENERGY FROM ANIMAL WASTE BY COMBINED DRYING, COMBUSTION AND HEAT RECOVERY"

Author   : H. Have

Contract Number: ESE-R-028-DK (G)

Duration   : 36 months   1 July 1980 - 30 June 1983

Total Budget   : Dkr. 1.886.220   CEC contribution 50%

Head of Project: Prof. T. Tougaard Pedersen

Contractor   : Jordbrugsteknisk Institut
                    Royal Veterinary and Agricultural University

Address   : Jordbrugsteknisk Institut
                    Royal Veterinary and Agricultural University
                    Rolighedsvej 23
                    DK-1958 Copenhagen V

Summary

The purpose of the project is to develop a combustion system which can produce low grade heat for space heating purposes from animal waste and other sources of wet biomass. The process comprises separated drying and combustion followed by heat recovery through flue gas cooling and water vapour condensation. Predictions of the performance have shown promising results.

After preliminary investigations concerning manure characteristics and methods of structurization, drying and condensation a full scale experimental plant comprising all the elements of the process have been constructed. So far testing of each of the elements are about to be completed with generally satisfactory results.

The testing is to be continued with an overall performance test.

1. Introduction

Large quantities of biomass (animal waste, energi crops etc.) are too wet for conversion by direct combustion in the traditional way. However, if drying and combustion are separated wet biomass may be burned and the heat can be recovered by condensation of water vapours in the flue gas.

The purpose of the project is to develop and test a system which make use of this principle for combustion of animal manure in order to produce thermal energy for space heating purposes or similar. A description of the process is given in section 3.

The project is carried out in cooperation with Vølund Miljøteknik A/S which is a company in the field of refuse burning equipment. Only few details of the development work is given in the following as a commercial utilization is under way.

2. Initial investigations.

The performance of the combustion system under realistic conditions has been predicted earlier (1). This calculation showed, that the system would be able to operate on biomass with up to approximately 82% moisture (wet base) without use of auxiliary fuel. The maximum temperature of the produced heat would be up to 80°C and the overall thermal efficiency 60 - 65%.

Moisture content, consistency and characteristics of manure in relation to drying was reported earlier. (2). Samples of different types of manure showed moisture contents ranging from about 50% w.b. for chicken manure to about 85 % for cattle and pig manure. The consistency depends on moisture content. At higher contents the manure is soft and adhere to any material it gets in touch with. During drying the strength and hardness increases until it, at lower moisture contents, becomes rather a strong, porous, and hard material. When drying up in contact with some other material it sticks strongly to it.

A small laboratory drier was constructed for preliminary drying investigations. It was equipped with a structurizer which formed the manure in aggregates of suitabel size. The form of these aggregates where then stabilized and maintained through drying. (2).

The scrubber system was investigated in another laboratory test (3). This test showed that it was possible to scrub and cool the flue gas with a scrubber designed according to the common theory of cooling towers. Also it showed that the scrubber is a very efficient heat exchanger with capacities in the range of 4 - 9 MW/m^3 packing column.

A preliminary energy balance (2) based on:

    a) operational electric energy consumption of a small scale feeding mechanism,
    b) estimated operational electric energy consumption of other auxiliary equipment,
    c) estimated nitrogen losses,

showed an output/input ratio about 9.

3. The full scale experimental incinerator system

A full scale experimental system for manure and similar products has been designed and constructed on basis of the results obtained from the preliminary investigations summarised above. An outline of the system is shown in figure 1, and the function is as follows:

Wet manure is metered from a buffer silo through a structurizer/feeder unit to the drier section. When dried a suitable amount of manure is fed to the burner through a stoker auger, while the surplus (only occuring at lower moisture contents of the wet manure) is taken out of the system. In practice this surplus may be used as solid fuel or as organic fertilizer.

The flue gas from the burner, is led to the drier and further on to the scrubber before released from the system. At the inlet to the drier the temperature is in the range 600 - 1000°C depending on conditions. In the drier it is cooled down to the range 100 - 125°C and in the scrubber further on to 35 - 45°C. Thus most of the heat energy is recovered in the scrubber, and the wash water is heated up to about 60 - 70°C. The heat is then transmitted to the cooling water in a plate type heat exchanger. The surplus of condensate is released from the system after cooling.

Dust and ash particles from drying and combustion are to a great extend retained in the wash water (condensate). Part of it is settled in a clarifier from which it is released as sludge and perhaps recirculated to the drier.

The gas burner fitted to the combustion unit is used for start up and for separate studies of drying. Alternatively the proces may be started up on dry biomass fed to the stoker auger between the drier and the burner.

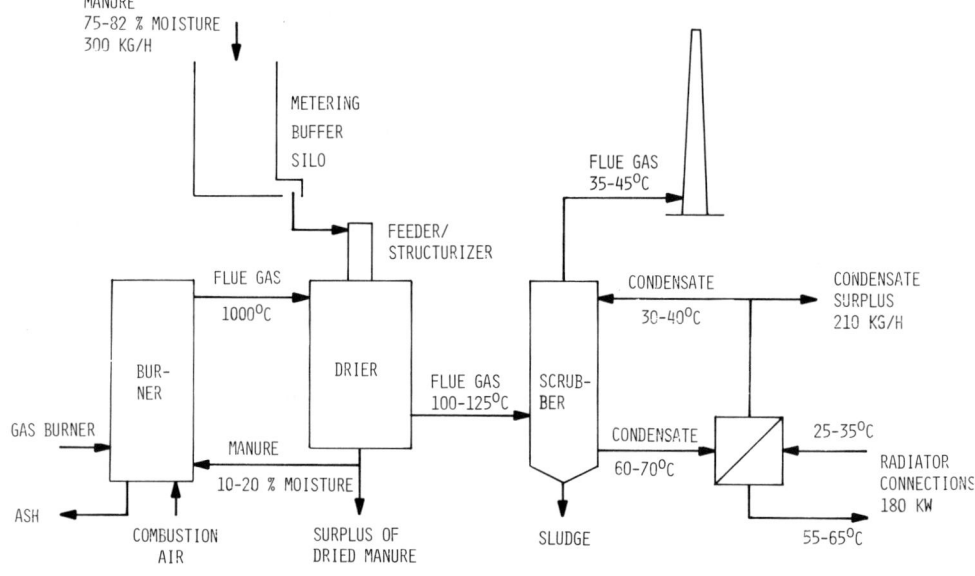

Figure 1. Flow sheet showing the experimental low grade waste incinerator system. (Patent pending).

The incinerator system is equipped with transducers and sensors necessary for setting up energy and mass balances in order to find the performance under various conditions.

The results obtained with the experimental unit until now may be summarized as follows:

The feeding/structurizing system is working satisfactorily with manure containing moisture in the range 55 to 85% v.b. and straw contents up to approximately 40% on dry matter basis. The drying system, which is new in concept, still needs some improvement. So far it has been operated at lower temperatures without excessive gasification of the material. The manure is dried into formstable, porous aggregates which are easy to handle and burn. The burner system appear to work satisfactorily with 1/3 of the combustion air supplied as primary air. Tempeatures up to $1.200^{\circ}C$ are uptained with yields between 100 and 300 kW. The scrubber system has shown good capacity of efficient heat transmission and removal of particles from the flue gas. No harmful sediments have been observed in the packing column. Also the heat exchanger is working well.

The testing is to be completed with an overall performance test and analysis of ash and condensate for contents of plant nutrients. Also the condensate and flue gas is to be analysed for possible contents of environmentaly harmful products.

4. Future work.

It is the intention to test the experimental unit with a number of other wet biomass sources, e.g. energy crops, peat, and some industrial waste products.

5. References

(1) Have, H. (1980):
    Extraction of Heat Energy from Animal Waste by Combined Drying, Combustion and Water Vapour Condensation. - Proceedings of the 1st Coordination Meeting of Contractors, 18/19 Sep. 1980, Amsterdam.

(2) Have, H. and Fritze, M. (1981):
    Heat energy from animal manure by combined drying, combustion and heat recovery. - Energy from Biomass, Solar Energy R & D in the European Community, Series E. Vol. 1. D. Reidel Publishing Company, Dordrecht, Holland.

(3) Nielsen, H.K. and Have, H. (1981):
    The use of gas scrubber for heat extraction from straw furnaces. - Energy from Biomass. Solar Energy R & D in the European Community, Series E, Vol. 1. D. Reidel Publishing Company, Dordrecht, Holland.

HIGH TEMPERATURE STRAW GRANULATION

Authors	: J. LUCAS, JF MOLLE
Contract numbers	: ESE/071/F
Duration	: 11 months 1 July 81 - 30 May 1982
Total budget	: 270 000 UC CEC 30 %
Head of project	: J. LUCAS
Contractor	: CEMAGREF
	Parc de Tourvoie
	92160 ANTONY (France)
Subcontractor	: ALSTHOM-ATLANTIQUE
	PROMILL

SUMMARY

Cereal straws biomass is widely avalaible for energy production purposes. It is particularly well adapted for domestic heating wich requires small-size and high density straw particles. However, conventional conditioning techniques have not yet made it possible to reduce the size and to increase the density of the straw bales while at the same time reduce costs. Therefore, the objective of the project is the study of a new high temperature straw granulation technique. Straw is over-dried and heated before being processed into pellets. An apron drier was selected. The flexibility of the process is the main reason for which such overdrier was selected to manufacture a mobile high temperature granulation unit.
This drier was built and fitted with a control system.
Tests are due to begin during the next phase of the project.

1.1 Description of the process

The equipment is carried by means of a truck trailer. During operations, it is placed at one end of a field where straw bales were stocked immediately after harvest. During the process, the bales are picked up and fed into a grinder for rough grinding. Ground straw is then fed into an overdrier and, finally, it is turned into pellets by a pelleting-press. Overdried straw becomes oily and brittle, which makes agglomeration considerably easier.

1.2 Overdrier technology

Every piece of straw must be heated up to 200°C. Higher temperatures should be avoided for fear of irreparable carbonization.

At first, an entrained bed drier had been tested. It was compact and highly efficient but it was difficult to control the length of time the straw remained in the drier, which resulted in overheating and carbonization. This type of drier was therefore discarded.

An apron drier was ten chozen because it allowed perfect control of the length of time the straw remained in the drier.

Unfortunately, heat and mass transfers were less efficient and the machine was therefore bulkier than in the previous case.

The operation heat-level is 250°C . Hot gases go through the straw layer from the bottom up. They are sucked up by a fan and (in the present state of the prototype) heated up by a fuel-oil burner before being forced back into the drier.

Particular attention was devoted to problems caused by volatile matters (corrosion and condensation on mobile parts) and to safety (risks of explosion in case of introduction of air).

1.3 Pressing

A conventional die-press was tested.

Condensation and polymerization of tar in the various openings of the machine when the press was not working made it difficult to set it in motion again. This is why the mobile unit was equipped with a variable compression press which can be emptied after each test.

1.4 Present state of the project.

Drying and compression trials were carried out on probatory models from March to July 1982. The tests led to a study, and the information was gathered up into a file drafting the manufacture of a mobile drying unit, form July to September 1982. The study of the unit automation ended up in December 1982. The pelletization prototype has been manufactured since the beginning of 1983 and has been mounted on a semi mounted trailer.

1.5 Prospects of the project.

As the mounting work is coming to an end, hopefully the trials on the field will take place in September, after stationary experimentation at the CEMAGREF. Studies will be carried out and drying units will be manufactured, especially for grain and pellet-fueled domestic heating. The automatized units will allow the concerned manufacturens to undestake the mass production of these products.

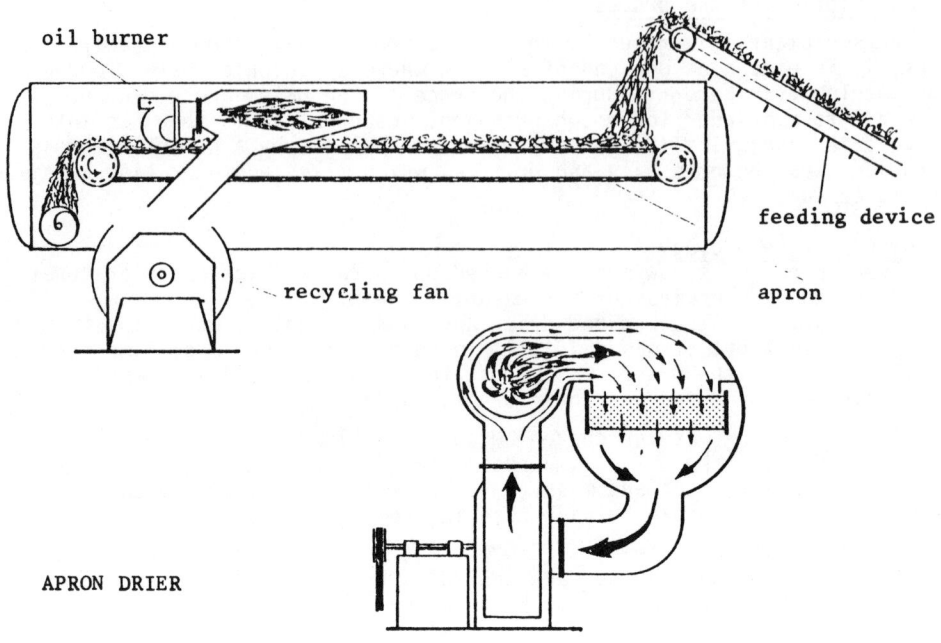

APRON DRIER

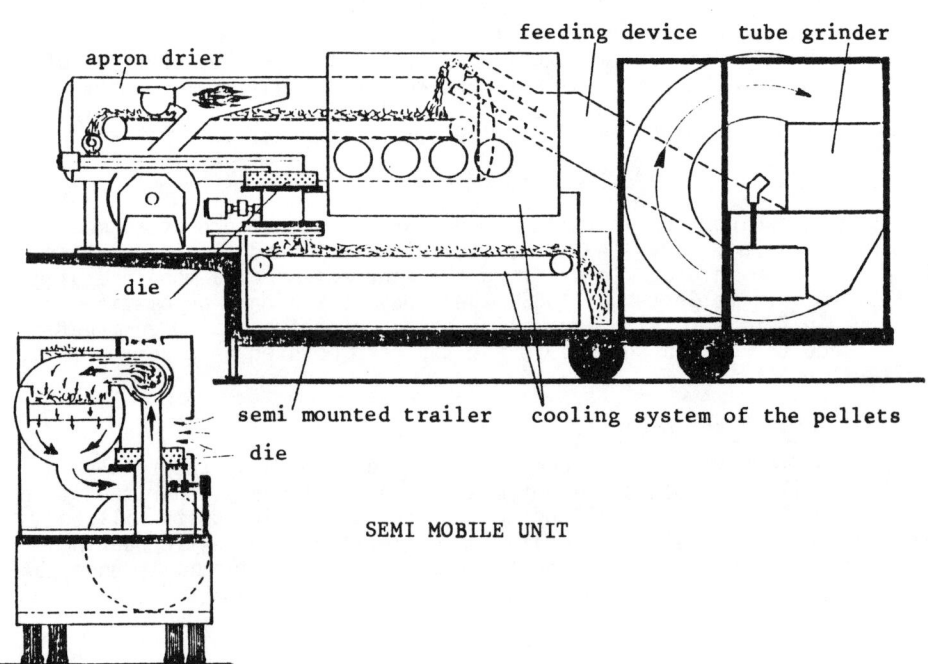

SEMI MOBILE UNIT

GENERAL VIEW

THE DRIER

THE BELT

THE RECYCLING FAN
THE OIL BURNER

THE PRESS

MOBILE PYROLYSIS PLANT

Author	:	S. Hauptmeyer, Messrs. FRITZ WERNER
Contract Number	:	ESE - R - 048 D
Duration	:	18 months; 1st October 1981 - 31st March 1983
Total Budget	:	621.290.-- DM
Head of Project	:	S. Hauptmeyer, Messrs. FRITZ WERNER
Contractor	:	FRITZ WERNER INDUSTRIE-AUSRÜSTUNGEN GMBH Department "New Technology"
Address	:	D-6222 Geisenheim Industriestraße

Summary:

Today mobile charcoal kilns made of sheet steel are applied in various styles e.g. the drum-kiln in Kenya. As a result, charcoal is produced and the pyrolysis oil is flared with the pyrolysis gas.

Compared to the charcoal kiln there is a higher charcoal output with the retorts which furthermore are of a better quality. In addition the pyrolysis oil is obtained. These positive aspects of the retorts must form the economical antipole for the high investments.

For the project at hand the fluidised-bed system was selected for the development of a mobile pyrolysis. The reason for this choice was the possibility to reach a high material throughput, to optimize the profitability and what is more, now also finely structured biomasses like reeds can be used for the pyrolysis.

Since the profitability of the plant limits the plant design immensely its prevailing development has not been very simple and is not yet completed.

Via the adjustable ring for gasification control with nozzles (17) air or oxygen are added to the gas streaming upwards for the after-cracking. Thus, the gas produced is cleared of higher hydrocarbon compounds. An additional device for the fluidisation (16) influences the after-cracking process positively.

After filtration and cooling the gas may be used in internal-combustion engines.

3.0 Utilization of the Pyrolysis Products
The utilization of the pyrolysis products strongly depends on the infra-structure of the country in question because refining is recommended.

3.1 Charcoal
The charcoal produced in the fluidised-bed is in small lumps or in the form of dust and unless it is used in a dust furnace, it must be briquetted for cooking or boiling purposes e.g. with starch as a binder. A thermal after-treatment is necessary when briquetting with tar oil. The charcoal produced at high temperatures can be broken down into different fractions by means of filtration. Thus, it can be used as low quality activated coal e.g. even for the carbonisation of steel.

3.2 Pyrolysis oil
The pyrolysis oil may be roughly separated in e.g. simple tar containers or the individual components are cracked in costly refineries e.g. in the phases methanol, methyl acetate, acetic acid, propiolic acid and butyric acid. The remaining oil is usually burnt for heating purposes. Because of the high portion of carbon, it burns with a bright flame and high radiant heat. As the oil is heavily loaded with the finest dust particles, it requires injection burners with large nozzles.

Paraffin and naphtalene are only deposited when the temperature drops. Since tar oil on the contrary to crude oil does not have any lubricating abilities, moving parts call for separate lubrication.

As the oil which has not been treated contains very high parts of acetic acid container, piping, valve and burner must be manufactured of stainless steel or copper or in the lower temperature range of plastic material. The fumes are neutral and much cleaner as those of crude oil because the latter contains a more or less large portion of sulphur. The pyrolysis oil is mixed with 15% water or heated to 50 - 75° C to obtain a better distribution by the oil burner. Wood tar differs from coaltar and lignite by its low content of ammonia compounds naphtalene and paraffin.

3.3 Gas
The gas has a low calorific value so that bottling is not worthwhile. Therefore, it is only used for a self-supporting energy supply of the plant.

1.0 Introduction
Though the mobile pyrolysis plant can easily be operated, the maintenance personnel must comply with certain requirements.

Due to the automatic operation, a mechanic and an electrician should shortly be available to repair possible malfunctions.

As with all pyrolysis processes the used biomass should have a 15% moisture content at the most. From this point of view the utilization of chopped straw is quite advantageous. However, the high ash content causes a great reduction of the effectiveness of the produced charcoal.

Materials up to a size of 25 x 25 x 5 mm can be applied in the fluidised-bed process used here.

2.0 Description of the Plant
A few items were altered in the plant conception which was introduced in May 1982.

Figure 1 shows the new system of the mobile pyrolysis.
The most substantial changes are as follows:
a. The gas is not cooled to approx. 80° but to approx. 115° C. Thus, the larger portion of the water content does not condensate. However, the calorific value of the gas drops.
b. The reactor is no longer heated externally by help of a double casing but by means of the flush gases according to the Reichert procedure. Thus, the thermal strain on the reactor casing is diminished.
c. The charcoal is no longer separated from the gas in the cyclone but extracted separately out of the reactor. Therefore, the rather unreliable cyclone is avoided.

Figure 2 shows the reactor which has been applied for a patent with its new solution for the charcoal extraction. The flush gas (21) enters the fluidised-bed chamber (10) with a slight excess pressure and a temperature of approx. 500° C through a lower grid (18) and produces with the fuel (20) a homogeneous fluidised-bed.

The height of the fluidised-bed chamber (10) is variable because of the adjustable hearth ring (11). Thus, an overflowing of the fluidised-bed into the ring chamber (15) can be influenced.

This trend of solid fuel separation is supported by the adjustable cover of the hearth (12) with its reversing of the gas stream.

The charcoal is discharged out of the ring chamber (15) and via a discharge opening with the sweeping auger (22).

The structure of the fuel (biomass) is maintained during the first phase of carbonisation. But during the second step (further residence in the fluidised-bed) the fuel particles suffer mechanical and thermal destruction before they leave the reactor vessel.

The separated coal is very homogeneous because it remained in a hot zone for a determined time. The pyrolysis products charcoal, pyrolysis oil and gas are produced at fluidised-bed temperatures between 400° and 600° C.

2.1 Operation at high Process Temperatures
When the fluidised-bed temperature is around 900° C a temperature which can be obtained e.g. by using pre-heated air as flush gas, then the charcoal contains a higher carbon content. The charcoal is now of better quality and it can be used as a low quality activated coal. However, no more pyrolysis oil is produced. The charcoal thus produced does not stick together and it can be discharged out of the reactor by chutes. Thus, the sweeping auger (22) can be relinquished.

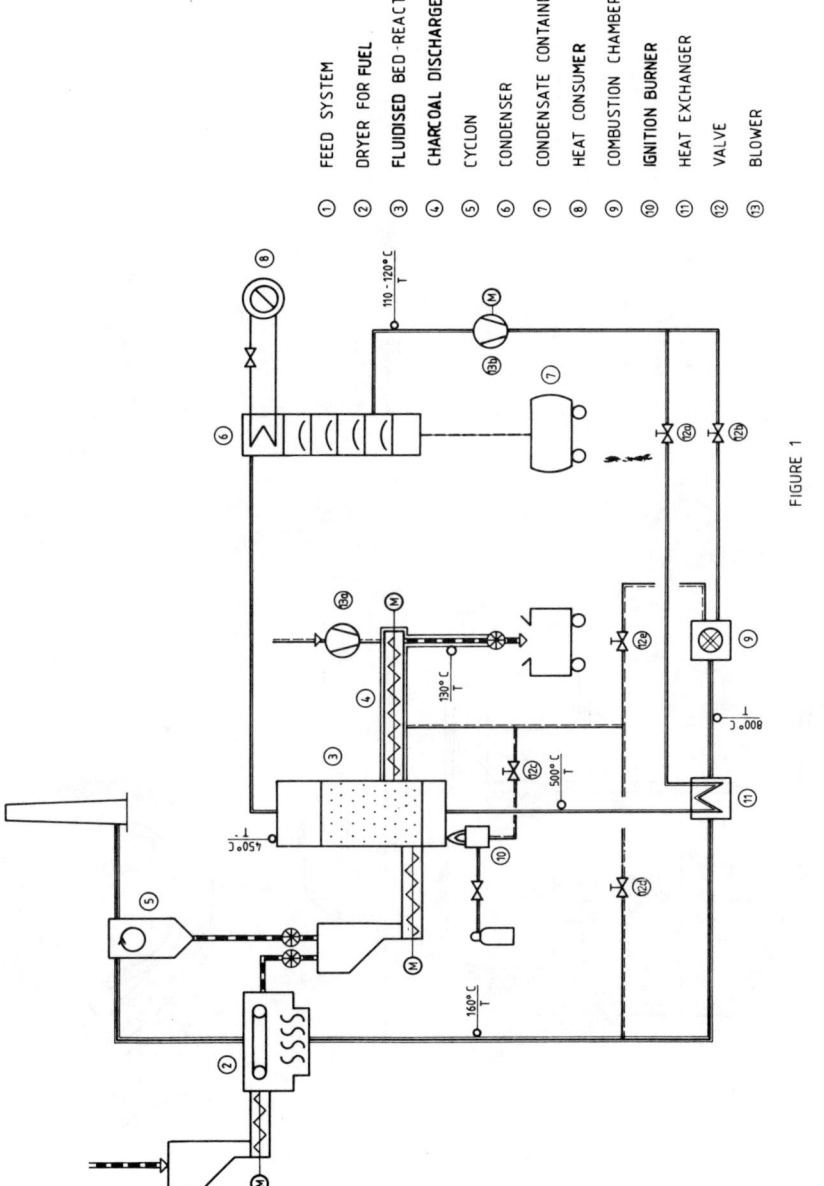

FIGURE 1   MOBILE PYROLYSIS PLANT

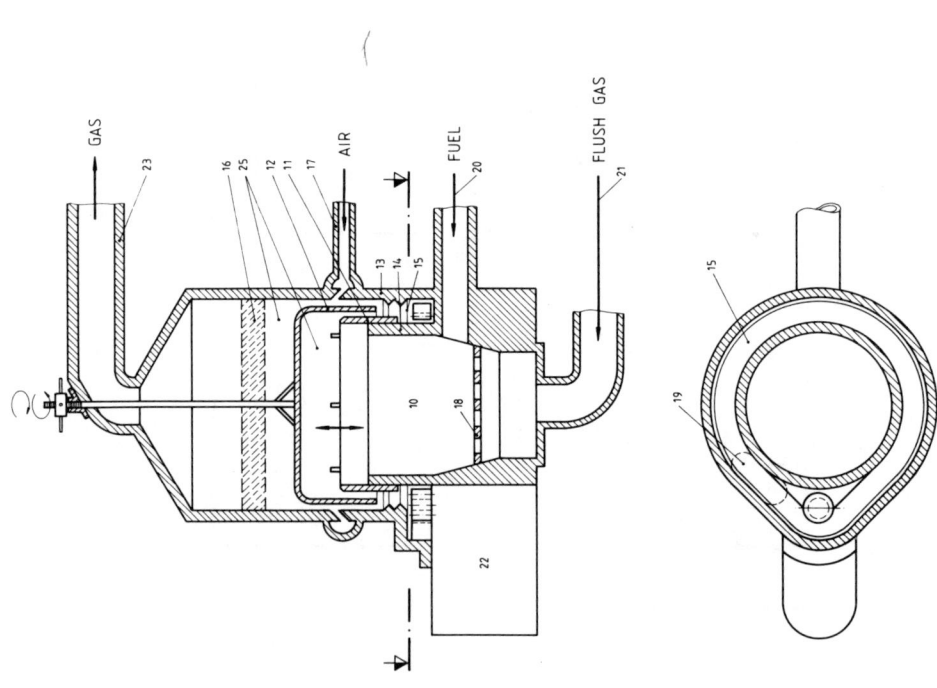

10 FLUIDISED-BED CHAMBER
11 HEARTH RING
12 COVER OF THE HEARTH
13 REACTOR OUTER CASING
14 HEARTH CASING
15 RING CHAMBER
16 DEVICE FOR THE FLUIDISATION
17 ADJUSTABLE RING FOR GASIFICATION CONTROL WITH NOZZLES
18 LOWER GRID
19 DISCHARGE OPENING
20 FUEL
21 FLUSH GAS
22 DRIVE OF THE SWEEPING AUGER
23 GAS PIPE
25 REACTOR CHAMBER

FIGURE 2

FLUIDISED-BED REACTOR
WITH CHARCOAL DISCHARGE

MODELING OF A FLUIDIZED BED WOOD GASIFIER

Authors   : F.G. VAN DEN AARSEN, A.A.C.M. BEENACKERS,
                     W.P.M. VAN SWAAIJ

Contract number   : ESE-R-029-NL

Duration   : 36 months   1 July 1980 - 30 June 1983

Total budget   : Dfl. 986,000.-   CEC contribution: Dfl. 304,000.-

Head of project   : Prof.dr.ir. W.P.M. van Swaaij,
                     Twente University of Technology

Contractor   : Twente University of Technology
                     Laboratory of Chemical Reaction engineering

Address   : P.O.Box 217
                     7500 AE   Enschede
                     The Netherlands

Summary

A mathematical model of a fluidized bed wood gasifier was developed, which incorporates the hydrodynamic features of a bubbling fluidized bed reactor.
Wood pyrolysis product yields, char gasification kinetics and bubble to dense phase mass transfer kinetics were separately determined and incorporated in the model. Model calculations are in good agreement with experimental findings.

Introduction
In a fluidized bed wood gasifier the following processes occur simultaneously:
- partial oxidation of the wood
- fast pyrolysis of the wood
- gasification of pyrolysis char
- conversion of pyrolysis tar.

The complexity of this integrated process asks for a mathematical model of the gasifier in order to correlate the effects of process variables on the product gas composition and the dimensions of the gasifier. Therefore a mathematical model of a fluidized bed wood gasifier was developed, which incorporates the hydrodynamic features of a bubbling fluidized bed reactor.

Gas-solid contacting in fluidized beds has received much attention over the passed decades (1,2). Our reactor model is an extended two phase fluid bed model incorporating Davidson's mass transfer correlation (3), Darton's bubble growth relation (4) and Werther's bubble rise velocity correlation (5). We made a separate study of mass transfer kinetics in a fluidized bed reactor applying particles with high minimum fluidization velocity as in fluidized bed gasifiers (6).

Figure 1 represents schematically some of the aspects involved in fluidized bed wood gasification.
In our model a separate oxidation zone is distinguished where part of the incoming wood is combusted under consumption of all oxygen supplied. The gas composition and quantity leaving this section, follows from thermodynamic considerations.

The extended two phase fluid bed model, including experimentally determined pyrolysis product yields, char gasification kinetics and bubble to dense phase mass transfer kinetics, is applied for calculating the pyrolysis/gasification section. In the next sections these subjects will be treated separately.

Thermodynamics
Several thermodynamic models exist to calculate a gasifier product gas composition on basis of fuel and air composition.

We make use of the Schläpfer (7) approach to calculate the required air for gasifying wood of a certain moisture content. The basic equations for these calculations are:
1. Componental balances:
   a. Carbon balance
   b. Oxygen balance
   c. Hydrogen balance
   d. Nitrogen balance
2. Enthalpy balance
3. Dalton's law
4. Ideal gas law
5. Homogeneous water shift equilibrium: $K(T) = \dfrac{(CO)(H_2O)}{(CO_2)(H_2)}$

Input data are:
- Equilibrium temperature = reactor temperature.
- Wood: composition, moisture content, heating value.
- Air: composition, moisture content, inlet temperature.
- Heat losses through reactor wall.
- Hydrocarbon content of the product gas.

Results of these calculations are:
- Equilibrium product gas composition and gas production volume (m^3 productgas/kg fuel).

- The air-wood ratio = required air (Nm^3) per kg wet fuel.
This air-wood ratio is required for calculating the composition of the gas leaving the oxidation zone. In this calculation it is assumed that part of the incoming wood is completely combusted in the oxidation zone and that all oxygen from the incoming air is consumed in this combustion process. The remainder of the wood is mixed up with the reactor contents above the oxidation zone and subsequently converted in the so called pyrolysis and gasification zone.

The height of the oxidation zone is set by the height of the wood inlet and depends on the fraction of wood that is required for consuming all the oxygen by combustion. This results in a height somewhere between the lower and the upper level of the wood entrance (between 9 and 21 cm).

Basic reactor model

The gasification-pyrolysis section of the fluidized bed reactor is modelled by an extended two phase model, taking into account chemical enhancement of mass transfer, analogous to gas-liquid systems (Danckwerts (8) and (Westerterp, Van Swaaij, Beenackers (9)) and through flow of the dense phase above minimum fluidization quantities (fig. 2).
The applicability of chemical enhancement of mass transfer in gas solid fluidized bed reactors has been suggested by van Swaaij and Zuiderweg (10) and was demonstrated by Werther (11).

The following assumptions are made:
1. Gas in bubble and dense phase flows in plug flow.
2. No solids in the bubble phase.
3. Interfacial area and bubble hold-up are a function of bed height.
4. The dense phase gas through flow is: $u_d = u_{mf} + C(u - u_{mf})$ with $0 \leq C \leq 1$.

The differential equations of the mass balances for stationary conditions with n-th order reaction, applying a simplifying solution for the film equation are:

$$(1-\gamma)\frac{dC_b}{d\zeta} - N_\alpha(C_d - C_b) + N_\alpha \frac{k\delta^2}{4 \, D}(C_b^n + C_d^n) + N_p(C_b - C_d) = 0$$

(bubble phase) (1)

and

$$\gamma\frac{dC_d}{d\zeta} + N_r C_d + N_\alpha(C_d - C_b) + N_\alpha \frac{k\delta^2}{4 \, D}(C_b^n + C_d^n) + N_p(C_d - C_p) = 0$$

(dense phase) (2)

These equations are solved numerically applying experimentally determined yields of the pyrolysis process, char gasification kinetics and local values for the mass transfer parameter as presented in the next sections.

In our model calculations we have estimated a reactor height required for completing the fast pyrolysis process. This height depends on the particle pyrolysis time and the solids mixing time. On basis of literature (12, 13) and experimental data (14) we derived an emperical expression for the pyrolysis height, in case biomass is fed near the bottom of a fluidized bed:

$H_{pyr} = 300 * d_p$ [m]

where,
d_p : characteristic dimension of a wood slab $\leq 3 * 10^{-3}$ m
H_{pyr} : pyrolysis height measured from the centre of the biomass inlet.

Further we assume in our model calculations the establishment of the homogeneous water gas shift equilibrium in both dense phase and bubble phase after each differential conversion step $\Delta \zeta$.

Fast pyrolysis of wood particles in a fluidized bed

Experimentally obtained product yields for the fast pyrolysis of 1 mm beech wood particles were reported previously (15). In table I the results are summarised.

Table I. Yields (g/g wet fuel) of fast pyrolysis of 1 mm beech wood (15% wt moisture)

Pyrolysistemp. °C	gas g/g	water g/g	tar g/g	char g/g
715	0.662	0.186	0.06	0.092
815	0.759	0.125	0.06	0.057
915	0.765	0.124	0.06	0.051

The elemental composition of the pyrolysis tar and char is presented in table II.

Tabele II. Elemental composition of tar and char from fast pyrolysis of 1 mm beech wood (15% wt moisture)

Pyrolysistemp. °C	tar	char
715	$CH_{0.81}O_{0.2}$	$CH_{0.25}O_{0.14}$
815	$CH_{0.75}O_{0.13}$	$CH_{0.21}O_{0.12}$
915	$CH_{1.0}O_{0.17}$	$CH_{0.14}O_{0.14}$

Figure 3 gives the dry pyrolysis gas composition. The experimental yields in hydrogen consequently showed a deficit, whereas for carbon and oxygen the elemental balance matched within 5%; for this reason we corrected the hydrogen content of the pyrolysis product gas to meet mass balance requirements.

Char gasification kinetics

Knowledge on the reactivity of char as it is present in a fluidized bed wood gasifier is necessary for proper modeling of this reactor. For several carbon dioxide concentrations and reaction temperatures we determined the gasification rate of char (originating from our pilot plant) in in a bench scale. Typical properties of this char are:
a_o = 271 m^2/g
S_o = 105 * 10^8 m^2/m^3 particle volume
ε_o = 0.69 m^3/m^3 particle volume
Figure 4 gives the dependency of reaction rate on carbon dioxide concentration, resulting in:

$$R'' = k'' c_{CO_2}^{0.83} \quad \text{mol C/m}^2 \cdot \text{s} \tag{3}$$

The 0.83 power indicates the inhibiting effect of CO_2 and thus suggests Langmuir Hinshelwood kinetics. Also CO will inhibit the reaction rate according to the surface reaction mechanism. The use of global kinetics however is practical and justified within the range of experimental conditions.
The dependency of reaction rate on temperature is represented in figure 5 by means of an Arrhenius plot, resulting in:

$$k'' = 7.2 \exp(-) \frac{166156}{RT} \quad \frac{\text{mol}}{\text{m}^2 \text{s}} \left(\frac{\text{mol}}{\text{m}^3}\right)^{-0.83} \tag{4}$$

Equations (3) en (4) are applied in the main reactor model.

Bubble to dense phase mass transfer kinetics

Mass transfer kinetics received much attention in literature and amongst those of the most frequently applied correlations in fluidized bed modeling is the one as presented by Davidson (1977). This correlation couples the diffusive exchange rate at the bubble-dense phase interface with the convective exchange rate between the bubble and dense phase. The contribution of the convective term, is relatively small for small particle systems, but increases with increasing minimum fluidization velocity. However, the validity of this relation has not been tested for large particle systems yet; we therefore investigated the mass transfer kinetics for such large particle systems, applying the catalytic oxidation of carbon monoxide as a model reaction. This enabled us to check the applicability of Davidson's mass transfer relation, Darton's bubble growth relation and Werther's bubble rise velocity in our reactor model.

Table III presents the local values of the mass transfer parameters as applied in our model. Figure 6 shows the experimental results of our mass transfer measurements, which indeed show an increasing contribution of the convective term with minimum fluidization velocity and an excellent agreement between mass transfer model and experimental results.

Table III. Mass transfer parameters

$$\alpha = P * u_{mf} + 0.91 \frac{D^{1/2} g^{1/4}}{D_e^{1/4}} \frac{\varepsilon_{mf}}{(1+\varepsilon_{mf})}$$

Davidson: $P = 1.19$

$$D_e(h) = 0.54(u-u_d)^{2/5}(h+4.0\sqrt{A_o}^{4/5}/g^{1/5}) \text{ for } h < h^*$$
$$\text{else } h = h^*$$

$$\varepsilon_b(h) = \frac{u - u_d}{u_A}$$

$$u_A(h) = c \sqrt{g \cdot D_e}$$
$$c = 0.6 \text{ for } D < 0.1 \text{ m}$$
$$c = 1.6 \, D^{0.4} \text{ for } 0.1 < D < 1 \text{m}$$
$$c = 1.6 \text{ for } D > 1 \text{ m}$$

$$a(h) = \frac{6}{D_e(h)} \varepsilon_b(h)$$

Performance of a fluidized bed wood gasifier: model and experimental results

Beech wood particles have been gasified at several conditions in our pilot plant fluidized bed gasifier. Details of this test rig were described in previous papers (15). Operating temperature, bed height and biomass feed rate were variables in these experiments, while gas production rate, gas composition, gas heating value and char hold-up were measured. We used these data for verification of our model calculations and some of the results are summarised in Table IV.

Table IV. Calculated en measured fluid bed gasifier data

	H_2	N_2	CO	CH_4	CO_2	C_2H_4	L.H.V.	$\emptyset_{wood}$	$\emptyset_{air}$	T	H	ε
	-dry gas composition mol%-						kJ/m^3	kg/hr	m^3/hr	°C	m	m^3/m^3
1a.calc.	12.3	50.8	17.4	3.7	15.0	0.87	5342	(22.4)*	(33.0)*	815	0.60	0.035
b.meas.**	14.4	46.5	18.0	3.6	16.7	0.75	5504	22.4	33.0	815	0.60	N.M.
2a calc.	12.2	51.3	17.0	3.6	14.9	0.86	5264	(25.6)*	(37.9)*	820	0.46	0.038
b meas.**	11.9	51.5	16.8	3.9	14.6	1.15	5568	25.6	37.9	820	0.46	0.032

* experimental data, as used in model calculations.
** cooled gas sampling probe installed.

The agreement between measured and calculated product gas composition is good.
The char hold-up in the reactor is an important parameter, which is dependent on char reactivity, temperature and bed height. The char hold-up that follows from our model calculations is compared to the measured values in figure 7; The experimental data are summarised in Table V.

Table V. Experimental data char hold-up

	$\emptyset_{wood}$ [kg/hr]	$\emptyset_{air}$ [m^3/hr]	T [°C]	H [m]	$\varepsilon_{measured}$ [m^3/m^3]	$\varepsilon_{calculated}$ [m^3/m^3]
1*	25.6	37.9	820	0.46	0.032	0.038
2	31.6	34.7	805	0.25	0.21	0.18
3	28.9	33.0	840	0.42	0.045	0.036

* Cooled gas sampling probe installed.

Conclusions

We developed a fluidized bed wood gasifier model based on:
- a two phase fluid bed reactor model incorporating Davidson's mass transfer relation, Darton's bubble growth relation and Werther's bubble rise velocity correlation
- a distinguishable oxidation zone fixed by thermodynamic considerations
- an empirical relation for the height required for pyrolysis
- experimentally obtained gasification kinetics
- experimentally obtained pyrolysis product yields.

Using this model we proved to be able to predict:
- the exit gas composition
- the char hold-up as function of the oparating variables (temperature, thoughput, bed height).

This model also enables us to predict concentration profiles along the reactor height; experimental verification of these calculations are still underway.

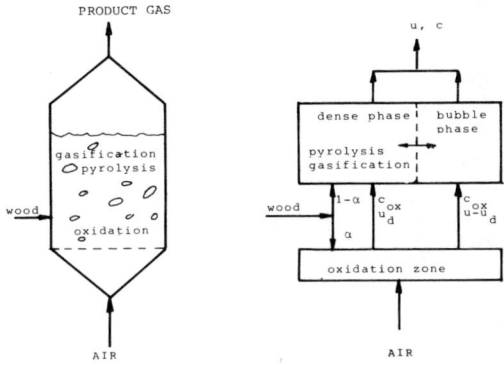

Figure 1: *Scheme of a fluidised wood gasifier*

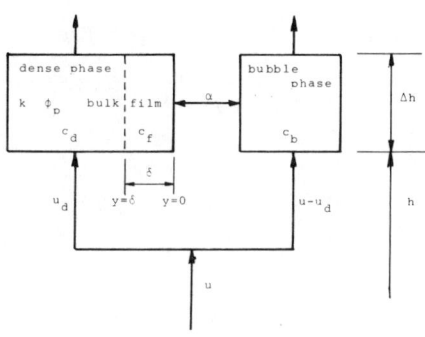

Figure 2: *Fluid bed reactor model*

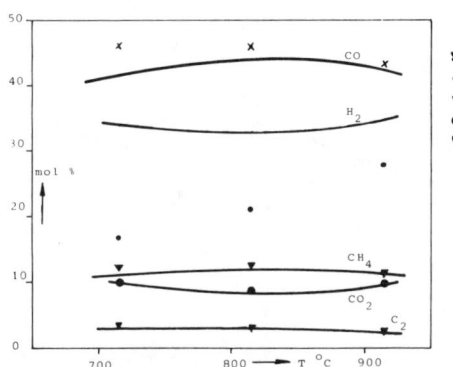

Figure 3: *Dry gas composition for pyrolysis of 1 mm beech wood particles (15 % wt moist)*

Figure 4: *Surface reaction rate of fluidized bed gasifier char as function of carbondioxide concentration* →

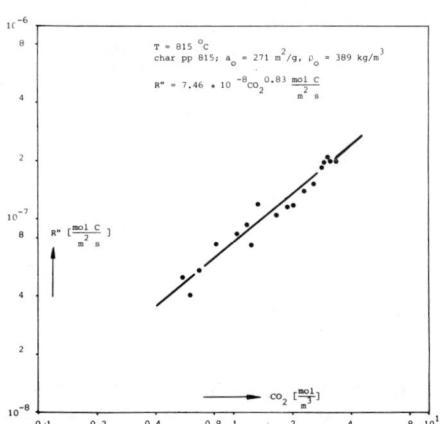

$T = 815\ ^\circ C$

char pp 815; $a_o = 271\ m^2/g$, $\rho_o = 389\ kg/m^3$

$R'' = 7.46 \cdot 10^{-8} c_{CO_2}^{0.83}\ \dfrac{mol\ C}{m^2\ s}$

$R''\ [\dfrac{mol\ C}{m^2\ s}]$

$CO_2\ [\dfrac{mol}{m^3}]$

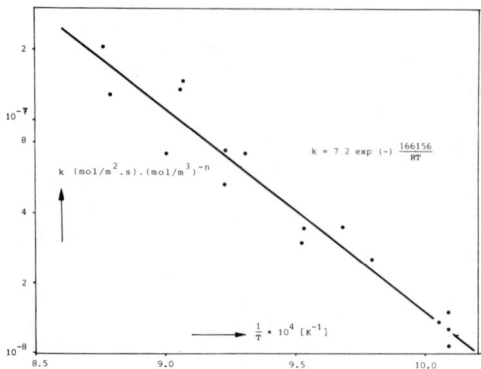

Figure 5: *Arrhenius plot for reaction rate constant of char-carbon dioxide reaction*

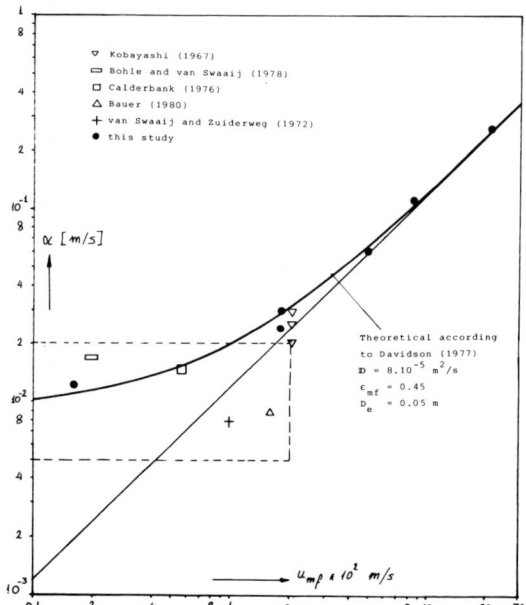

Figure 6: *Mass transfer coefficient as function of minimum fluidisation velocity*

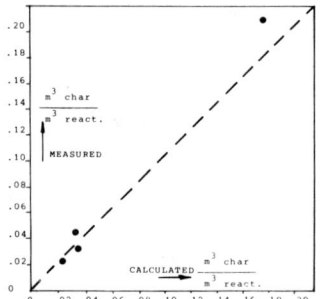

Figure 7: *Comparison of measured and calculated values of the char hold-up in a fluid bed wood gasifier*

- 262 -

LIST OF SYMBOLS.

A_o	area of distribution plate per orifice	[m^2]
a	specific surface area of bubbles	[m^2/m^3]
a_o	initial specific surface area of solids	[m^2/g]
α	bubble to dense phase mass transfer coefficient	[m/s]
α	air to fuel ratio	[m^3/kg]
c_{ox}	gas concentration in oxidation zone	[mol/m^3]
c_d	gas concentration in dense phase	[mol/m^3]
c_b	gas concentration in bubble phase	[mol/m^3]
c_f	gas concentration in film	[mol/m^3]
C	constant	[--]
γ	dense phase throughflow ratio $=(\frac{u_d}{u})$	[--]
d_p	particle dimension	[m]
D_e	effective bubble diameter	[m]
δ	film thickness $=(\mathbb{D}/\alpha)$	[m]
$\mathbb{D}$	gas diffusion coefficient	[m^2/s]
ε_{mf}	gas hold up at minimum fluiditation	[m^3/m^3]
ε_o	initial particle porosity	[m^3/m^3]
ε_b	gas hold up in bubble phase	[m^3/m^3]
ϕ_p	Volumetric production rate of pyrolysis gases	[$m^3/m^3.s$]
g	gravitational acceleration (=9.81)	[m/s^2]
h	height coordinate	[m]
H	height of bed	[m]
K	equilibrium constant	[--]
k	reaction rate constant in equation (1),(2)	[$(\frac{mol}{m^3})^{1-n}.s^{-1}$]
N_α	number of mass transfer units $(=\frac{\alpha aH}{u})$	[--]
N_p	number of pyrolysis units $(=\frac{\phi_p H}{u})$	[--]
N_r	number of reaction units $(=\frac{k(1-\varepsilon_b-a\delta)c^{(n-1)}H}{u})$	[--]
n	reaction order	[--]
P	constant	[--]
R	gas constant (= 8.314)	[$\frac{J}{mol.K}$]
R''	surface reaction rate	[$mol/m^2.s$]
S_o	initial specific surface area of solids	[m^2/m^3]
u	superficial gas velocity	[m/s]
u_d	superficial gas velocity in the dense phase	[m/s]

u_b superficial gas velocity in the bubble phase [m/s]
u_A bubble rise velocity [m/s]
ζ dimensionless height (=h/H) [-]

LITERATURE.

1. van Swaaij, W.P.M.; in "Fluidization", Davidson and Harrison (eds.), to be published
2. Wen, C.T.; in "Residence time distribution theory in chemical engineering"; A. Pethö and R.D. Noble (eds.); Verlag Chemie-Weinheim (1982)
3. Davidson, J.F. et. al.; in "Chemical Reactor Theory, a Review"; L. Lapidus and N.R. Amundson (eds.); Prentice Hall, Eaglewood Cliffs, N.J. (1977)
4. Darton, R.C.; Trans. Inst. Chem. Engnrs. 55 (1977) 274
5. Werther, J.; Chem. Ing. Tech. 50 (1978) 11, 850-860
6. van den Aarsen, F.G. et. al.; "Bubble to dense phase mass transfer kinetics in gas solid fluidised beds"; to be published
7. Schläpfer, P. and Tobler, J.; "Theoretische und praktische Untersuchungen über den Betrieb von Motorfahrzeugen mit Holzgas", Bern 1933 - Berichte nr. 3 (Separatabdruck aus "Der Motorlastwagen")
8. Danckwerts, P.V.; "Gas-Liquid Reactions"; Mc Graw-Hill-London (1970)
9. Westerterp, K.R., van Swaaij, W.P.M. and Beenackers, A.A.C.M.; "Chemical Reactor Design and Operation"; Wiley (1984)
10. van Swaaij, W.P.M. and Zuiderweg, F.J.; Proc. 5th Europ. Symp. on Chem. React. Eng.; B9-25 - B9-35; Elseviers Publ.; Amsterdam (1972)
11. Werther, J.; Chem. Eng. Sci., 35 (1980) 372-379
12. Nienow, A.W. et. al.; Am. Inst. of Chem. Eng. Symp. Series, 74 (1978) 176, 45
13. Masson, H. et. al.; Paper presented at Symp. "Materials and Energy from Refuse", Antwerpen, Oct. 1981
14. van den Aarsen, F.G.; Ph.D. Thesis, to be published
15. van den Aarsen, F.G., Beenackers A.A.C.M. and van Swaaij, W.P.M.; in "Energy from Biomass"; Vol. 3; Grassi, G. and Palz, W. (eds.), Reidel Publishing Company (1982)

BIOMASS GASIFICATION - PROGRAMME STATUS

Contract No.	:	E-S-E-R-032-UK(N)
Duration	:	41 months. Aug. 1983 - Dec. 1983
Total Budget	:	£322,000. CEC contribution £91,400
Head of Project	:	H. T. Wilson, Foster Wheeler Power Products Limited
Contractor	:	Foster Wheeler Power Products Limited
Address	:	Greater London House, Hampstead Road, London, NW1 7QN.

SUMMARY

A gasification test facility has been constructed. The design will permit evaluation of different reactor types on a direct comparative basis. The system is capable of operation at pressures up to 30 bars and temperatures of $1200^{o}C$. The main design/construction problems which arose in the course of the programme have been resolved. The data from this area will be of considerable value in construction of commercial scale gasification plant - not necessarily restricted to Biomass. Complimentary theoretical studies in relation to chemical equilibrium and pyrolysis resulted in the development of a suite of versatile computer programmes which have been used to predict performance on a range of commercial gasifier reactors. The correlation between predicted performance and actual performance is very high for specific reactor types. From examination of the relationships between feed form condition vs. reactor selection it is now possible to match a range of feed materials to specific reactor types and anticipated products. Development of a partial oxidation reactor has opened up the possibility of modifying any gasifier output to expand operational capabilities by design or retrofit. The main experimental programme has been extended due to the technical design difficulties experienced and the associated necessity to re-finance the remainder of the programme.

1. Overview

As part of the EEC Energy from Biomass programme, Foster Wheeler Power Products Limited proposed the construction and operation of a versatile gasification test facility within relevant system parameters could be evaluated on a comparative basis. This test facility has now been built (Fig. I). In view of the interest in pressurised gasification for the production of synthesis gas followed by catalytic conversion to liquid fuels, the facility was designed to operate at pressures up to 30 bar and temperatures of up to 1200°C with a range of atmospheres.

Translating the objectives of the programme into hardware proved to be singularly difficult with numerous engineering problems requiring resolution. In perspective, these problems which were met and overcome are implicit in the design of any gasification plant and their resolution thus provides a measurable consolidation and advance of knowledge in this area.

Parallel theoretical studies undertaken by Foster Wheeler Power Products Limited in the field of chemical equilibrium and pyrolysis resulted in the development of a suite of computer programmes offering a powerful technique to examine performance of any gasification facility under a wide range of operating conditions. In reactor systems which operate at or close to equilibrium, the correlation between predicted performance and actual performance is very high. Also in systems where sufficient data has been correlated in relation to the pyrolysis reactions, a combined pyrolysis-equilibrium model has been demonstrated as extremely valuable in predicting performance (I).

Arising from the combination of practical engineering experience and theoretical approach, a design was evolved for a Partial Oxidation Reactor. This has been built and awaits testing (Fig. II). Its commercial operation as a design feature or in a plant retrofit situation will permit modification to any gasifier system output and could provide a means of enhancing operational flexibility.

2. Key Design Considerations

The test plant was constructed generally in accordance with ASME VIII Division 1 codes with the associated design constraints. Reactor material selection was based on provision of an interior two-layer refractory lining to drop the maximum working temperature at the surface to 300°C. The design was optimised on the basis of shape vs. diameter and minimum wall thickness, also projected reactor configurations arising from changes in the internal refractory. Material selected for vessel construction was a fully killed aluminium carbon-manganese steel to BS 1501-224 Gr. 32 B. This is a low alloy carbon steel which offers optimum material performance characteristic in relation to the anticipated working conditions.

A large number of references were reviewed in order to ascertain the relevance of Hydrogen Embrittlement and Stress Corrosion cracking to rig components. Above 220°C hydrogen atoms permeating carbon steel can gradually react with iron carbide in the steel to form methane. The effect is twofold :
    (a) decarburisation of the steel with resultant loss of strength and,
    (b) fissuring of the steel by the pressure of methane formed at the grain boundaries.

In certain circumstances hydrogen damage can be severe. Material selection for high pressure gasifiers application should be made with reference to the well known Nelson curves (II) which define acceptable limits of temperature and hydrogen partial pressure for low-alloy steels.

The test plant service life in terms of creep to rupture failure of extreme temperature components (particularly the steam superheater arrangement) was determined in accordance with ASME VIII and manufacturer's data.

A high temperature thermal 'sleeve' design was developed to allow penetration of superheater tubes into the furnace vessel forming part of the plant. Superheater tube temperature Ca $1000^{\circ}C$ - max. furnace vessel surface temperature $350^{\circ}C$. Superheater material was selected as Incoloy 800 H. The superheater tube design was complex and iterative due to the experimental requirements.

The feed system valves were subjected to Non-Destructive Testing utilising a dye penetrant technique. This resulted in the valves being initially rejected due to body cracks which could have caused catastrophic failure under the projected operational conditions.

As part of the development programme, the feed system arrangement was initially modelled in full scale. Three grades of wood chip were passed through the model. It was found that fine chips, i.e. $>$ 6 mm, and stranded chip $>$ 3 cm, gave rise to feeding problems due to bridging in the selected triple valve arrangement. Normal feed will be 1-2 cm Hacker Chips which passed freely. The triple valve arrangement was selected on the basis of a requirement for continuous batch operation and safe venting of trapped (interspace) product gases. The lower valve acts additionally as a thermal shield.

3. Present Status

The test facility has been partially commissioned. Due to the programme delays experienced because of technical problems, the project timetable has been rearranged. The Partial Oxidation Reactor will be installed prior to hot commissioning of the first Reactor now in place - Cross Flow. A detailed design for the next reactor has been prepared - Down Draught. A further reactor is under consideration - Fluidised or Entrained Bed.

An experimental programme for the first two reactor systems has been evolved.

Because of the delays and associated costs it has proven necessary to re-finance the remainder of the programme. This is presently being negotiated.

It is projected that the remainder of the programme will be finalised by July 1984.

REFERENCES

(1) J. Gibbins and H. T. Wilson
"Equilibrium Modelling a Cheap and Efficient Aid to Gasifier System Design".
I.Chem E. Symposium Series No. 78. London, 12-15 October 1982.

(2) API Publication 941
"Steels for Hydrogen Service at Elevated Temperatures".
Second Edition 1971.

FIGURE I - End View of the Gasification Test Facility

FIGURE II - Partial oxidation reactor awaiting installation

HYDROGEN ABSORPTION IN LaNi$_5$ DISPERSED IN LIQUID

Authors	: K.J. PTASINSKI, A.A.C.M. BEENACKERS, W.P.M. VAN SWAAIJ
Contract number	: ESE-R-062-NL
Duration	: 24 months 1 January 1982 - 31 December 1983
Total Budget	: DFL 589.569,-- CEC contribution 50%
Head of project	: Dr.Ir. A.A.C.M. Beenackers
Contractors	: Twente University of Technology
Address	: Twente University of Technology P.O. Box 217 7500 AE ENSCHEDE, the Netherlands

Summary

A new process is proposed for hydrogen absorption in a slurry containing finely dispersed particles of metal hydride in an inert liquid. The method has several attractive features such as easy continuous operation, improved heat transfer to metal particles and reduction of problems known to occur in discontinuous fixed beds, which is the conventional way of operation. The proposed process is demonstrated for the system: hydrogen - LaNi$_5$ - silicone oil. The experiments have shown that hydrogen can be successfully absorbed in the slurry with sufficient rates and without poisoning interference between metal hydride and the liquid. A flexible, continuous pilot plant presently under construction is discussed briefly.

1. Introduction

In the last few years recovery of hydrogen from off-gases or mixed process-gas streams has attracted much attention because of increasing cost of natural gas. It also has been shown [1] that hydrogen recovery from producer gas obtained by air gasification can be a key process for a feasible "methanol from biomass" route.
The methods of hydrogen separation which are used in practice include cryogenic separation, pressure swing adsorption or membrane separation. A relatively new method which received some interest recently [2,3] is the application of metal hydrides as hydrogen absorbents. The standard way of this operation is the pressure swing method in a packed bed which is a discontinuous process and therefore a serious disadvantage. The poor heat transfer properties of packed beds of hydrides, formation of fine powders which leads to blocking of the beds and large forces on the reactor wall during absorption are other disadvantages.
In order to overcome these drawbacks a new process has been proposed [1] using metal hydrides dispersed in an inert liquid. This way a continuous operation can be obtained with good heat transfer properties, low pressure drop and no or reduced problems with embrittlement.
In this paper the proposed process is experimentally demonstrated for the system: $LaNi_5$ - silicone oil. $LaNi_5$ hydride has been selected as it has several advantages over other hydrides. It has a large absorption capacity and high absorption and desorption rates at relatively low temperatures. $LaNi_5$ is also easily activated and has a relatively small sensitivity for impurities

2. Experimental

2.1. Apparatus and Procedure

A schematic diagram of the experimental apparatus is shown in Fig.1. A batch stirred-cell reactor of a 0.5 dm^3 volume (Fig.2) was used for absorption experiments. The reactor was immersed in a water bath in order to maintain isothermal conditions. Hydrogen was supplied from a pressure vessel A of 1 dm^3 volume through a pressure regulator in order to keep constant pressure of hydrogen in the reactor. After having the absorption completed, hydrogen was then removed from the reactor by reducing reactor pressure. The volume of hydrogen removed was also measured to check the mass balance. All parts of the apparatus in contact with hydrogen were of stainless steel 316. The slurry temperature in the reactor and the hydrogen temperature in the vessels were measured by thermocouples. Precise Bourdon gages were used to measure hydrogen pressure in both the reactor and the vessels.
Utrahigh purity hydrogen of ≥ 99.999% purity (Matheson) was used in all experiments. $LaNi_5$ alloy was supplied by INCO Europe Ltd. Alloy was activated in the reactor before introducing the liquid, following standard activation procedures [4]. After obtaining the activated alloy a pressure-composition equilibrium relationship at $25^\circ C$ was determined, by a P-V-T method (Fig. 3). The sample of alloy was also subjected to several hydriding-dehydriding cycles in order to establish the maximum capacity and the absorption rate. Thereafter, silicone oil (type Baysilone PD-5) was introduced into the reactor. This liquid was outgassed under vacuum before it was mixed with the hydride. Hydrogen solubility in the liquid was determined by a P-V-T method (Fig.4) using another, but similar set-up.

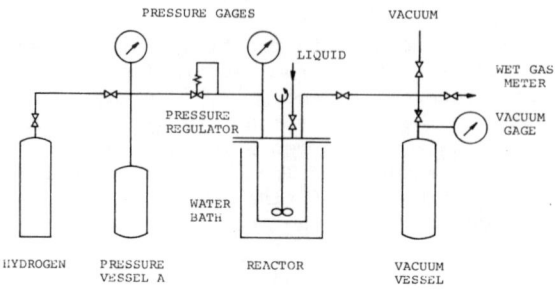
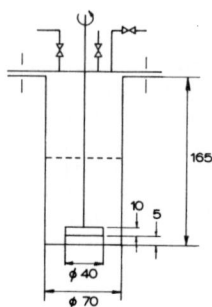

Fig.1 Schematic diagram of the experimental apparatus.

Fig.2 Stirred-cell reactor used for absorption experiments.

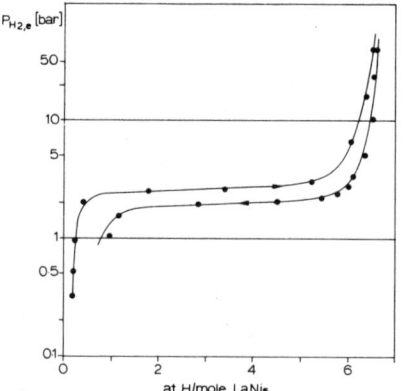
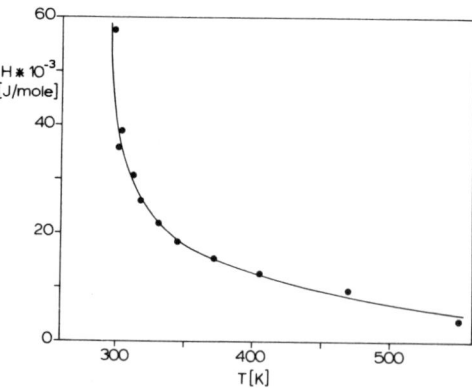

Fig.3 Pressure-composition isotherm at 25°C for $LaNi_5$.

Fig.4 Solubility of hydrogen in silicone oil PD-5.

2.2. Experimental results

The experiments were carried out within the following range of parameters: temperature 25-55°C, hydrogen pressure 6-46 bar, stirring speed in the reactor 166-1200 RPM. In all experiments a slurry was used of 0.101 kg LaNi$_5$ and 0.255 kg liquid (slurry concentration 28.4 wt per cent). The amount of H$_2$ absorbed in LaNi$_5$ was calculated from the P-V-T readings of pressure vessel A and the reactor. A measure for the progress of the hydrogen absorption is the relative degree of LaNi$_5$ conversion X, that is the ratio of hydrogen amount absorbed by LaNi$_5$ at time, relative to the maximum hydrogen amount that can be absorbed at the given temperature and pressure (calculated from Fig. 3). Figures 5 to 8 show the relationships between the relative degree of LaNi$_5$ conversion X and the absorption time. It can be seen that in all cases the conversion reaches almost unity with the relationships X(t) being linear in the two phase alloy-hydride region. Fig. 5 and 6 show the results for different values of stirrer rotation speed at 25°C and p_{H2} = 26 bar. The absorption rate increases with an increase of mixing efficiency and relatively high rates are obtained for the higher stirrer speeds. Fig. 7 shows the effect of hydrogen pressure on the conversion rates for the experiments at 25°C and 400 RPM. As expected, absorption rates are strongly influenced by hydrogen pressure. Fig. 8 shows the effect of temperature on the absorption rates at a constant hydrogen pressure of 26 bar and a constant n = 400 RPM. The temperature has a positive effect on the absorption rate, while furthermore it should be noticed, that the equilibrium hydrogen pressure changes with temperature. The specific rate of absorption R, was calculated from the slopes of the X(t) lines and expressed as kmoles of hydrogen absorbed per unit of time and unit of surface area of the liquid in the reactor, assuming the slurry surface to be flat. It should be noticed that this assumption is only correct for stirring speeds ≤ 400 RPM. R was found to be proportional to the difference between the hydrogen pressure in the gas phase p_{H2} and the hydride equilibrium pressure $p_{e,H2}$, which is shown in Fig. 9. Therefore the absorption rate can be expressed per unit of $\Delta p = p_{H2} - p_{e,H2}$ as $r = R/\Delta p$ [kmoles H$_2$/s.m^2.bar], being dependent only of temperature and stirring speed. Fig. 10 shows the influence of stirring speed on r for a constant temperature of 25°C. It can be observed that the absorption rate strongly increases with increasing the mixing efficiency, especially for n > 400 RPM. This can be explained by an increase of the interfacial gas-liquid surface area for this range of rotation. As mentioned before, visual observation showed that below 400 RPM the liquid surface in the reactor was flat whereas for n > 400 RPM gas bubles penetrate into the liquid. Absorption times for the higher stirring speeds were almost of the same order of magnitude as obtained for dry hydrides. In Fig. 11 the influence of temperature on the absorption rate r is shown for n = 400 RPM. The equilibrium hydrogen pressure $p_{e,H2}$ that exists in Δp_{H2} was calculated from the value of the equilibrium pressure at 25°C (Fig. 3) and the heat of formation of LaNi$_5$, ΔH = -31 KJ/mol H$_2$ [5]. One of the important properties of hydrides is the sensitivity to impurities. Fig. 12 shows the change of the final relative degree of conversion of LaNi$_5$ as a function of the number of absorption/desorption cycles. It is shown that after the first cycles where the absorption capacity decreased to about 0.95 of that for fresh dry hydrides it further remains constant. This slight decrease of capacity is comparable with that caused by presence of water in hydrogen [6]. Indeed some traces of water could be detected in the oil. The kinetics of absorption didn't change with cycling number as follows from the comparison X(t) lines in Fig. 5 for absorption no 11 (line marked ▯-403 RPM) and absorption no 1 (line marked 0-400 RPM).

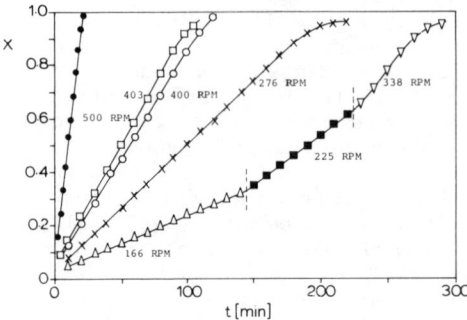

Fig.5 LaNi$_5$ conversion vs absorption time for various stirrer speeds (T = 25°C, p_{H2} = 26 bar).

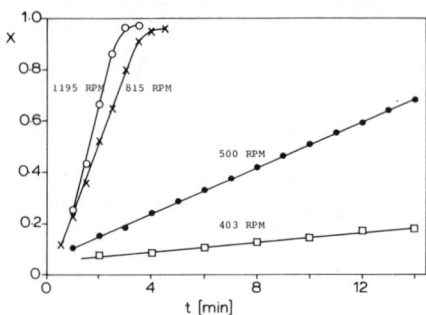

Fig.6 LaNi$_5$ conversion vs absorption time for various stirrer speeds (T = 25°C, p_{H2} = 26 bar).

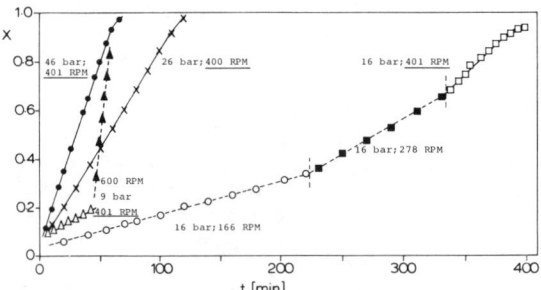

Fig.7 LaNi$_5$ conversion vs absorption time for various hydrogen pressures (T = 25°C).

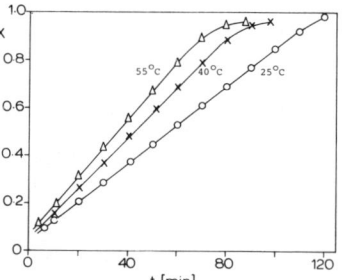

Fig.8 LaNi$_5$ conversion vs absorption time at various temperatures (p_{H2} = 26 bar; n = 400 RPM).

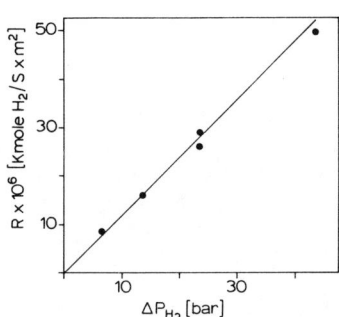

Fig.9 Variation of absorption rate R with pressure difference of hydrogen: $p_{H2}-p_{e,H2}$ (T = 25°C; n = 400 RPM).

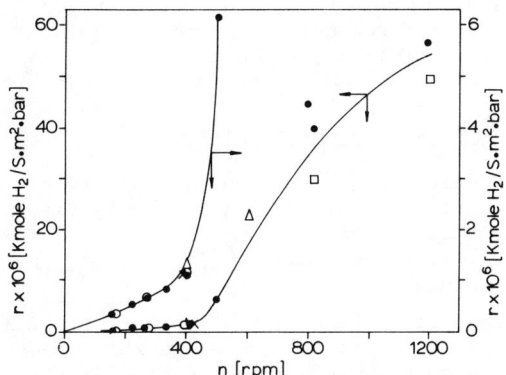

Fig.10 Effect of stirrer speed on absorption rate (T = 25°C, Δp_{H2}: □ -3.4, △ -6.4, ○ -13.4, ● -23.4, X -43.4 bar).

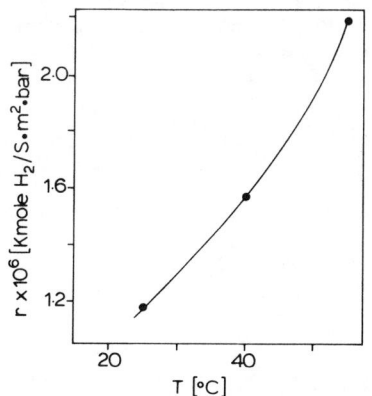

Fig.11 Effect of temperature on absorption rate (p_{H2} = 26 bar, n = 400 RPM).

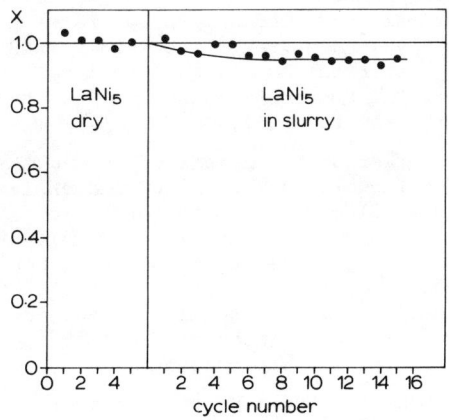

Fig.12 Final conversion of $LaNi_5$ vs number of absorption cycles.

2.3. Conclusions

The experiments showed that hydrogen can be succesfully absorbed in $LaNi_5$ finely dispersed in an inert liquid. The high absorption rates and the absence of interference between $LaNi_5$ and the liquid indicate that the investigated system is promising for technical applications.

3. Continuous pilot plant

Knowledge on continuous gas-slurry reactors as published in open literature is rather limited, reason why we designed and now have under construction a continuous pilot plant for studying hydriding behaviour of various alloy-oil slurries under conditions close to industrial practice in various contactors.

The pilot plant basically consists of 3 major sections (see Figure 13)
- a continuous hydrogen absorbing reactor (1)
- a continuous hydrogen desorbing reactor (6)
- a slurry storage section (7).

The hydrogen loaded liquid slurry continuously flows from the bottom of the absorber to the top of the desorber where the hydrogen is desorbed. Decharged slurry continously flows from the desorber to the top of the absorber in which it is to be reloaded with hydrogen. Storage tank 9 serves as a container for make-up of fresh slurry and for storing the process slurry during shut-down periods. A hydrogen containing gasflow enters the absorber at the bottom while the uncharged offgas leaves the absorber at the top. To reduce operating costs the main part of this offgas is made up with hydrogen and then recycled to the absorber by means of a compressor (2). Installment of cooling and heating coils in absorber and desorber, respectively, allows for maintaining, alloy specific, optimal temperatures for both the absorption and desorption process. Instead of isothermal reactor operation also a more or less adiabatic type of operation is possible by reducing the heat supply or withdrawal in the reactors. This allows for e.g. an optimal increased absorption rate in the bottom part of the absorber at a relatively high temperature in combination with realizing a large hydrogen absorption ratio by maintaining a relatively low temperature in the top section of the absorber. Such a way of operation asks for extra cooling outside the reactors (8).

The absorber unit consists of an autoclave in which a great variety of column contactors can be tested successively, e.g.
a) plate columns: - up to three sieve trays
- up to three valve trays
- up to three bubble cap trays
b) bubble columns
c) packed columns: various packings including cooling coils acting as a packing and each operated either countercurrently or cocurrently.

To change from one type of contactor to another only asks for replacing the internals of the autoclave which is a relatively simple procedure.

The actual temperature profile in the absorber can be recorded continuously.

The research program will concentrate particularly on absorber characteristics while the desorber (a stirred tank reactor) merely serves to allow for continuous reactor operation with continuous reuse of the same alloy-oil slurry. This way operating costs are reduced while simultaneously infor-

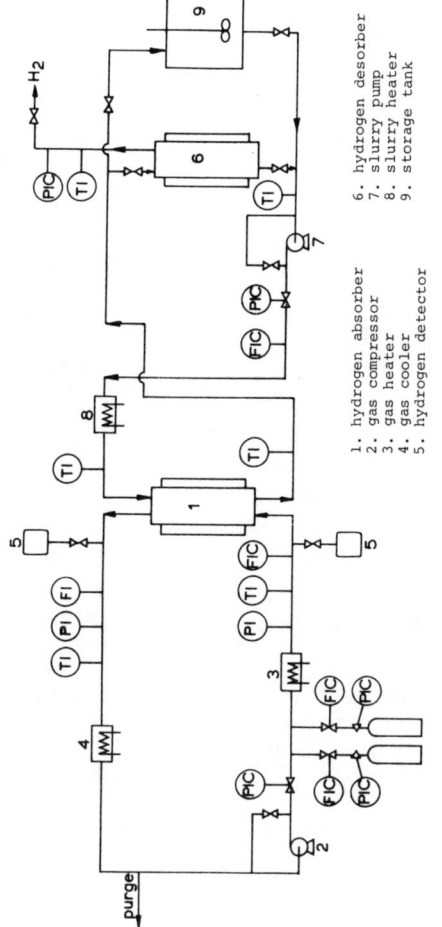

Fig.13 Experimental hydrogen separation and recovery pilot plant unit.

1. hydrogen absorber
2. gas compressor
3. gas heater
4. gas cooler
5. hydrogen detector
6. hydrogen desorber
7. slurry pump
8. slurry heater
9. storage tank

mation on any slurry deactivation with time will be obtained.

Main points to be investigated:
- mass transfer characteristics of various contactors with several slurries as a function of gas and slurry load (mass transfer coefficients, specific contact area)
- flow behaviour of the suspension (stable flow or problems with solid deposition on plates or coils)
- slurry deactivation behaviour
- selection of optimal reactor type.

Process conditions in pilot plant

temperature:	$20 \leq T \leq 300\,°C$
pressure:	$1 \leq p \leq 35$ bar
slurry flowrate:	$0 \leq \phi_L \leq 50*10^{-6}$ m^3/s
gas flowrate:	$0 \leq \phi_G \leq 0.1$ Nm^3/s
internal absorber diameter:	$D \leq 0.08$ m
reactor height:	$H \leq 0.75$ m

Symbols

H	-	Henry's coefficient	J/mole
n	-	stirring speed	RPM
p_{H2}	-	hydrogen pressure	bar
$p_{e,H2}$	-	equilibrium hydrogen pressure	bar
R	-	specific absorption rate	kmole H_2/s.m^2
r	-	$R/(p_{H2} - p_{e,H2})$	kmole H_2/s.m^2.bar
t	-	time	min
T	-	temperature	°C or K
X	-	relative degree of LaNi$_5$ conversion	-

References

[1] Beenackers A.A.C.M. and van Swaaij W.P.M. in: W. Palz and G. Grassi (Eds.) Energy from Biomass, Reidel, Dordrecht 1982, pp 201-206
[2] Cholera V. and Gidaspow D., Inst. Chem. Eng. Symp. Ser. 1978, 54, 21
[3] Kaplan L.J., Chem. Eng. (N.Y.) 1982, 89, 34
[4] Buschow K.H.J. and van Mal H.H., J. Less - Common Metals 1972, 29, 203
[5] Swartzendruber L.J., Carter G.C., Kahan D.J., Read M.E. and Manning J.R., Adv. Hydrogen Energy 1979, 1, 1973
[6] Sandrock G.D. and Goodell P.D., J. Less - Common Met. 1980, 73, 161

CATALYTIC LIQUEFACTION OF WOOD MATERIAL

Authors	: A. BURTON, D. DE ZUTTER, G. PONCELET, P. GRANGE, B. DELMON
Contract number	: ESE-019-B
Duration	: 36 months 1 July 1980 - 30 June 1983
Total budget	: 10.027.624 BF CEE contribution : 100.000 uc
Head of project	: Prof. B. DELMON, Université Catholique de Louvain
Contractor	: Groupe de Physico-Chimie Minérale et de Catalyse
Address	: Place Croix du Sud 1 B-1348 Louvain-la-Neuve (Belgium)

SUMMARY

Total liquefaction of wood is achieved in the presence of a catalyst, after 30 minutes of reaction at 400°C, under 120 bar H_2 (working pressure) and with a solvent/wood ratio of 5/1.

Three catalytic systems have been compared: Ni-Mo catalysts exhibit the best selectivities towards phenol and cresols.

Comparative liquefaction tests on barks performed in similar operating conditions yield lower conversions.

Preliminary results of asphaltenes hydrotreatment are also reported.

1. Introduction

The rational exploitation of biomasses, which consist of elaborated lignocellulosic structures, cannot be confined to the mere production of liquid fuels: through its controlled degradation, biomass may lead to the production of chemicals with high added value.

From this point of view, the catalytic liquefaction appears more and more as a key process among those which are considered for the valorization of lignocellulosic materials (1).

One possibility is the direct catalytic hydroliquefaction of biomass, which avoids the hydrolysis step and yields principally liquid fuels and, to a lower extent, phenols, which contribute to the valorization of the lignin content.

An industrial scheme which integrates both selective hydrolysis of carbohydrates into hexoses and pentoses and catalytic hydrogenation of the lignin content maximizes the valorization of the various wood constituents.

The pay-off of the investments for a catalytic liquefaction process might be improved considerably by using highly selective catalysts, capable of depolymerizing the lignin residues of the paper industry.

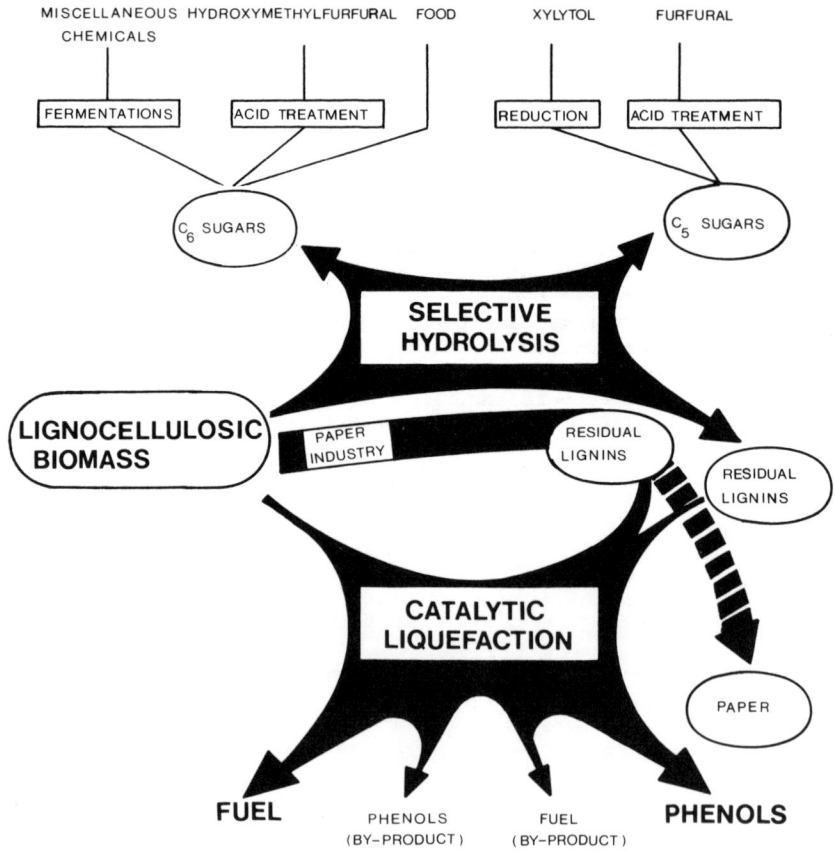

2. Main achievements of the project

2.1. Literature survey
Internal report (2)
(not published)

2.2. Analysis

- Global evaluation of the efficiency of the reaction
  - direct extraction of the phenols from the crude
  - fractionated distillation

  CEE report, Copenhagen (3)
  M.S. Thesis (4)
  (not published)

- Setting of a methodology for refined analysis
  - vacuum distillation
  - alkaline extraction
  - discrimination asphaltenes-maltenes

  CEE report, Brussels (5)

- Identification of the main constituents of the fraction by GC/MS

  CEE report, Berlin (1)

- Preliminary analyses of the gas phase
- Preliminary technological study on the light fraction
  - heating capacity
  - distribution as a function of the boiling points
  - sulphur content

  This report

2.3. Systematic investigation of the experimental parameters

- Preliminary tests in microreactor
  - screening of the solvents
  - choice of the experimental conditions

  CEE report, Copenhagen (3)

- Preliminary runs with a 1 liter vessel
  - reproduction of some runs done in microreactor

  M.S. Thesis (4)
  (not published)

- Systematic investigation of various parameters and selection of the experimental conditions leading to a total liquefaction

  CEE report, Brussels (5)

- Evidencing the importance of the contact between gas-wood-solvent-catalyst

  This report

2.4. Catalysis

- Screening of different catalysts
- Reproducibility of the catalyst activity
- Improving the selectivity towards the production of phenols by the choice of the catalyst
- Control of the duration of the catalyst activity

  CEE report, Copenhagen (3)
  Final report (in preparation)
  This report

2.5. Behaviour of different ligneous materials

- Hard and soft wood
- Cellulose, hydrolyzed lignin
- Bark

  CEE report, Copenhagen (3)
  This report

2.6. Synthesis of the results
Final report (in preparation)

3. New results (since CEE report, Brussels, 1982)

3.1. Influence of the reaction temperature

Figures 1 and 2 show the evolution of the light, phenolic, neutral, aqueous and heavy fractions as a function of temperature.

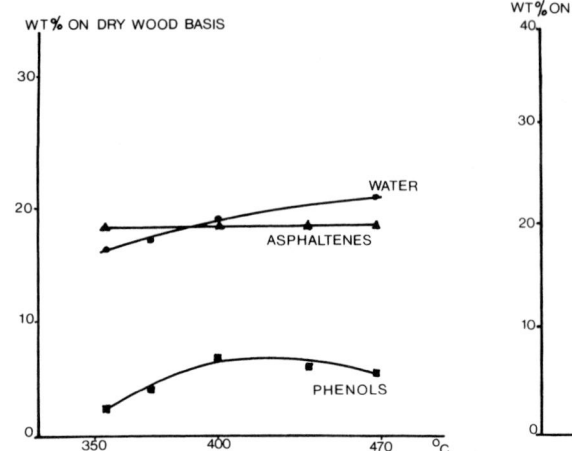
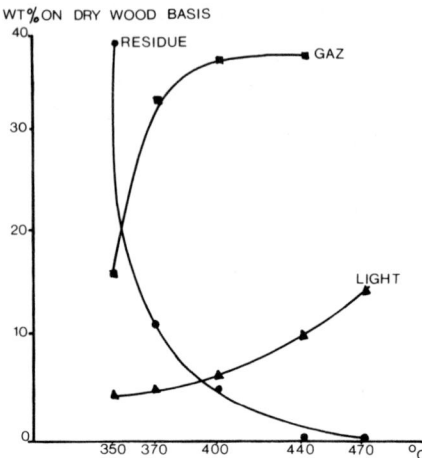

Experiments were carried out at different reaction temperatures using a beech/tetralin ratio of 3, and an initial hydrogen pressure of 40 bar. The catalyst was CoMo and the reaction time, 30 min.

The different fractions obtained from the crudes, using the analytical methodology described earlier (1), are given in Table 1.

TABLE 1 : Yields of the different fractions (in weight % on a dry wood basis)

Run	T°C	Gas fract.*	Light fract.	H_2O	Phenolic fract.(Tot.)	Neutrals	Asphalt.	Residue
64	350	15.8	4.5	16.8	2.6	3	18.3	39
68	370	33.4	5	17.6	4.3	6.1	21.6	11
43	400	37	6.5	19	6.5	5.7	19.4	5
69	440	46	9.6	20	5.7	7.1	12	0
69b	440	37.1	10.1	18.2	6.3	8.5	19.8	0
69t	440	27.8	11.6	20.8	4.9	8	21.9	0
66	470	43	14.9	18.2	5.5	7.7	9.5	1.1
66b	470	31	14.4	21.7	5.8	7.3	19.3	0.6

* In this Table, as in the following ones, the gas fraction is obtained by difference.

In a few runs, we observed a poor reproducibility concerning the asphaltenes content (runs 66 and 69 in Table 1). For the other fractions, the small differences were within the experimental accuracy.

Problems in the preparation of the catalyst and/or the viscosity of the slurry were at first suspected. The first possibility could be discarded from experiments carried out in similar conditions using model molecules. On the other hand, increasing the solvent/wood ratio (5/1 instead of 3/1), i.e. improving the rheological properties of the wood-catalyst-solvent slurry and hence, the agitation efficiency and the contact between the components, led to better reproducibility, as can be seen in Table 2.

TABLE 2. Yield of the different fractions (%) obtained from poplar *

Run	Gas fract.	Light fract.	H_2O	Penolic fract. (tot.)	Neutrals	Asphaltenes	Residue
83	30.5	6.6	23.6	5.8	9.3	21.8	2.4
87	38.2	6.4	21.6	5.8	8.0	20.0	0.0

* Operation conditions : T = 400°C; solvent/wood ratio : 5/1; $P^i_{H_2}$ = 40 bar; reaction time : 30 min.

3.2. Hydroliquefaction of different lignin materials: hard and soft wood, raw and treated pine bark, cellulose

Different materials were reacted in the 1 liter vessel under hydrogen pressure in the presence of tetralin and a CoMo catalyst. The liquid phase collected was separated into different fractions, as in Table 1.
The runs performed on the hard and soft woods, were done in the following conditions :
solvent/wood   : 3/1
catalyst   : CoMo
$P^i_{H_2}$: 40 bar
temperature   : 400°C
reaction time   : 15 min
The percentages (wt %) of the different fractions are given in Table 3. They are expressed on the basis of dry wood.

TABLE 3. Liquefaction of hard and soft wood: yields (wt %) of the different fractions (gas fraction is obtained by difference)

Type of wood	Gas fract.	Light fract.	H_2O	Phenolic fraction			Neutrals	Asphalt.	Residue
				I	+ II	= Tot.			
Beech	44.5	6	20	1.4	3.3	4.7	5.8	17	2
Pine tree	39.3	7	16.3	3.4	1.8	5.2	4.5	8.7	19
Poplar	27.2	7	26.8	1.0	3.9	4.9	5.0	20.4	8.6

Under similar conditions, beech and poplar are more readily transformed than pine tree, for which the conversion (liquefaction rate) reaches only 80%. The conversion (or liquefaction rate) is defined as :

$$\frac{\text{wt dry wood - wt residue}}{\text{wt dry wood}} \times 100$$

Beside, we can expect that pine tree, which has a higher lignin content, will produce a greater amount of phenols (at total conversion) than deciduous trees assuming that the lignin of deciduous trees and of conifers react in the same way in the presence of catalysts.
For beech and poplar, there is a difference in the distribution of the monophenols, as can be seen hereafter.

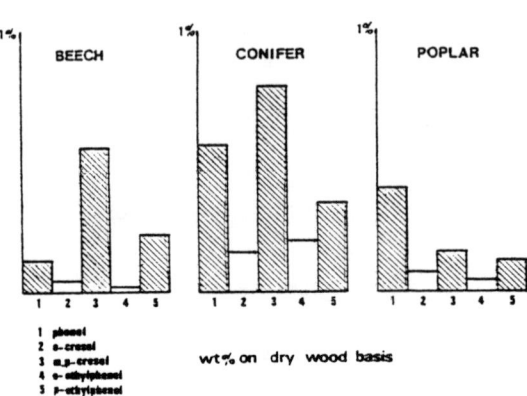

MAJOR PHENOLICS REPARTITION

BEECH CONIFER POPLAR

1 phenol
2 o-cresol
3 m,p-cresol
4 o-ethylphenol
5 p-ethylphenol

wt% on dry wood basis

It thus appears that the choice of the wood biomass will have an influence on the nature of the phenols obtained. In spite of the fact that their liquefaction is only partial, conifers lead to higher fractions of phenols 1, 2, 3, 4 and 5, and the amounts of phenol and cresols (m and p) are almost identical.

3.3. Liquefaction of other biomasses

Pine bark has been hydroliquefied in the following conditions :

solvent/bark	(wt by wt)	: 5/1
catalyst		: CoMo
$P^i_{H_2}$		: 60 bar
temperature		: 400°C
reaction time		: 30 min
adjuvent		: 1 g

Barks have a high lignin content (30 - 44%). They are presently poorly exploited and constitute an important waste of the paper and forest industry (6).

The analysis of the liquefaction products gave the following results :

TABLE 4. Amount (wt %) of the different fractions obtained from pine bark

Nature of biomass	Gas fract.	Light fract.	H_2O	Phenolic fraction			Neutrals	Asphalt.	Res.
				I	+ II	= Tot.			
Pine bark	12.6	4.8	22	2.3	3.6	5.9	6.3	37.6	10.8
Pine bark*	18.8	4.9	18.4	3.3	1.2	4.5	3.0	28.0	22.4
Cellulose	22.1	10.6	33.2	0.6	1.3	1.9	7.4	24.8	0.0
Poplar	35.1	8.9	21.0	2.1	7.7	9.8	4.2	18.7	2.3

* extracted with ethanol-water.

The barks are characterized by a conversion that is appreciably lower than the other substrates. The yields in phenols are low and do not correspond to the high lignin content present in the starting material.

The lower yields obtained from the treated bark may be explained by the removal of the simple polyphenols during the ethanol-water extraction (6).

The gas-liquid chromatographic analysis of the fractions coming from the hydrogenation of bark shows the presence of different products, which have not been yet identified.

The texture and the aggregation state of the barks are less homogeneous than that of wood flour.

This fact combined with the presence of several compounds which are specific to bark (suberine, cork) may be at the origin of the lower efficiency of the catalyst (6).

Cellulose is completely transformed (conversion = 100 %, yield of oil soluble in benzene : 45 %), but produces much water and asphaltenic compounds.

The presence of an important asphaltenic fraction in the product of hydrogenolysis of wood seems to indicate that the majority of these products are coming from cellulose, which implies several dehydrogenation reactions.

3.4. Hydrogenation of condensed aromatic structures (asphaltenes)

The asphaltenic fraction has been submitted to two successive hydrogenations (7). The operating conditions are given in Table 5.

TABLE 5. Hydrogenation of asphaltenes

Run	Weight of asphaltenes (g)	Weight of tetralin (g)	Temperature °C	$P^i_{H_2}$ (bar)	Time (min)	Weight of catalyst (g)
73	70	150	400	40	30	3.5
76	44	155	400	40	30	3.5

The yields of the different fractions are given in Table 6.

TABLE 6. Hydrogenation of asphaltenes : amounts of the different fractions (wt %)

Run	Gas fract.	Light fract.	H_2O	Phenolic fraction I	+ II	= Tot.	Neutrals	Asphaltenes
73	17	5.1	4.5	1.4	3.7	5.1	4.4	64
76	?	2.5	2.3	1.4	2.8	4.2	6.4	87

The total liquid products in run 76 exceeds 100 %. Two possible reasons may be invoked :
- the yield of the unreacted asphaltenes may be overestimated because of the presence of the solvent in the structure. However, this fraction has been lyophilized until constant weight.
- it may be that during the analytical procedure, they have been oxidized at the contact with the atmosphere.

The second explanation seems more realistic when considering the oxygen content obtained by the elemental analysis. Because of time shortage and of the priority of the goals, we have not attempted to set up an experimental procedure allowing to avoid the oxidation of the asphaltenes.

The elemental analysis of asphaltenes (method of Pregl-Dumas) gave the following composition :

C	H	N	O (by difference)
77.7	6.9	0.5	14.9
78.7	7.2	0.2	13.7

The partial conversion of the asphaltenes cannot be due to a rapid catalyst deactivation, since each run was done with a fresh catalyst.

The NMR analysis of this fraction has shown the absence of α-O-H and β-O-H ether bonds, which excludes the presence of nondepolymerized lignin fragments (8).

3.5. Influence of the type of the catalyst

Three different hydrotreatment catalysts (r = 0.34) have been used (9). The experiments were carried out in the following conditions :

solvent/wood (poplar)   : 5/1 (wt by wt)
catalyst   : 7 % (wt)
temperature   : 400°C
$P^i_{H_2}$: 60 bar
reaction time   : 30 min

TABLE 7. Influence of the catalyst composition on the liquefaction of poplar

Catalyst	Gas fract.	Light fract.	H_2O	Phenolic fraction I	+ II	= Tot.	Neutrals	Asphaltenes	Residue
NiMo	32.2	9	24	2.8	2.8	5.6	7.2	22	0
NiW	24.2	6.2	23.4	1.3	3.7	5.0	7.6	18.2	15.2
CoMo	39	6.4	21.2	1.4	4.4	5.8	8.0	20	0

With the NiMo and CoMo catalysts, the transformation is complete, while the NiW catalyst leads to a lower conversion, the total phenolic fraction, however, being similar to that obtained with the two former catalysts.

GCMS analyses of the phenolic fraction indicated that the selectivity for pure phenol (non substituted) is higher with NiMo than with CoMo catalyst, as shown in Table 8. This Table gives also the comparative values obtained from two other liquefaction processes of residual lignin. Clearly, CoMo and especially NiMo are more efficient catalysts.

TABLE 8. Yields of phenol from poplar wood[xx] and from lignin hydrogenolysis

	Poplar liquefaction CoMo	NiMo	Lignol process (10)	Nogushi process (11)
Phenol	3.9[x]	4.7[x]	2.5[x]	3.0[x]

[x] wt % based on organic content of lignin.
[xx] lignin content of poplar wood : 23.5 %

4. Conclusions

Hydroliquefaction is and will stay a key option among other possible ones that are considered for the valorisation of biomass, provided it will not be only focussed on the production of liquid fuels.

The simultaneous production of liquid fuels and chemicals with high added value, as demonstrated in this research programme, is quite feasible through the use of unsupported CoMo catalysts.

However, the yields of the combustible fraction as well as the phenols so far obtained can still be improved.

Indeed, the experimental results have shown that the global yield of the phenolic fraction and the selectivity (limited number of chemicals: phenol, cresols, ethylphenols) could be ameliorated by using adequate catalytic systems. Future research should be focussed on the optimization of these two aspects.

In order to increase the yield of the combustible fraction, it will be necessary to crack the heavy compounds produced during liquefaction, or to inhibit their formation.

This goal could be reached by means of bifunctional catalytic systems.

Beside, liquefaction done on purified substances (cellulose, lignins, model molecules) will enable to gain a better knowledge of the complex mechanisms involved during the reaction and, as a consequence, to help in the optimisation of the process.

The best performing catalysts will be tested on more complex matter (wood, industrial wastes) in batch reactor first, and in a continuous pilot reactor afterwards. Once this goal will be reached, the economical feasibility of the process could be established.

Acknowledgements

This research programme is also financially supported by the Service de la Programmation de la Politique Scientifique (Belgium).

We thank Professor E. de Hoffmann, Director of the Mass Spectrometry Laboratory of Louvain University, for running the GCMS analyses.

The authors are indebted to Professor J.M. Dereppe and coworkers (UCL) for the NMR study of the asphaltenes.

We gratefully acknowledge Dr. R. Marutzky for providing the samples of bark, (Fraunhofer-Institute for Wood Research - WKI, W. Germany.)

We thank Dr. R. Cahen for the determination of the calorific values (Labofina s.a., Belgium).

References

(1) A. Burton et al., Energy from Biomass, Proceedings of the Int. Conf. (2nd EC) on Biomass, Berlin, (P. Chartier, A. Strub and G. Schlesser, Eds.) Applied Science Publishers, 935, 1982.
(2) Internal Report: Wood Liquefaction: bibliography, 1980, M.Van Horenbeeck.
(3) B. Delmon et al., Energy from Biomass, Proceedings of the EC Contractor's Meeting, Copenhagen, (P. Chartier, W. Palz, Eds.) Reidel Publ. Co, 209, 1981.
(4) J.F. Lambert, Mémoire de fin d'études, Faculté des Sciences Agronomiques, Université Catholique de Louvain, 1982.

(5) A. Burton et al., Energy from Biomass, Proceedings of the EC Contractor's Meeting, Brussels (Serie E, Vol. 3), D. Reidel Publishing Co, 207, 1982.
(6) B. Dix and R. Marutzky, Fraunhofer-Institute for Wood Research-WKI, D-3300 Braunschweig, personal communication.
(7) R.B. Long, The concept of asphaltenes, Preprints, Div. of Petr. Chem. ACS, $\underline{24}$,(4), 891, 1979.
(8) Prof. J.M. Dereppe and coworkers, Internal Report.
(9) P. Grange, Catalytic Rev. Sci. Eng.,$\underline{21}$(1), 135, 1980.
(10) H.J. Parkhurst, Preprints, Div. Petr. Chem. ACS, $\underline{25}$, 657, 1980.
(11) D.W. Goheen, Adv. Chem. Ser., $\underline{59}$, 205, 1966.

STUDY ON THE PYROLYSIS OF AGRICULTURAL WASTES

Authors : A.LUCCHESI, G.MASCHIO

Contract number : ESE-R-082-I-(S)

Duration : 6 months   1 August 1982 - 31 January 1983

Total budget : Lit. 10.000.000

Head of project : A.Lucchesi, Istituto di Chimica Generale, Facoltà di Ingegneria, Pisa

Contractor : Centro Ricerca Industriale Tecnologia Avanzata (C.R.I.T.A.)

Address : C.R.I.T.A.
I-56100 PISA, lungarno Sonnino 20

Summary

The state of the art of research and of industrial application on pyrolysis of biomass in Italy for the production of charcoal and oil has been pointed out. Trade market forecasts on pyrolysis products has been investigated.The utilization of charcoal as solid pulverized fuel can be easily appreciated: the charcoal is better than coal from ecologic point of view because sul - phur content is very low and its ashes are likely to be a good fertilizer. The oil produced by pyrolysis of biomass depends strongly on the kind of biomass: resinous material give an yield of oil higher than that straw,lu- cerne or wood. On the conversion behaviour has been based the process and reactor design. As a main conclusion it appears that the pyrolysis gas ena- bles the system to save energy because the direct heat exchange between the combustion products of such gas and biomass improves heat balance.The resul ts of the study lead to the conclusion that pyrolysis efficiency is higher than combustion efficiency by a coefficient about 1.3(with biomass pre-hea- ted system).It is resulted from our tests that it is possible to decrease strongly the acidity of the oil avoiding the condensation of the water con - tained in the pyrolysis gas.

INTRODUCTION

Before examining the results of the experimental tests, it is useful to recall the most important stages of the process of biomass pyrolysis.

At lower temperature (below 150°C), biomass is dried, with evident decrease in moisture content. On heating at higher temperatures (170-270°C), the process involves evolution of H_2O, CO_2, CO and volatile organic compounds (prepyrolysis). Then, at about 280°C, biomass pyrolysis provides a variety of gaseous and volatile organic products; in the range from 280 to 500°C, reactions occur with always faster rates, providing other products.

At still higher temperature, 800°C, char, produced by pyrolysis, undergoes gasification reaction, which begins at about 650°C if the catalytic action of the ashes, contained in lignocellulosic materials, takes place.

Biomass pyrolysis provides char, oil and aqueous fraction and gaseous products; their yields and properties depend on the chemical composition of biomass, temperature and process conditions (1).

Table I show the composition of the lignocellulosic materials used in the tests which will be discussed later.

TABLE I - Analysis and heating values of biomass.

Biomass	Hemicellulose %	Cellulose %	Lignin %	Extractives %	Ash %	H.H.V. kJ/kg	L.H.V. biomass
olive-husks	21.1	22.2	45.0	8.1	3.6	20777	19225
straw	24.0	40.0	21.0	8.0	8.1	18948	17326
lucerne	13.7	45.5	21.3	10.6	7.6	19100	17680
pine-cone	13.8	35.1	43.0	7.3	0.8	21034	19225
wood	19.4	47.5	24.0	7.5	1.6	18739	17096

The different constituents of biomass, cellulose, hemicellulose, lignin, display different thermal properties, which can be observed on thermogravimetry of biomass and its components (Fig.1).

The hemicellulose are the less stable components so that they pyrolyze under mild conditions, the cellulose at 300-370°C, while lignin is not completly degradated at 450°C. Based on the above considerations it is useful to know the different composition of biomass to understand the results of the experiments.

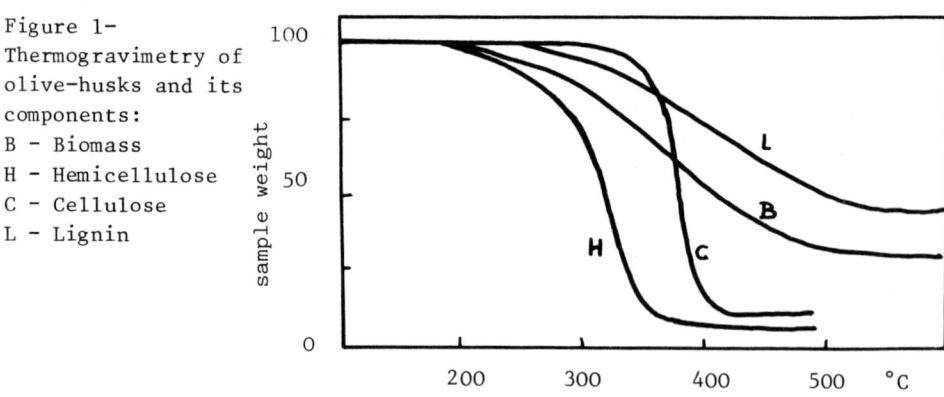

Figure 1- Thermogravimetry of olive-husks and its components:
B - Biomass
H - Hemicellulose
C - Cellulose
L - Lignin

EXPERIMENTAL RESULTS

The products, obtained by pyrolysis of biomass, can be divided into charcoal, aqueous fraction, oil fraction and gas.

Their yields depend on chemical composition of biomass, type of reactor, and operative variables, among which temperature is the most important one.

Some experimental results about the yields of the products of the investigated biomass are reported in Table II (2).

TABLE II - Yields of pyrolysis products of some biomass.

	Temperature (°C)	300	350	350	400	400	500	500	550
S T R A W	Preheating weight loss (CO_2 + H_2O)	-	-	-	-	16.7	-	20.0	-
	char	31.0	-	-	30.8	30.8	30.0	30.0	-
	oil fraction	6.8	-	-	37.5	23.3	19.0	3.3	-
	aqueous fraction	31.0	-	-					
	gas	29.2	-	-	31.7	29.2	51.0	46.7	-
O L I V E H U S K S	Preheating weight loss (CO_2 + H_2O)	-	-	11.5	-	11.9	-	-	-
	char	-	37.0	31.8	30.0	33.0	26.9	-	20.0
	oil fraction	-	12.0	9.3	11.0	8.5	8.7		7.0
	aqueous fraction	-	32.7	19.0	31.0	22.5	23.0		27.1
	gas	-	18.3	28.4	28.0	24.1	41.4		46.0
W O O D	Preheating weight loss (CO_2 + H_2O)	-	-	30.0	-	33.0	-	28.5	25.6
	char	-	-	24.0	21.0	21.0	-	19.2	18.2
	oil fraction	-	-	8.5	7.1	10.1	-	10.5	11.0
	aqueous fraction	-	-	16.5	47.4	19.0	-	18.5	17.1
	gas	-	-	21.1	24.6	17.1	-	23.8	29.0

As it possible to note from these data increasing temperature above 500°C, the yields of char, aqueous fraction ond oil progressively decrease, while the gaseous products greatly increase.

In order to maximize char and oil fraction, which are the most promising products from an energetic point of view, the best operative temperature range is 400-500°C.

In addition to the above mentioned isothermal experiments the possibility of operating with variable temperature was investigated.

To evaluate the feasibility of a pyrolysis plant the most interesting experimental data are the yields of products, their properties, the rates of reaction at different temperatures.

The rate of reaction of pyrolysis is assumed to be a first order reaction.

Table III summarizes the experimental kinetic results at different temperatures: rates of reaction are expressed as % converted biomass/min.

TABLE III - Kinetic results of pyrolysis runs carried out at different temperatures (% converted biomass/min).

Biomass	temperature of pyrolysis (°C)					
	300	350	400	450	500	550
straw	4.3	12.4	18.7	-	24.0	-
olive husks	-	3.9	8.3	11.5	13.1	13.6
pine cone	5.5	-	10.1	12.0	16.2	25.0
wood	-	-	7.0	-	15.3	-
dried straw	-	18.3	23.1	-	35.5	-
dried olive-husks	-	8.3	16.7	-	-	-
dried wood	-	12.5	14.7	-	30.1	35.0

From these data it can be noted that the rates of pyrolysis reaction with dried biomass are about twice as fast than those with corresponding wet materials. This fact points out that the preliminary biomass drying is important because it allows faster rates of reaction and, consequently, smaller reactor sizes.

It seems possible to try to relate chemical composition of lignocellulosic materials to the thermodynamic and kinetic data of pyrolysis conversion.

To investigate this subject of pyrolysis, which is quite important in our opinions, a specific study should be developed.

CHARACTERIZATION AND APPLICATION OF PYROLYSIS PRODUCTS

Charcoal

The pyrolysis process can be operated to produce charcoal with a volatile content ranging from 5 to 25%.

As shown in Table IV the low heating values of the charcoal range from 22,000 to 28,000 kJ/kg. For some feed materials, such as wheat straw, the ash content of the char (17%) can be high and consequently, its low heating values is less than that for a char of low ash content (2-3% for wood charcoal).

The bulk density of char from the tested biomass is in the range from 150 to 300 kg/m^3.

The charcoal from lignocellulosic waste has essentially a very low content of sulphur and, consequently, it does not emit any sulphur oxide emission when burned. The charcoal can be used as a substitute for pulverized coal and has value as a diluent for sulphur containing fuels.

Moreover, it has been reported that blends of number six fuel oil (the heaviest fuel oil) and a mixture of low volatile charcoal and pyrolytic oil was prepared to produce a slurry containing 30% charcoal. The slurry was fired in a, oil fired, packaged firetube boiler. The flame stability in all tests was reported as excellent; the NO_x emission were lower that those obtained from firing a coal-oil slurry, and the SO_x emission were proportionate to the concentration of sulphur in the slurry(3).

In addition, it is interesting to show some remarkable properties of the char obtained by the pyrolysis of biomass because it has good values of surface area and consequently shows adsorbent properties toward dyes, so that it is possible to use this char as a substitute of the activated carbon in the treatment of some polluting industrial waste-waters.

TABLE IV - Chemical analysis and heating values of charcoal.

Charcoal from	C	H	O	N	Ash	H.H.V.	L.H.V.
	% by weight					kJ/Kg	
wheat straw	66.4	2.7	11.1	0.6	17.1	25176	23897
olive-husks	64.7	5.2	11.6	2.4	7.5	27897	26982
wood	72.2	3.0	17.4	0.3	2.6	28382	27391
lucerne	61.6	2.9	13.8	2.6	16.2	22835	22045
pine cone	82.4	2.7	10.6	0.2	1.7	30656	29660

Pyrolysis oils

The oils produced by pyrolysis of biomass are organic substances with a wide spectrum of compounds. The oils have carbon and hydrogen contents higher that those of primary biomass ones, and their oxygen content is substantially lower that biomass one. The ash content is obviously very low (see Table V).

The oils contain essentially no sulphur. The boiling range of the components varies from the low boiling volatile compounds to the very high boiling ones.

The oils are heat sensitive and begin to decompose when subjected to temperature above 150°C; they contain reactive components and their viscosity increases on exposure to air.

The low heating values of the oils, separated from water by a centrifuge, range from 21,000 kJ/kg (wood) to 28,400 kJ/kg (olive-husks). The density of the pyrolysis oils ranges from 0.93 to 1.05 g/cm^3 as compared with fuel oil densities of 0.80 to 0.85 g/cm^3.

The oils are acidic, and this characteristic must be taken into account in handling and processing the oil; they are soluble in organic polar solvents, such as acetone, but only slightly soluble on nonpolar solvents, such as heptane and gasoil. The viscosity of the oils approximately ranges from fuel oil N°4 to fuel N°6. The oil has been burned satisfactorily in combustion tests conducted by Tech-Air Corporation (4); the oil from the pilot-plant of Tech-Air Co. has been sold for use as a fuel in a cement kiln, a power boiler and lime kiln. It has been fired directly and as a 20% blend with a N.6 fuel oil. The water content in the oils lower their viscosity and, therefore, improves their handling properties, but it appreciably reduces their low heating values.

The pyrolysis oils contain a wide variety of organic compounds and hence are a source of materials for chemical applications. The oil contains approximately 20% phenolic compounds, which have potential for the production of resins. The Environmental Protection Agency has supported research and development programs to yields products for industrial chemical applications and therefore, enhance the economic value of the oils.

TABLE V - Chemical analysis and heating values of pyrolysis oils.

oil from	C	H	O	N	Ash	H.H.V.	L.H.V.
	% by weight					kJ/kg oil	
wheat straw	-	-	-	-	-	26806	24474
olive husks	64.5	7.7	19.9	3.1	0.3	30405	28403
wood	52.3	6.7	38.4	0.2	0.1	22614	20992

Gas

The pyrolysis gas consist mainly of hydrogen, carbon dioxide, carbon monoxide, methane, water, small amounts of low molecular weight hydrocarbons and vapours of organic compounds.

The relative amount of charcoal, oil and gas depend upon the conditions of operation, on rising in pyrolysis temperature corresponds an increase of the gaseous phase. The heating values of the gas phase increase with increasing temperature and range from to 10,500-14,600 kJ/Nm^3. The pyrolysis gases have such caracteristics that it is necessary to utilize them close to the pyrolysis plant as a fuel for supplying the heat required for the pyrolysis process or for other utilities.

In spite of the not high heating value of pyrolysis gases, it is possible to reach an high temperature of flame in combustion as compared with the more common fuels.

MASS AND ENERGY BALANCE FOR A PYROLYSIS PROCESS

On the basis of experimental results it is possible to evaluate the process efficiencies and to develop the most suitable one. A fixed bed reactor has been selected in order to maximize the yield of products (char and oil), the overall process has been designed to be simple and flexible en - ough (5).

Two methods of heating have been proposed: recycle of the pyrolysis gas indirectly heated or mixed with combustion gas obtained from combustion of a quote of pyrolysis gas.

The preliminarly drying reduces the energy required for the reaction unity and, hence, its size. By this previous separation and avoiding the condensation of water produced in the process, cooling the pyrolysis gas at about $90^{o}C$, we obtained a oil practically free of water, so that it is possible to lower the pyrolytic oil corrosive aggressiveness. The following balances have been developed on the basis of data obtained from pyrolysis of olive husks. The heat available to preheat feed from ambient temperature up to $200^{o}C$ and to dry it is about 600 kJ/kg.

It is difficult to predict pyrolysis reaction heat; data available in literature suggest that the energy required for the process ranges from 850-1050 kJ/kg (6).

On the basis of the experimental data have extimated that the heat of reaction for our biomass is about 650 kJ/kg at $400^{o}C$. In figure 2 is shown the mass-flow diagram for the process. The off-gas stream from the reactor is passed through the cooling section to condense the oil fraction.

The pyrolysis gas from the scrubber-condenser is separated in two stream; a portion of the gas is used as a fuel at the burner and then mixed with the remainder in a proper unity. The hot stream obtained from this operation is sent to the reactor and to the dryer to supply the necessary heat.

To compare the performances of the different methods of supply the heat to the reactor, in Table VI are summarized the results of the heat and material balances, the efficiencies of the process, NTE, and the ratio between pyrolysis and combustion efficiencies, η.

TABLE VI

Process	Char %	Oil %	N.T.E*	η
Continuous with indirectly heating	33	9	0.63	1.25
Continuous with directly heating	33	9	0.63	1.36
Batch	27	10	0.52	1.15

*N.T.E. = $\dfrac{\text{heating value of products - process energy}}{\text{heating value of feed}}$

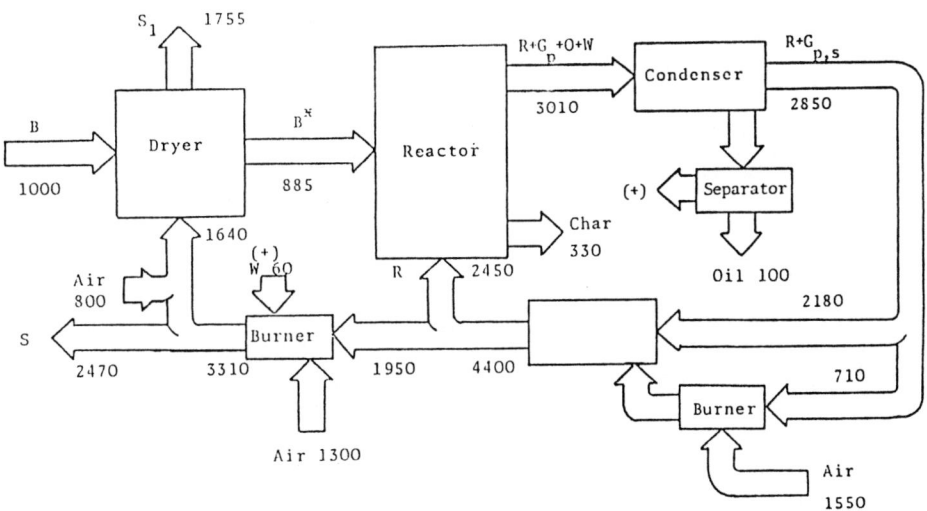

Fig. 2 - Mass-flow diagram for the pyrolysis process (Basis on 1000 kg/hr of wet biomass).

Table VI shows that pyrolysis processes are more efficient than combustion one from an energetic point of view; moreover it is to be pointed out pyrolysis products can be storaged and transported more economically than biomass itself.

The continuous operating system is more efficient, it involves more homogeneous products qualities, larger automated unities and minor energetic losses respect to the batch operation.

REFERENCES

1. Shafizadeh,F., J.Analytical Applied Pyrolysis, $\underline{3}$, 283-305 (1982).

2. Lucchesi,A., Maschio,G., Proceedings 2nd E.C. Conference. "Energy from Biomass", 1109-1111. Berlin 1982.

3. Knight, J.A., "Progress in Biomass Conversion", Vol.1. Academic Press Inc. New York, 1979.

4. Sofer,S.S., Zaborsky,O.R., "Biomass Conversion Processes for Energy and Fuels". Plenum Publishing Corp., New York, 1981.

5. Maschio,G., Tesi di laurea in Ingegneria Chimica, Univ. Pisa (1978).

6. Shafizadeh,F., AIChE Symposium Series, 74, 76-82 (1978).

BIOMASS CONVERSION (BIOLOGICAL ROUTES)

The anaerobic digestion of farm wastes and energy crops

Anaerobic filter digestion of agricultural wastes

Two-phase process for the anaerobic digestion of organic wastes yielding methane and compost

The feasibility of thermophilic anaerobic digestion for the generation of methane from organic wastes

Alcoholic fermentation : improvement of the technology based on physiological phenomena

Enzymatic saccharification of native cellulose : Effect of product inhibition and biomass pretreatment

Enzymic liquefaction and saccharification of agricultural biomass

The Anaerobic Digestion of Farm Wastes and Energy Crops

Authors	: D. A. Stafford, S.P. Etheridge, D. A. Hughes, U. E. A. Leroff.
Contract Number	: ESE-R-023UK
Duration	: July 1980-June 1983
Total Budget	: £ 100,000 EEC Contribution 40%
Head of Project	: Dr. D. A. Stafford
Contractor	: University College, Cardiff.
Address	: Department of Microbiology University College Newport Road Cardiff, Wales, UK.

Summary

The present EEC Contract is almost completed and the Anaerobic Digester Research Unit at Cleppa Park has been fully established. Four pilot plant reactors have been constructed and evaluated under semi-continuous operation with pig waste. Optimum performance parameters have been determined and a summary of results obtained so far are given. Crops grown specifically for digestion include, Forage Pea and Italian Rye Grass, Barley and Augusta Rye Grass and Taronda Stubble Turnip. The latter could not be harvested fully due to poor weather conditions. Full laboratory analysis has been undertaken during the digestion of these crops at laboratory and pilot scale. The potential advantages of maceration have been demonstrated but full scale maceration of crop material has been difficult to achieve. Experimental data has been obtained from trials performed with the unit presently installed. Feedstocks of 1% T.S. have been produced when crop material has been macerated with water and this has been digested at 30 day retention times. Experiments to evaluate the use of digester effluent as a fertiliser have been continuing at Walnut Tree Farm, Suffolk. The Trials are now in their third year and preliminary data is published here for the first time.

FIG. 1. PILOT PLANT DIGESTERS AT CLEPPA PARK, PRELIMINARY RESULTS

Digester names	Plug flow	Plug flow	Contact	Hydraulic	Filter
Design type	Plug flow	Plug flow	Conventional High Rate Contact	Novel (UCC)	Anaerobic filter
Operating volume (m³)	30	30	30	2	1.5
Material of construction	Fibre Glass (Helically wound)	Fibre Glass (Helically wound)	Mild Steel	Fibre Glass (Helically wound)	Plastic (ex-orange-juice container)
On-line since	June 1981 (2yrs)	June 1981 (2yrs)	November 1982 (½ yr)	June 1980 (3 yrs)	Spring 1982 (1 yr)
Loading frequency	Daily	Daily	Hourly	Daily	Daily
Waste substrate used	Pig	Pig	Pig, limited crop	Pig	Pig
Situation installed	Below ground	Below ground	Above ground	Above ground	Above ground
Trials undertaken HRT (days)					
60	0.009		0.385	0.646	-
30	0.051		0.469	0.169	0.351
20	0.324		-	0.430	-
15	0.555		0.713	-	0.751
lower	-		-	0.578-0.59	0.257-0.753
Temp. 35	✓		✓	✓	✓
lower	✓		-	-	-
Cost (£/m³) (approx)	300		1200	500	50

Introduction
The first major achievement of this contract was the establishment of a purpose built research facility on a 'green field' site owned by University College, Cardiff (Fig.9.). The Anaerobic Digester Research Unit is fully equipped and facilities include electrical points, drainage channels, water points for hose connection, silage storage clamps and weighing scales for loads of up to 150 Kg of crop material. A Portacabin on site comprises a small Office with telephone for direct link to the University Computer, and a fully equipped laboratory. All analyses are performed on site.
In Fig. 1. the details of construction of the four pilot digesters are outlined as well as operating procedures and some preliminary experimental achievements.

High Rate Contact Digester
This is a $30m^3$ (Fig.2.), fully automatic conventional digester which was to be directly compared with the Plugflow Digester. Valuable data has been obtained from the digestion of pig waste when operated in the mesophylic range (in practice 33-37 ℃). The digester has mostly been operated under gas mixed conditions, except during winter periods when the liquid ring gas recirculator is prone to freezing. Hydraulic retention times studied so far have included 60,30,20, and 15 days. Limited trials have been made with macerated crop material at a 30 day retention time. The feed characteristics are a 1% T.S. (approximately) mixture of Barley macerated in water. Although this is not representative of an expected feed substrate to a crop digester, results obtained in the time available have shown that a yield of 0.438 m^3/Kg V.S. is possible at these loading rates.

Plugflow Digester
This digester has been operated so far at retention times from 60-15 days. Results have indicated that the performance of this digester compares well with the Contact digester. The simplicity of this design has meant very little down time and its low aspect ratio enables its installation in many locations without unsightly consequences (Fig.4,5). Commercial experience could reduce installation times to one day and would therefore present no extra problems over traditional designs which must be insulated and commissioned further on site.
Heat transfer studies of this digester are another aspect for which the Research Unit would be ideally suited, with facilities for temperature probes at several points along and across the digester. This would directly compare with work carried out at Cornell University, USA, on a similar Plugflow system.

The Hydraulic Digester
This $2m^3$ digester (Fig.3) is a novel design conceived and developed at the University which has proved itself versatile and efficient. Retention times as low as 3.3 days have been reached giving gas yields of 0.578 m^3/ Kg V.S. The suitability for patenting this design is presently receiving attention and it is hoped to publish more details in the final report. This digester has operated entirely in the mesophylic range on a pig waste substrate. Heating mechanisms used have included direct steam injection and external recirculation through a water bath.

Fig.2. Contact Digester

1. Macerator   2. Contact Digester   3. Contact Feed Tank
4. Effluent Chimney   5. Effluent Tank

Fig.3. Hydraulic and Filter Digesters

1. Feed Tank   2. Hydraulic Digester   3. Effluent Tank
4. Anaerobic Filter   5. Heat Exchanger

Fig.4 & 5. Plug Flow Digester

1. Plug Flow Digester   2. Influent End
3. Emergency Gas Relief   4. Effluent End
5. Heating Pipes   6. Boiler
7. Gas Holder   8. Macerator

The Anaerobic Filter
This 1.5m³ unit has worked for approximately one year (Fig.3) and has operated at retenion times from 30-7 days. It has always been operated at 35°C and entirely on pig waste. At 15 days retention time reductions in C.O.D. of 70% have been achieved and at 7 days retention time there have been reductions of 50-60% C.O.D. Results of experiments with Anaerobic Filters from University College, Galway are presently being compared with those recorded at Cleppa Park in order to optimise their operation and design.

Data Management
All analyses and operating results are stored and manipulated by a computing facility at the University; a GEC 4090 which was supplied by the Science Research Council and forms part of the SRC Network around the country. The language used for the management of Cleppa Park data is the GEC 4090 version of APL.
Weekly data files are input for all digesters and form the 'raw' data subsequent parameters, such as Gas Yield, C.O.D. loading rate and reduction factors are routinely calculated. These files can then be printed for hard copy storage or used to compile graphs which may represent up to 52 data files. Statistical data is also calculated when the graphs are compiled and Means, Standard Deviations, Maxima and Minima are determined for all data provided.

Catch Crop Studies
Various crop material substrates have been studied and one major part of these studies is the evaluation of the effect of maceration on different crop substrates. Laboratory scale digesters have been used to confirm the rapid increase in gas yield obtainable by pre-maceration (Fig.6,7).
Attempts to evaluate crop substrates at full scale started in 1982 when a field of Forage Pea mixed with Italian Rye Grass was sown after consultation with colleagues at Reading University. This was harvested later that year, and 'Big bag' ensiled, which involved rolling large bales of whole crop material and packaging each bale into a large black plastic sack. One particular advantage with this system was that each bag could be opened separately when required for feeding to the digesters.
During 1982 a second crop was grown, a mixture of Barley and Augusta Rye Grass. This time the substrate was 'Fine-Chop' harvested to an average length of about 3-4 cm.

350m³ Digester, Suffolk
A full scale digester at Walnut Tree Farm, Chediston, Suffolk was monitored intensively and now general data is available covering a three year period. Comprehensive data is still awaited from the manufacturer and this data will be compared with information already gathered from Cleppa Park when it is made available.
Effluent from this digester has been evaluated for its ability to replace commercial fertiliser when used on grass leys. This work has been carried out during the summer months in 1981 and 1982. The Fertiliser Trials for 1983 are currently sponsored by the Ministry of Agriculture. The results will be published at the end of the 1983 trials so that all data can be evaluated simultaneously. The results for the 1982 Trials are given however (Fig.8.).

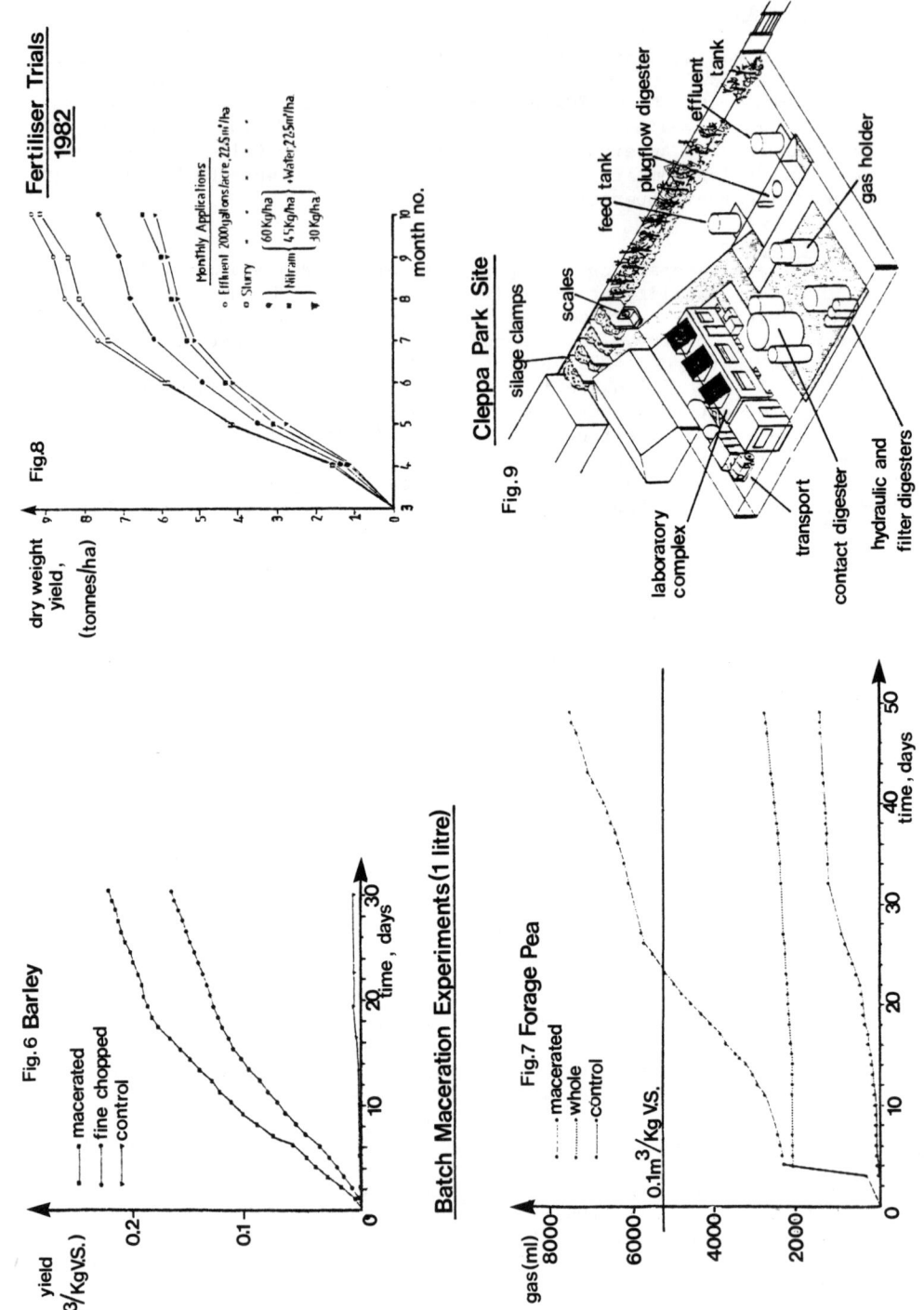

The Future and Conclusions
   The continuation of this project is essential if crop digestion data is to be obtained at this scale in the U.K., as it has in some other countries notably New Zealand and Germany. The research unit is in an ideal position to continue such work, considering the amount of stored crop material available for digestion, as well as the two 30m^3 digesters which have been specifically designed for the digestion of crop material, either in admixture with pig and other wastes, or alone. Plugflow systems have been shown to be fully reliable in operation and performs as well as, if not better than the Contact system. The simplicity and cheapness of construction coupled with low running costs (no mixing necessary) recommend plugflow digesters for the treatment of pig waste.
   The high efficiencies of Hydraulic and Filter digesters operating at low retention times (3 days) are preferred to high rate Contact systems. The latter operate at 10-15 days retention times and capital costs preclude their use if the technological know-how of the two former systems is available.

References

Stafford D.A., Hughes D.E. and Etheridge S.P.,'Anaerobic Digestion of Farm Wastes and Energy Crops'.In Proceedings of EEC Contractors Meeting, Copenhagen (1981).

Stafford D.A. and Etheridge S.P.,'Farm Wastes, Energy Production and the Economics of Farm Anaerobic Digesters'.In Anaerobic Digestion, Elsevier Biomedical Press,pp 255-268,(1982).

Stafford D.A. and Etheridge S.P.,'Research in the U.K.: Energy Farms and Anaerobic Digestion',Helios 13, Solar Energy Unit, University College, Cardiff,(1981).

Stafford D.A., Hughes D.E. and Etheridge S.P.,'Anaerobic Digestion of Farm Wastes and Energy Crops'.In Proceedings of EEC Contractors Meeting, Brussels (1982).

Stafford D.A., Etheridge S.P. and Hughes D.A.,'Anaerobic Digestion of Farm Wastes and Energy Crops',Biomass 82, Berlin, (1982).

Stafford D.A. and Etheridge S.P.,'The Hydraulic Digester',BABA Digest, November 1982.

Stafford D.A. and Etheridge S.P.,'The Anaerobic Digestion of Industrial Wastes, Farm Wastes and Sewage Sludges. An Assesment of Design Types and Performances',In Effluent in the Treatment Process Industries,I.Chem.E. Symposium Series No. 77,pp 141-148,(1983).

Stafford D.A.,'The Effects of Mixing and Volatile Fatty Acid Concentrations on Anaerobic Digester Performance',Biomass 2,pp 43-45,(1982).

Turner J.,Stafford D.A.,Hughes D.E. and Clarkson J.,'The Reduction of Three Plant Pathogens in Anaerobic Digesters', Agricultural Wastes 6,pp 1-11,(1983).

ANAEROBIC FILTER DIGESTION OF AGRICULTURAL WASTES

Authors	: A. WILKIE, M. BARRY, P.J. REYNOLDS, N. O'KELLY and E. COLLERAN
Contract Number	: ESE-R-024-EIR
Duration	: 36 months 1 July 1980 - 30 June 1983
Total Budget	: £133,050 CEC contribution: £66,525
Head of Project	: Dr. Emer Colleran
Contractor	: University College, Galway
Address	: Department of Microbiology, University College, Galway, Ireland.

Summary

Further studies on the development of a two-phase process for the anaerobic digestion of animal slurries and solid agricultural residues were carried out. The process envisages linking a Phase I hydrolysis reactor to a Phase II methanation reactor of the retained biomass design. The anaerobic filter was shown to be highly suited to two-phase digestion processes by a series of digestibility trials at various loading rates and retention times and using grass and wheat straw hydrolysate and silage wastewater as influent feed. Studies with different support matrices revealed that a fired clay matrix allowed more rapid start-up and higher operational efficiency than plastic, coral or mussel shell support materials. The effect of varying the matrix configuration from random packed to channelled and of downflow versus upflow feed modes on the start-up phase of anaerobic filter digestion of pig slurry was also investigated.

In general, two-phase digestion of solid residues such as grass, straw and cabbage leaves allowed CH_4 yields which were comparable with those reported in the literature for 30-day retention periods in batch or continuous single-stage digesters. The development of appropriate seed cellulolytic cultures for the hydrolysis reactor was identified as meriting detailed further study.

1.1 Introduction
       The random packed support matrix bed and the upward feed flow
characteristic of the anaerobic filter precludes its use for single-phase
digestion of wastes containing a high content of suspended material or
solid agricultural residues such as straw, grass, etc. Considerable
attention has focussed recently, however, on the application of a two-
phase digestion system to biogas production from animal slurries and solid
residues (1,2). A two-phase approach, in which the hydrolytic and fermentative phase is carried out in a separate batch reactor which is linked to
an upflow anaerobic filter for rapid biomethanation of the liquefied first
phase products, has been under study in Galway throughout the duration of
this contract (3,4,5). The two-phase process has been applied to the
digestion of pig slurry and to a variety of crop residues including straw,
grass, cabbage leaves and some industrial organic residues such as paper
wastes, horse chestnut residue, etc.
       Hydrolysis of the polymeric components of waste residues is the major
rate-limiting step in solid substrate digestion (6) and the lengthy substrate retention times required for effective hydrolysis in conventional
single phase digesters is prohibitive from an economic standpoint. Phase
separation allows separate optimisation of the hydrolytic and methanogenic
phases of digestion and independent variation of both loading rate and
substrate retention time in the two reactors, thereby affording better
overall process control.

1.2 Description of the Process
       Two-phase digestion of pig slurry utilises the existing farm-practice
of keeping slurry in holding tanks for quite lengthy time periods prior
to eventual disposal by land-spreading. Hydrolysis and liquefaction of
the solid material occurs naturally at ambient temperatures in the holding
tanks and gravity settlement or, preferably, screen separation using conventional farm slurry separators, yields a liquid fraction with a low content of suspended solids and which contains 70-90% of the total COD of the
waste (7,8). The supernatant fraction is readily digested with 65-80%
COD removal efficiency by a 3-day hydraulic retention time in an anaerobic
filter at 30-35°C (3,7).
       The two-phase process utilised in this study for solid residue diges-
tion has been previously described (4,5). It consists of a batch reactor
which is loaded with the solid substrate and to which pig slurry supernatant is added to provide the desired solids concentration (4-15% range).
The soluble hydrolysis and fermentation products are removed from the
first reactor by a liquid displacement system in which effluent from the
linked anaerobic filter is passed to the first phase reactor, thereby
displacing the desired liquid volume containing the liquefied material.
The liquid displacement cycle used is dictated by the nature of the substrate undergoing liquefaction, the total solids content of the reactor
and the rate of fermentation acid production within the reactor. Biomethanation of the hydrolysis and fermentation products in the anaerobic
filter proceeds with a high COD conversion efficiency of 80-90%.
       The results of laboratory scale studies on filter digestion of pig
slurry and of two-phase digestibility trials of straw and grass residues
have been previously reported (4,5). The present report is concerned with
(i) a 3 m^3 pilot scale study of filter digestion of pig slurry;
(ii) evaluation of a variety of filter support media and feed flow modes
at laboratory scale; (iii) assessment of the suitability of the anaerobic
filter as the methanation reaction in two-phase solid substrate digestion

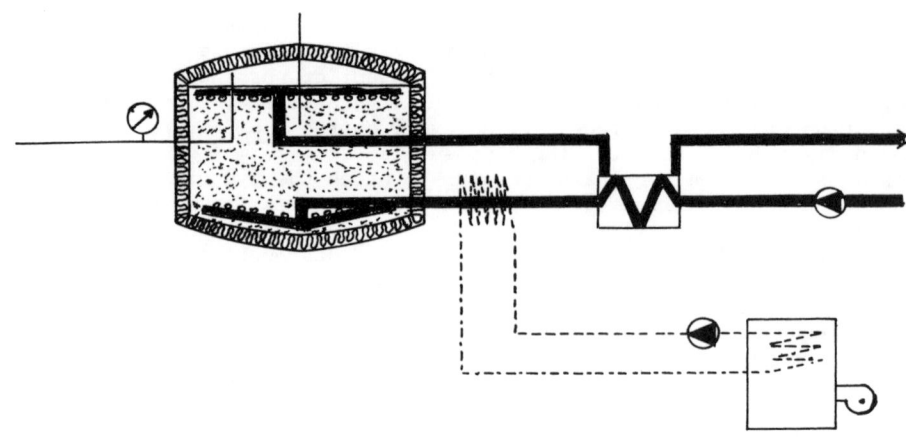

Fig. 1: Schematic diagram of a 3.5 m^3 upflow anaerobic filter for pig slurry digestion

and (iv) optimisation studies and additional digestibility trials of solid substrates in the Phase I hydrolysis reactor.

2. Materials and Methods

Four identical anaerobic filters of 18 L void volume were constructed as previously described (7) and packed with clay, coral, mussel shell and plastic ring support media. Digestibility of Phase I liquified material was carried out in a limestone matrix filter of 8.4 L active volume and a clay particle filter of 12.4 L active volume. The pilot scale slurry digester had a total volume of 3.5 m^3, was filled with a plastic ring matrix of 94% porosity and operated at a range of temperatures from 25-35°C using an external heat-exchanger. All analytical measurements were performed according to Standard Methods.

3. Results and Discussion

3.1 Pilot Scale study of Anaerobic Filter Digestion of Pig Slurry

A schematic diagram of the 3.5 m^3 anaerobic filter installed in the piggery of the Agricultural College at Athenry, Co. Galway is presented in Fig. 1. The digester was packed with plastic rings of 3.8-5 cm in diameter and seeded with anaerobic sludge from the bottom of a slurry holding tank on the farm. The slurry supernatant available for digestion was considerably diluted by wash and rain-water. The COD strength varied between 13.2 and 26 g/L during the initial phase of the trial period. The digester was started up in an upflow feed mode at a hydraulic retention time of 6 days and an operating temperature of 25°C. An average COD removal of 66% was obtained at loading rates ranging from 2.16 - 4.32 kg COD.m^{-3}.d^{-1} and with the CH$_4$ productivity and yield shown in Table 1.

The hydraulic retention time was subsequently reduced to 3 days and the temperature was maintained at 25°C. The influent slurry COD averaged 25.2 g/L giving a loading rate of 8.4 kg COD.m^{-3}.d^{-1}. A significant drop in COD removal and in methane yield per kg COD introduced to the reactor was noted when steady state removal efficiencies were obtained (Table 1). It is apparent from the results obtained that the COD volumetric load applied at this temperature and hydraulic retention time exceeds the capacity of the filter unit. The operating temperature was consequently increased to 33-35°C while maintaining the same organic loading and hydraulic retention time. The results obtained to date indicate that the COD removal efficiency increased significantly and ongoing studies at steady state will reveal the full extent of the increase. It is intended to further reduce the hydraulic retention time at 33-35°C to 2.5 and 2 days to ascertain the maximum loading rate and minimum retention which the filter unit can sustain. The reactor is designed to accommodate upflow and downflow feed modes and final experiments to compare the operational characteristics at a variety of loading rates in downflow feed mode will also be carried out.

Table 1: Pilot Scale Anaerobic Filter Digestion of Pig Slurry

HRT (days)	Loading Rate (kg COD.m^{-3}.d^{-1})	COD removal efficiency %	Volumetric CH_4 productivity (m^3.m^{-3}.d^{-1})	CH_4 yield (m^3.kg^{-1} COD_o)
6	2.16-4.32	66	0.82-1.04	0.22
3	8.4	52	1.47	0.172

3.2 Support Matrix effects in anaerobic filter digestion

Considerable differences between support matrices have been reported by van den Berg for downflow fixed film reactors with respect to the rate of attached biofilm development, the density and uniformity of film formed and the ease of sloughing off by hydraulic flow or biogas bubble generation within the film (3). As reported previously (5), the start-up phase for four laboratory-scale anaerobic filters containing fired clay, plastic rings, mussel shells and coral support materials was compared using a pig slurry supernatant feed. The operating temperature was 35°C and the initial hydraulic retention time was 6 days. The fired clay matrix provided the most rapid start-up and reached maximum COD removal efficiency by day 30 (5). When steady state removal was attained by all four filters, the COD removal and conversion efficiency to CH_4 was found to be marginally better with the clay and plastic filters (73% COD removal) than with the coral (70%) and mussel shell units (69%) under identical COD loading rates and operating conditions. The loading rate to the four filters was increased stepwise by decreasing the retention time from 6 to 3, 2 and 1 days. Table 2 compares the results obtained at 2 day hydraulic retention time for the four filter units. The clay and coral filters clearly allow more efficient conversion of the waste at this high loading and short retention time than the mussel shell or plastic matrix units. It is difficult to relate these differences in performance to differences in either matrix porosity or in matrix surface area. In channelled downflow reactors, overall performance depends solely on the total amount of attached film and hence on the surface to volume ratio of the support material (3).

Table 2: Effect of Support Material on Anaerobic Filter Performance
2 day HRT

Support Matrix	Clay	Coral	Mussel Shells	Plastic
Porosity	69	71	80	94
Surface Area per unit volume occupied $m^2 \cdot m^{-3}$	119	490	161	179
Loading Rate:				
$kg\ COD_t \cdot m^{-3} \cdot d^{-1}$	13.4	13.4	13.4	13.4
$kg\ COD_s \cdot m^{-3} \cdot d^{-1}$	12.7	12.7	12.7	12.7
% Removal Efficiency:				
COD_t	58.3	57.7	56.0	47.3
COD_s	67.3	61.3	60	50.7
% CH_4 in Biogas	87	88	88	88
Volumetric CH_4 productivity $m^3 \cdot CH_4 \cdot m^{-3} \cdot d^{-1}$	3.27	3.22	2.83	2.58
COD yield $m^3 CH_4 \cdot kg^{-1} COD_o$	0.255	0.240	0.22	0.194

By contrast, the studies of Young and Dahab with varying support media in upflow filters showed that the ability of the media to entrap and prevent washout of biological solids is a more important parameter than the unit surface area of the matrix (7). It is implicit in Young and Dahab's findings that the porosity of a matrix is of major importance in the performance of upflow anaerobic filters. In the present study, the plastic matrix with a porosity of 94% exhibited the poorest performance. Although the coral matrix had by far the largest surface to volume ratio, its performance was slightly poorer than that of the clay filter. The roughness of the clay surface and physico-chemical interactions such as electrostatic attraction or the leaching of essential inorganic nutrients may be a contributory factor to the superior performance of the clay matrix.

3.3 Effect of varying the Substrate Flow Direction

Virtually all of the studies carried out on random packed anaerobic filters have utilised an upward flow of substrate through the submerged matrix bed. This results in a plug-flow feed pattern in the reactor but may also lead to accumulation of influent suspended solids as well as flocculated or granular biological solids in the lower sections of the

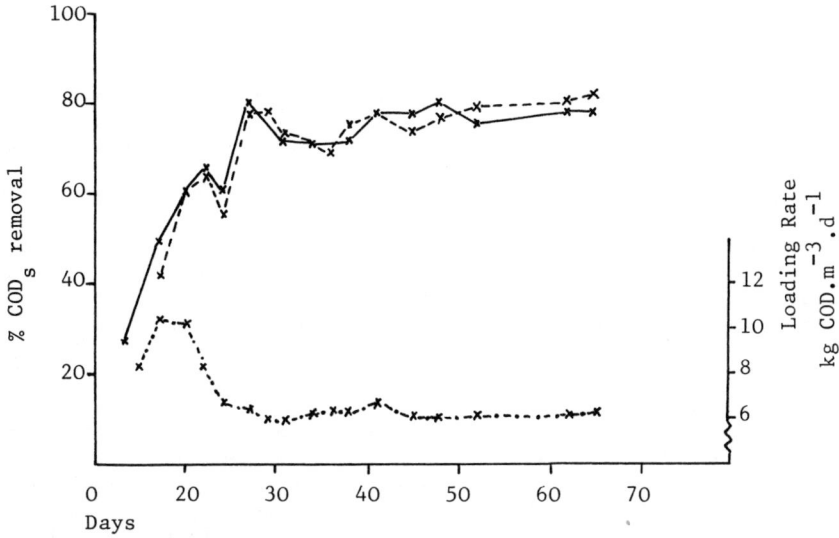

Fig. 2 Effect of Substrate Flow direction on start-up of random-packed clay filters
($COD_{soluble}$ removal efficiency: downflow ———; upflow ‐ ‐ ‐ ‐ ; COD loading rate ·—·—·—·)

filter. Solids build-up may cause channelling and short-circuiting which must be relieved by a back-wash or solids wasting procedure. By contrast, the studies of van den Berg and co-workers with channelled support media have utilised a down-flow mode of substrate distribution.

Four filters were constructed of equal empty bed volume (18 L) and containing a fired clay support. Two of the filters were packed on a random basis with clay particles of 4 cm length and 2.5 cm width and the other two were packed with a solid clay column containing vertical channels. One filter of each type was operated in upflow mode and the other with downflow feed distribution. The increase in soluble COD removal efficiency during the start-up phase of the random packed filters is illustrated in Fig. 2. The pig slurry input feed varied from an initial high level of 24-25 g/L to 18-19 g/L during the final 40 days of the start-up phase as illustrated by the total COD loading rate shown in Fig. 2. Very little difference was noted between the upflow and downflow filters with respect to the increase in COD removal efficiency and the yield of CH_4 during the start-up phase. Fig. 2 illustrates the soluble COD removal since analysis of total COD removal in the initial period after start-up is obscured by washout of seed sludge from the filters. At steady state operation on a 3 day retention time, there was no significant difference between the operational efficiencies of the two random-packed filters.

Analysis of the distribution of volatile suspended solids in the matrix beds indicated that the bulk of the biological solids were in the bottom section of the upflow filter in agreement with the observation that the COD removal was occurring chiefly in this section of the matrix bed.

By contrast, the distribution of biological solids was more uniform throughout the downflow matrix bed. The VSS and F_{420} content of the liquid samples from the top three ports were virtually identical, with a lower level present in the bottom port sample. COD removal in the downflow filter was carried out largely in the upper half of the filter. No clogging problems were presented by the downflow filter whereas some blocking of feed pipes and the feed dispersion plate in the upflow filter by influent suspended solids did occur. Experiments are currently in hand to evaluate the performance of both filters at higher loading rates and lower retention times with slurry and milk waste feeds.

The start-up phase of the channelled filters is currently under study. Preliminary results suggest that the downflow channelled unit exhibits a far longer start-up phase than the upflow unit. This is to be expected since the upflow unit retains some suspended biological solids which contribute to the COD removal efficiency of the reactor whereas the downflow filter is entirely dependent on the build-up of an active attached biofilm.

3.4 Suitability of the anaerobic filter for two-phase digestion of solid substrates

The two-stage digestion process developed during this contract involves linking a Phase I hydrolysis reactor to a suitable high capacity methanation reactor which will convert the hydrolysis products to biogas at high efficiency. The hydrolysate liquor from the Phase I reactor, when operated in batch mode, varies with respect to COD strength, total organic acid (TOA) content and pH during the batch reaction. It was considered essential, therefore, to fully investigate the flexibility of the anaerobic filter with respect to these three parameters. For this purpose, the hydrolysates from grass clippings and from wheat straw and effluent wastewater from a silage pit were used as feedstock for both a granite chip and a clay particle matrix filter. Silage effluent was used because of its close similarity to Phase I reactor effluent and its availability in large quantities and at varying strengths.

Fig. 3 illustrates the COD content of the influent and effluent from the granite filter treating silage wastewaters which ranged in COD strength from 23.8 to 42.2 g/L. The volumetric COD load ranged from 7.8 to 14.2 kg $COD.m^{-3}.d^{-1}$. It is evident from Fig. 4 that the efficiency of COD conversion to CH_4 was independent of the COD loading rate within the range tested and at the hydraulic retention time utilised (3 days). Reduction of the retention time to 2 days, while maintaining the COD load rate within the range listed above, did not noticeably alter the COD removal efficiency. Further studies to evaluate the operational efficiency at 1 day HRT and at higher loading rates are currently in hand.

Digestion of grass hydrolysate liquor at influent COD concentrations ranging from 12-45 $g.L^{-1}$ and under constantly varying COD loading conditions revealed that the filter effluent tolerated feed fluctuations without any acclimatisation and an average efficiency of COD removal of 84% was maintained throughout the trial. The TOA_1 content of the grass hydrolysate liquor varied between 4.2 and 10 $g.L^{-1}$ and an average TOA removal efficiency of 95% was maintained despite the wide fluctuation in influent acidity. High tolerance to fluctuation in the pH of the influent liquor was also noted. No decrease in filter efficiency was noted when the pH of the silage wastewater feed was maintained at 5.2-5.6 for a 20 day period. Shorter-term trials at influent pH values between 4 and 5 yielded similar results.

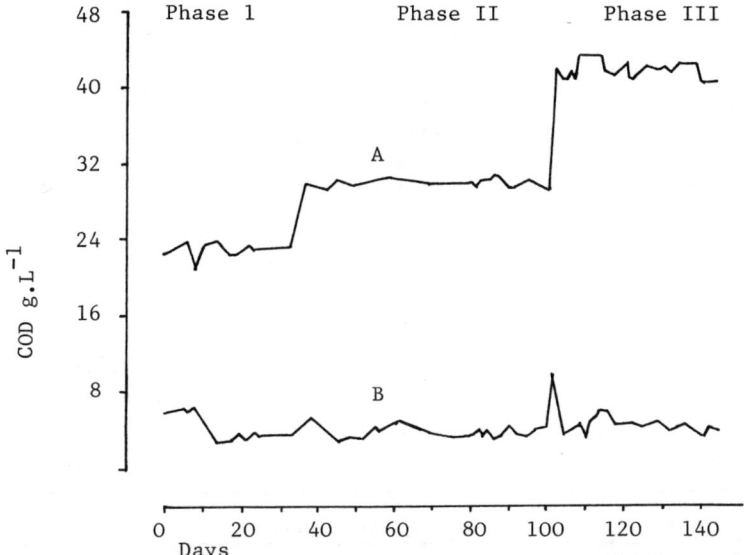

Fig. 3. COD content of feed and effluent during digestion of silage wastewater.

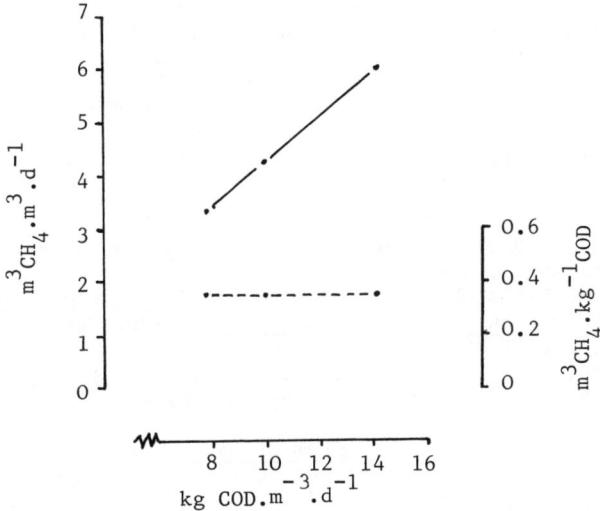

Fig. 4. Effect of COD loading on the volumetric rate of methane (STP) production (———) and on the rate of methane production per kilogramme of COD added (-------).

Table 2: Methane Yield from Two-Phase Digestion of solid substrates

Substrate	Retention time (days)		$m^3 CH_4 \cdot kg^{-1} VS$
	Hydrolysis reactor	Filter	
Straw (roughly chopped)	18	3	0.16-0.18
Straw (NaOH treated)	18	3	0.26
Grass (clippings)	8	3	0.25-0.33
Horse chestnut residue	28	3	0.17-0.23
Cabbage Leaves	16	3	0.30-0.35

3.5 **Substrate Digestibility and optimisation studies of the Phase I hydrolysis reactor**

Digestibility trials on straw, grass clippings, cabbage leaves and horse-chestnut residue using the two-phase system described previously (4,5) were also carried out with the results shown in Table 2.

The total solids concentration of the hydrolysis reactor was varied from 2-10% and a variety of liquid displacement regimes evaluated. Operating conditions which minimised CH_4 production in the hydrolysis reactor were identified. The evolution of hydrogen from the hydrolysis reactor was monitored and tests are ongoing to establish the quantity of hydrogen produced under different conditions. Initial experiments on seeding the reactor with defined isolates from mixed seed cultures suggest that the hydrogen output can be controlled by the use of appropriate cellulolytic seed strains.

References

1. Rijkens, B.A. (1983) in "Energy from Biomass", p. 572-580, Applied Science Publishers, London.
2. Ghosh, S. and Klass, D.L. (1976) in Clean Fuels from Biomass, Sewage and Urban Refuse and Agricultural Wastes, 27-30, Published by the Institute of Gas Technology, Chicago, U.S.A.
3. Colleran, E., Barry, M. and Wilkie, A. (1982) in Energy from Biomass and Wastes VI, 443-481; Published by the Institute of Gas Technology, Chicago, U.S.A.
4. Colleran, E. (1981) in Energy from Biomass, Series E. Vol 1, Edited by P. Chartier and W. Palz, D. Reidel Publ. Co., Holland.
5. Barry-Concannon, M., Wilkie, A., Faherty, G. and Colleran, E. (1982) in Energy from Biomass, Series E, Vol. 3, Edited by P. Chartier and W. Palz, D. Reidel Publ. Co., Holland.
6. Kotze, P.G., (1968) Water Research, 3, 545-558
7. Colleran, E., Barry, M. and Wilkie, A. (1982) Process Biochem, March-April, 12-17.
8. Brondeau, P., de la Farge, B. and Heduit, M. (1982) Genie Rural, Janvier-Fevrier, 5-10.

"TWO-PHASE PROCESS FOR THE ANAEROBIC DIGESTION OF ORGANIC WASTES YIELDING METHANE AND COMPOST"

Authors	: G. Hofenk, S.J.J.Lips, B.A.Rijkens, J.W.Voetberg
Contract Number	: ESE-R-040-N (N)
Duration	: 36 months 1 July 1980 - 30 June 1983
Total Budget	: Fl. 820.000 CEC Contribution : Fl. 410.000
Head of Project	: Drs. Berend A. Rijkens, IBVL
Contractor	: Institute for Storage and Processing of Agricultural Produce (IBVL)
Address	: Institute for Storage and Processing of Agricultural Produce (IBVL) Bornsesteeg 57 P.O. Box 18 6700 AA WAGENINGEN The Netherlands

Summary

The aim of the project was the development of the novel two phase process for the anaerobic digestion of solid organic wastes, producing biogas and compost. A pilot plant should be erected, based on laboratory results. The technological and economical feasibility should be determined.
After one year of laboratory studies, the first experimental liquefaction reactor and the second stage methane reactor were erected in 1981. One year later an improved liquefaction reactor of 300 m^3 with five compartments was built.
Full-scale experiments with tomato-plant wastes from glasshouses were completed and a feasibility study was based upon the acquired technological parameters and other information.
Preliminary experiments were carried out with solid waste from a beet-sugar factory and with the mechanically classified organic fraction of municipal waste.
The experiments will be continued.
In the laboratory digestibility studies with different kinds of wastes are in progress, to gather information on the suitability to anaerobic processing on a technical scale.
The promising results caused already quite some spin-off in respect of further R & D work ordered and funded by industry and governmental organisations.
This process may turn out to be a good alternative for dumping or incineration of perishable solid organic wastes.

1. Introduction

The anaerobic digestion process for solid wastes, as developed by the IBVL, is a two-stage process in which the wastes are percolated in a liquefaction/acidification reactor(R1), while the percolate containing the dissolved organic matter is fed to a methane reactor (R2)(UASB-type). The treated effluent of the R2 is recirculated to the R1. Figure 1 and 2 show a schematic view of the principle of the process and of the realization in practice.

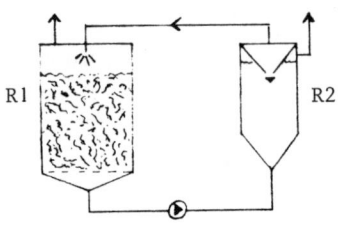
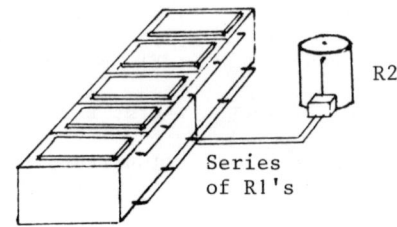

Fig.1. Schematic view of the principle of the IBVL-process

Fig.2. View of a full-scale plant

In the laboratory the process has been investigated as well as the digestibility of various wastes like waste-onions, sugar-beet pulp, sugar-beet wastes, cannery wastes, extracted poppy-seed hulls, tomato-plant wastes, potato-wastes, grass clippings, the organic fraction of municipal solid waste and wastes from public gardens. Results of these experiments have been used as design criteria for the construction of a pilot plant that has been erected at Delft in cooperation with the firm of "Zegwaard BV". The first prototype of the R1(100 m^3) is dismantled now and replaced by a 300 m^3 unit, consisting of 5 compartments of 60 m^3 each. The R2 is a 10 m^3 UASB reactor. Experiments with tomato-plant wastes have been carried out, now municipal waste is being digested.

2. Results of the experiments

2.1 Laboratory results

The rate of the decay is mainly controlled by the liquefaction rate of the solid matter in the R1. The liquefaction rate increases when the solid matter is comminuted by chopping or crushing, but also depends on the kind of organic matter and its composition. Woody material and strongly lignified cellulose are hydrolyzed slowly and will leave a larger residue; lignin is not digestible under anaerobic conditions and will protect part of the cellulosic matter against decay. Thus the breakdown starts at a high rate with the readily digestible part of the plant material and generally decreases after 10 to 14 days to a much lower rate, when mainly the lignocellulosic fraction is left. At the inflextion point the R1 generally has become selfsupporting in respect of its methane flora, that has been developed to an extent capable of converting all the dissolved organic matter (mainly v.f.a.) into biogas. At this point 40 to 80% of the organic matter will have been digested and the residue has become free of odours, so the residue can be discharged from the R1.

The effect of temperature on the rate of liquefaction is studied at 3,10,19,25,32,37,45 and 55°C. The highest rates were obtained at 32 and

37°C, below 10°C hardly any biological activity could be observed over a period of 30 days.

A comparison between the process-conditions at mesophylic and thermophylic conditions lead to the conclusion that especially the methane flora in the R2 at the thermophylic temperature is vulnerable. The sludge tends to be very fine and doesn't show good settling properties, so the sludge is washed out continuously. Further study on thermophylic digestion, however, showed that with glucose as a substrate a granular sludge with good settling properties and high activity (more than 20 kg COD/m^3.day) could be obtained. A thermophylic anaerobic filter, fed with a mixture of acetate, propionate, butyrate and glucose, showed a steady high capacity of more than 30 kg COD/m^3.day.

The methane flora (mesophylic) in the R2 never showed adaptation problems when switched to feed solutions from a different waste material.

Problems with the granular sludge of the methane bacteria only arose with too high concentrations of ammonia and alkalinity, especially when these two coincide. In that case the granular sludge desintegrades and is washed out. Tomato-plants tend to be very alkaline and also sometimes show an extremely high content of nitrogen. In practice this will mean that the recirculating water, accumulating ammonia and alkalinity, will have to be diluted with fresh water.

2.2 Pilot-plant results

Because the wastes require a retention time of several weeks in the liquefaction reactor, a cheap R1 on pilot-scale was designed as described previously (2). A cross-section of this R1 is shown in fig.3.

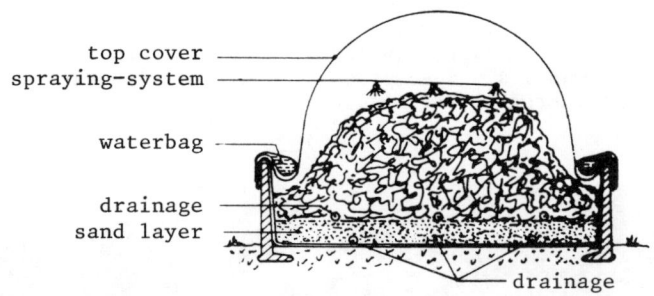

Fig.3 : Cross-section of the first R1.

This R1 can contain 100 m^3 of wastes and has been used for the digestion of tomato-plant wastes; the R2 has a volume of 10 m^3.

A series of 6 experiments have been carried out and from these experiments it was concluded that in a period of 2-3 weeks a vs-reduction of 40% could be obtained with a concomitant reduction in volume of 55%:
- Chopping of the wastes has a positive effect on the liquefaction rate while the density of the waste has increased by a factor of 2.7
- The H_2S-concentrations in the produced gasses are rather high: up to 4% for the R1 and 1% for the R2
- The wastes are high in calcium($\pm$5% of the TS) and other buffering compounds, so the pH tends to be high. The COD in the percolate of R1 is highly acidified (more than 90%).

- In all experiments CH_4 was produced in the R1, but could not be quantified because of leakage of the plastic cover.

After the digestion period the residue was odourless;an additional period of about 2 months was necessary for the aerobic after-composting of the residue.

The experiments with tomato-plant wastes were interfered by two major problems caused by the quality of the charge used.These problems were:
- Inhibition by NH_3,due to a high nitrogen-content of the wastes.During the digestion of one particular charge of tomato-plants the ammonia-concentration in the liquid rose to 4000 mg/l while the pH rose to more than 8, so that the concentration of free NH_3 exceeded the toxicity limit and no activity of the methane bacteria could be observed anymore. This problem can be solved by dilution with water, but this will result in a larger amount of waste water to be treated.
- Scaling due to a high calcium content of the wastes occurred especially in the heat exchangers and in the effluent-piping of the R2.Decalcifying or regular cleaning of the piping may be necessary.

The experiments with the first R1 revealed some serious drawbacks of this design:
- Leakages in the plastic cover prevented to draft a proper massbalance
- The piling height of the wastes was limited to 2 m
- The design didn't allow for experiments under fully submerged conditions.
- The sand-layer at the bottom clogged rapidly and couldn't be used for draining.
- This R1 was not insulated.

To overcome these drawbacks and because of the need for more R1's to bring about a more constant loading rate of the R2, a new R1 has been built of a totally different design. This new R1 consists of a concrete body, divided into 5 compartments of 60 m^3 each, which can be operated separately. A steel cover provided with water locks is placed on each compartment. The whole R1 is completely insulated. Figure 4 shows a picture of the new R1(photograph below).

With this new R1 experiments have been carried out with sugar-beet wastes.

The above-mentioned problems with tomato-plant wastes didn't arise in these experiments. After a week 70% of the vs added was converted. The produced biogas had a CH_4-content of about 75%. The production of methane took place for about 70% in the R2 and for about 30% in the R1. A loading rate of the R2 of 47 kg COD/m^3.day could be attained with a removal efficiency of more than 95%. Influent-concentrations for the R2 up to 40,000 mg COD/l were no problem, while buffering of the influent was not necessary. The reduction in volume of the wastes after digestion amounted to 45%. The residue, however, was very difficult to dewater and didn't get heated during the aerobic after-composting.

2.3 Aerobic after-composting

The residue discharged from the R1 after the main part of the biogas has been collected, is sufficiently free of odours to be exposed to open air. Woody or stalk material is still pretty tough, because the lignocellulosic structure is hardly attacked. So, though the material is already half-composted it has not yet obtained the right chemical and physical structure of a compost. After being exposed to air in a pile, the material rapidly starts to heat up by aerobic decay. Temperatures up to 75°C are obtained within ten days, whereupon the temperature decreases gradually. After one to two months the material has obtained a blackish colour, the woody parts can be crumbled easily, it has an earthen smell and the water content is lowered to about 45%(if shielded from rain). The final compost cannot be distinguished from a compost produced along a completely aerobic route. The high temperature has strongly reduced (plant) pathogenes and killed seeds. The quality of the compost is still under research.

3. Feasibility study

Several runs with tomato-plant wastes in the pilot plant provided the basic information for the calculations concerning a full scale processing plant of 50,000 t/annum capacity. The figures and parameters obtained and used for the calculations are:
- 50,000 t/annum tomato-plant wastes from the glasshouses in the area of Delft
- dry matter content (d.m.) 25% ≡ 12,500 t/annum
- organic d.m.(of the total raw material) 20% ≡ 10,000 t/annum
- residence time of the solids in the batch reactor R1   2 weeks
- anaerobically digested, 40% of the organic d.m. ≡ 4,000 t/annum
- CH_4 production   1,500,000 m^3/annum (gross)
- biogas production (70% CH_4) 2,000,000 m^3/annum (gross)
- biogas required for process heating 300,000 m^3/annum
- 33% of the biogas is produced by the R1
- 67% of the biogas is produced by the R2
- hydraulic retention time in the R2 : 8-10 h.
- conversion capacity of the R2 : 10 kg COD/m^3.day
- surplus water (to be discharged) 300 l/t waste ≡ 15,000 m^3/annum
- COD concentration in surplus water 2000 mg COD/l.
- ammonia concentration in surplus water 2000 mg NH_3/l.
- density of chopped tomato-plants 480 kg/m^3
- compost production (after 2 months of additional aerobic after-composting) 8,000 t/annum ≈ 14,000 m^3/annum.

Other assumptions or stated parameters:
- plant operation continuously and the year round
- biogas required for plant operation 15% of the gross production
- chopping costs Dfl. 10/t (crushing will be cheaper, but the apparatus

has to be developed).
- loading and internal transportation by shovels.

Not incorporated in the calculations are:
- external transportation costs
- office, managing and sales costs
- starting up costs
- taxes and profit
- gas purification and transportation
- indirect investments (not within battery limits)

Investments:
- liquefaction reactors (R1's)   Dfl.   1,200,000
- methane reactors (R2's)   Dfl.   400,000
- ancillary equipment   Dfl.   500,000
- erection, piping and instrumentation Dfl.   1,000,000
- property(land), roads, paving etc.   Dfl.   1,000,000
                               Total   Dfl.   4,100,000

Costs:

	Total	Per ton of waste
depreciation and interest	Dfl. 600,000	Dfl. 12
maintenance	Dfl. 200,000	Dfl. 4
chopping	Dfl. 500,000	Dfl. 10
loading, internal transportation	Dfl. 500,000	Dfl. 10
plant personnel and staff	Dfl. 100,000	Dfl. 2
aerobic treatment of surplus water	Dfl. 150,000	Dfl. 3
energy	Dfl. 50,000	Dfl. 1
aerobic after-composting	Dfl. 300,000	Dfl. 6
Total	Dfl. 2,400,000	Dfl. 48

Sales:
biogas (Dfl. 0.33/m^3)	Dfl. 600,000	Dfl. 12
compost (Dfl. 10/t)	Dfl. 80,000	Dfl. 1,6
Total	Dfl. 680,000	Dfl. 13,6

Net costs for 50,000 t/annum wastes, Dfl. 1,720,000, per ton of waste
                                                           Dfl. 34.40

Alternatives for disposal:
- dumping (controlled dumps), per ton of waste Dfl. 25-40
- incineration   , per ton of waste Dfl. 120

   A preliminary study concerning the anaerobic digestion of the organic fraction, mechanically separated from municipal waste by sifting, resulted in similar costs per ton of raw material or calculated for tomato-plant wastes.
   The calculations indicate that anaerobic digestion is definitely cheaper than incineration and tends to be in the same cost range as dumping. Since the dumping costs are expected to rise significantly and since anaerobic digestion has hardly any environmental disadvantage, this way of processing wastes shows good prospects for the future.

4. Indications for improvements and future R & D

   The anaerobic digestion of all kinds of organic wastes by this two-phase process is technically considered to be feasible. The problems we met can be solved by technical means, but they may for certain kinds of raw material lead to additional costs.
   Improvements lowering the costs may be found in the handling of the solids, the development of novel types of reactors R1 and the use of alternative types of reactors R2, already being developed for the treatment

of waste water. For readily digestible raw materials like fruit- and vegetable wastes, an alternative may be found in the single phase slurry process.

More research has to be carried out to develop reliable ways to monitor and controll the process, in order to prevent damage of the methane flora and to maintain a high efficiency.

Finally, more knowledge should be gathered about the microbiology and chemistry of anaerobic processes, about the analytical and structural composition of various kinds of plant material and about the qualities of the compost in respect of its direct effect on plant production and its long term effect on the soil structure.

5. Prospects for the implementation of fullsized processing plants

The promising results of the two-phase anaerobic digestion of solid organic matter has already attracted the attention of research, industry and authorities responsible to energy- and environmental policies.

Due to the steadily rising costs of organic waste disposal and to the increasing environmental awareness, the anaerobic digestion has come up as an alternative for dumping (landfilling) and incineration.

Push to study the possibilities has been exerted already by:
- Firms serving in the transportation and disposal of wastes from horticulture and of municipal waste (the firm of "Zegwaard BV" at Delft joined the research and partially financed the erection of the pilot plant at Delft together with the Netherland Ministry of the Environment and Public Health)
- Industries producing large amounts of perishable wastes like the beet-sugar industry (CSM) and the potato starch industry (AVEBE), who joined a research project together with IBVL on one- and two-stage processing of their wastes.
- The Ministry of Economic Affairs is pushing, through their subsidiary NEOM (Netherlands Energy Development Company) the utilization of wastes for energy and ordered the IBVL to make an inventory of the available wastes from agriculture and agro-industry.
- The same Ministry funded a preliminary investigation by the IBVL of the possibility to produce biogas and compost from the troublesome low grade grass clippings to be cleared for road verges.
- Provincial authorities responsible to waste management are confronted with the problem to find new locations for dumping of municipal waste. In the planning made several years ago generally incinerators are forseen. Now, because of doubt about their environmental implications, a movement is growing to reconsider the plans and to study the possibilities of the novel two-stage anaerobic digestion process. A large engineering firm is engaged in making this analysis.
- A variant of the two-stage process may be a relatively cheap solution for the disposal of surplus tomatoes and similar horticulture produce, causing environmental problems. A research project by the IBVL has been ordered and financed by the Central Bureau of Horticulture Auctions in the Netherlands.

In the Netherlands, but also in the EC, there may be a good future for medium- or large-sized local plants processing various solid organic wastes, possibly together with the organic fraction of municipal waste into biogas and compost.

A preriquisite for this approach will be the existence of a market for the compost; the utilization of the biogas will not be a real problem.

6. Acknowledgement

IBVL wishes to emphasize that the developments described above are the results of research carried out under two contracts, granted by the DG XII-, Solar Energy Programme.

7. References

1. Rijkens, B.A. and Voetberg, J.W. (1981). Energy from Biomass, vol.1, 121-125.
2. Hofenk, G., Rijkens, B.A. and Voetberg, J.W. (1982). Energy from Biomass, vol.3, 232-237.
3. Barry-Concannon, M., Wilkie, A.M., Faherty, G. and Colleran, E.(1982). Energy from Biomass, vol. 3, 224-231.
4. Hamer, G. (1982). Biotechnology and Bioengineering, vol.14, 511-531.

THE FEASIBILITY OF THERMOPHILIC ANAEROBIC DIGESTION FOR

THE GENERATION OF METHANE FROM ORGANIC WASTES

Authors   : W.M.Wiegant & G.Lettinga

Contract number : ESE-R-039

Duration   : 36 months, June 1980 - June 1983

Total budget   : 180,000 units of account (50 % EC-contribution)

Head of project : Dr. G.Lettinga

Contractor   : State Agricultural University

Address   : Salverdaplein 10
                   6701 DB Wageningen
                   Holland

Summary

In this fourth report on the feasibility of thermophilic (45 - 65 °C) anaerobic digestion for the generation of methane from organic wastes the results of experiments with high rate thermophilic digestion systems are presented. The high rate system used is the upflow anaerobic sludge bed (UASB) reactor, and the temperature used is 55 °C. For good performance of this type of reactor, good settling properties of the thermophilic sludge are important. So, first of all the impact of using substrates consisting mainly of volatile fatty acids on granular sludge cultivated on sucrose was studied. No deterioration of the granular sludge was observed. Adaptation to high levels of ammonia was observed to proceed in a way very similar to that reported for mesophilic digestion. Finally the performance of thermophilic UASB-reactors was investigated using practical wastewaters as the substrates. Two wastewaters were used, of which one exerted a clearly disintegrating effect on the granular sludge. Nevertheless, extremely high loading rates were achieved. It is concluded that application in practice is very well possible for wastewaters discharged at temperatures exceeding 50 °C.

1.1 Introduction

Until recently, little attention has been paid to thermophilic (45 - 65 °C) anaerobic digestion processes. Most of the recent reports concern systems without biomass retention. In this project thermophilic digestion of chiefly dissolved wastes is investigated in systems with a high biomass retention. These systems offer the possibility to achieve high methane generation rates and good treatment efficiencies, combined with low reactor volumes. This project deals with (i) the possibility to start and maintain thermophilic high rate systems, (ii) the capacity of these systems in treating various types of wastewaters and (iii) the comparison of thermophilic and mesophilic digestion processes in this respect.

Thermophilic digestion can be started up quite easily by using an inoculum containing mesophilic methanogenic activity, *e.g.* cow manure, digested sewage sludge or sludge from mesophilic high rate reactors. Thermophilic digestion is most satisfactorily started up by bringing the inoculum directly to the desired temperature (1). Slowly increasing the temperature from the mesophilic to the thermophilic temperature range is not profitable, as most methanogens do not survive the increase in temperature (2). Both completely mixed reactors (CSTRs) and reactors with a high biomass retention, like the upflow anaerobic sludge bed (UASB) reactor can easily be started up by bringing the inoculum immediately to thermophilic conditions.

The performance of the UASB-process is greatly enhanced when the biomass is growing in a granular form (3). Granulation proceeds readily with simple sugars like glucose and sucrose as substrates, starting with fresh cow manure as inoculum. Upon granulation high organic loading rates, up to 45 kg COD/m^3.day were achieved with a concomitant high conversion efficiency of over 80 % (4). However, in using volatile fatty acid solutions as substrates little if any granulation occurs and in using a granular mesophilic seed sludge even a disintegration of sludge granules takes place (4).

The results presented in the previous reports concerned the digestion of simple substrates and the effect of the addition of a single toxicant to a dilute substrate. The main objectives were to assess (i) the potential of the thermophilic UASB-process or (ii) the ability of thermophilic digestion processes to withstand high levels of ammonia ($NH_3 + NH_4^+$), which was chosen because it occurs in many wastes and wastewaters in high concentrations.

The aim of the investigations described in this report was first of all to determine whether a substrate consisting mainly of volatile fatty acids has a detrimental effect on granular sludge cultivated on sucrose. More insight in this matter is of great importance, because the nature of most wastewaters is such, that only a relatively small amount of acetogenic biomass can be cultivated. Furthermore research has been devoted to the question whether adaptation to high toxicant levels occurs. Ammonia was used as a relevant toxicant. Finally the performance of the UASB-process was investigated using practical wastewaters as the substrates. Wastewaters which are discharged at high temperatures, *e.g.* those of alcohol distilleries and vegetable canning and potato processing industries, often contain high levels of both COD and toxic compounds. Two of these wastewaters were used for this purpose, one from a potato processing industry and one from an alcohol distillery.

Table I. Results of experiments with thermophilic granular sludge with volatile fatty acids as the main substrates.

feed	acids[a] + sucrose	acetate + yeast extract	
concentration	8.02	1.41	kg COD/m^3
VFA	6.48	1.35	kg COD/m^3
non-VFA	1.54	0.06	kg COD/m^3
hydraulic retention time	5.7	2.1	hrs.
organic loading rate	35	16.0	kg COD/m^3.day
duration	105	31	days
treatment efficiency[b]	84.5	98.9	%
conversion efficiency[c]	82.8	91.5	%
biomass in effluent	0.010	ND[d]	kg COD/m^3

a: the acids were acetic, propionic and butyric in a 1:1:1 (w/v) ratio
b: (1 - filtrated effluent-COD/influent-COD) x 100 %
c: (methane-COD/influent-COD) x 100 %
d: ND, not determined

1.2 Materials and methods

The materials and methods used were the same as described previously (1,4). Substrates consisting mainly of volatile fatty acids and of two wastewaters in varying dilutions were fed to UASB-reactors in which the sludge bed occupied less than 40 % of the reactor volume. The granular seed sludge used in the experiments was cultivated in the way described previously (4). The COD of a wastewater from a potato processing industry amounted 25 g/kg. Approx. 10 g/kg of this COD was in the form of suspended solids. The wastewater was stored anaerobically at 20 °C. The second wastewater was concentrated beet vinasse, a waste product from alcohol distilleries. After dilution to its original volume the COD of the waste was 125 g/kg, the ash content was 3.3 % (w/v) and the Kjehldal nitrogen content was approx. 6.1 kg N/m^3. Sulphate levels were rather low (approx. 0.5 kg/m^3). These values are in good agreement with those found by Robertiello for concentrated beet vinasse, taking the dilution into account (5). This re-diluted concentrated vinasse will be further referred to as "vinasse".

An experiment in which the adaptation to high ammonia levels was studied was carried out in a semi-continuously fed digester with a retention time of 5.5 days. The feed consisted of acetate (9.4 kg COD/m^3) and trace elements and mineral salts as described in (1), with addition of 2.5 kg NH$_4$Cl-N/m^3 and 0.03 kg/m^3 Na$_2$SO$_4$. To this solution 40 ml/l of fresh cow manure was added, which resulted in a VSS concentration in the feed of 0.024 %.

All experiments were carried out at 55 °C.

1.3 Results

experiments with volatile fatty acids as the main substrates

No deterioration of the granular sludge was observed in using volatile fatty acids as the main substrate for prolonged periods of time, although the appearance of the sludge altered somewhat. The sludge granules

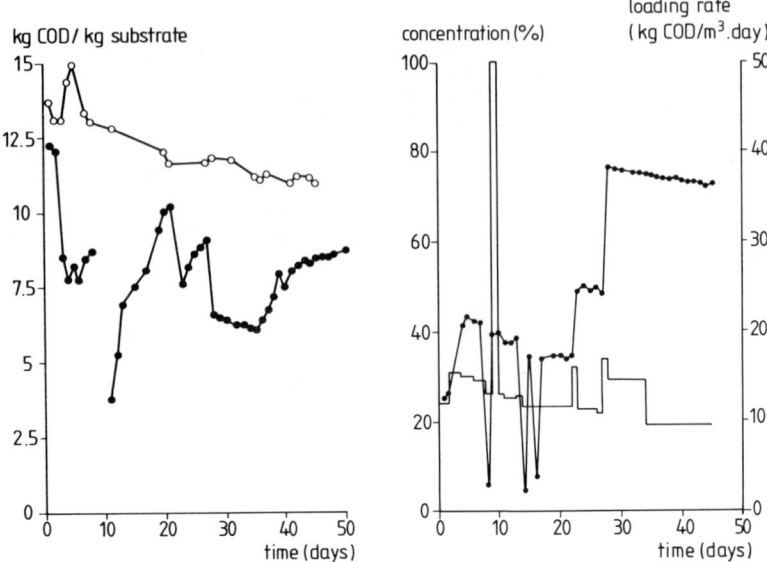

Fig. 1A (left). The amount of methane (●), and CH$_4$-COD + VFA-COD (o), expressed in kg COD/kg undiluted substrate as a function of time during the treatment of wastewater from a potato processing factory.
Fig. 1B (right). Applied organic loading rates (●) and the feed concentrations (given in percentage of undiluted wastewater) as a function of time in the same experiment.

decreased slightly in diameter and their colour changed from yellowish to grey. Relevant characteristics of the performance of the reactors are given in Table I.

experiments with wastewater from a potato processing industry

The main objective of the experiments was to determine the loading capacity of the system treating potato processing waste. The results of the experiments are shown in Fig. 1A and 1B. From the results it is clear that a substantial part of the waste is not degraded. The amount of acidified COD, defined as the sum of methane-COD (evolved and in the effluent) and VFA-COD, is slowly decreasing. This probably is a result of the storage of the substrate. Apparently the methane conversion of the substrate is strongly influenced by the concentration applied as can be seen by comparing Fig. 1A with 1B. In feeding undiluted substrate (day 9) a strong decrease in the methane production is apparent. This effect seems to be reversible. An increase in the loading rate to 38 kg COD/m^3.day (day 29) leads to a sharp decrease in the conversion to methane at a substrate concentration of 29 %. A slight further decrease in the methane conversion occurs thereafter. The methane conversion recovered distinctly when a 19 % substrate concentration was applied (from day 35 onwards).

experiments with vinasse as feed

The experiments were made with a granular thermophilic seed sludge cultivated on sucrose as substrate. Much care was paid to the adaptation of the

Table II. The process performance of a thermophilic UASB-reactor treating vinasse.

loading rate	concentration (% dilution)	treatment[a] efficiency	$\dfrac{\text{VFA-COD}}{\text{COD}_{in}}$	potential treatment efficiency
kg COD/m^3.d	%	%	%	%
17.2	21.0	61.9	21.2	83.1
25.6	14.8	59.0	19.7	78.7
25.5	19.6	64.9	17.4	82.3
24.8	22.3	56.8	20.1	76.9
25.3	24.9	52.6	27.1	79.7
38.6	23.6	58.8	23.5	82.3
83.6	14.6	59.6	22.5	82.1
98.3[c]	12.9	58.9		

a: (1 - filtrated effluent-COD/influent-COD) x 100 %
b: (1 - (CODfilt.eff. - VFA-COD/influent-COD) x 100 %
c: at this loading rate sludge washout occurred within a few hours after the loading rate was raised to this value.

seed sludge to the vinasse substrate, *i.e.* the substrate was gradually shifted from sucrose (approx. 2 %) to vinasse (20 % w/v). The methane generation rate was kept constant during this phase. During the shift of substrate the granular appearance of the sludge was gradually lost, *i.e.* the diameter of the granules decreased from 1-3 mm to less than 1 mm. The results of this experiment will be described in detail elsewhere (6). After completion of the adaptation the system was investigated for its loading potential and for the effect of different vinasse concentrations. The results are summarized in Table II. As with the potato processing wastewater, the methane conversion is strongly dependent on the concentration of vinasse in the feed. This is clearly shown in Fig. 2.

adaptation to high ammonia levels

In a semi-continuously fed digester with 9.4 kg COD/m^3 of acetate and fresh cow manure (0.024 % VSS) as feed, the $NH_3 + NH_4^+$ concentration was raised in one step from 2.5 to 4.0 kg N/m^3. The pH was maintained at 7.3 to 7.5. As appears from the results shown in Fig. 3 the adaptation proceeds quite slowly.

1.4 Discussion

In accordance with the preceding reports this report deals with the experimental results of investigations concerning the assessment of the feasibility of thermophilic anaerobic digestion for the generation of methane from organic wastes (1,2,3). Part of the results obtained will be described in separate publications (6,7). Now, after three years of research, the potential of thermophilic digestion will be evaluated on the basis of the presently available insight.

Thermophilic digestion as a process for the generation of methane has found little application throughout the world. For many years the process has been considered as being too sensitive to slight environmental changes

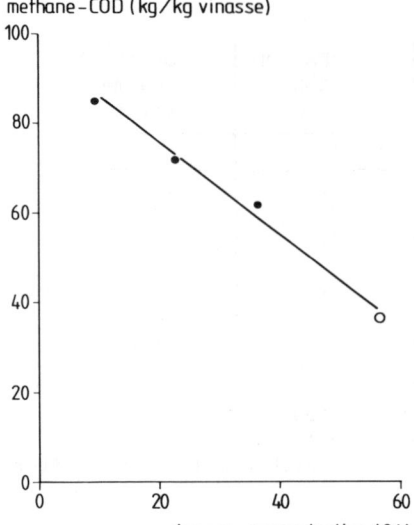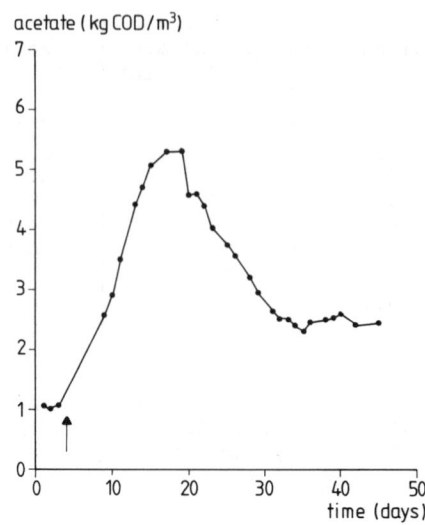

Fig. 2 (left). Methane production in relation to the vinasse concentration as determined in completely mixed reactors without biomass retention at 55 °C. Retention times were 18.2 days (●) and 33.3 days (o).

Fig. 3 (right). Adaptation of a reactor digesting acetate (9.4 kg COD/m^3) and diluted manure (0.024 % VSS) to an instantaneous increase in the ammonia concentration from 2.5 to 4.0 kg N/m^3. The moment of the increase is indicated with an arrow. In the figure the effluent-concentrations of acetate are given.

such as changes in pH and temperature to make the process attractive for practical application. Moreover, the process is very susceptible to a number of compounds and in addition the start up is considered as being difficult and time consuming. Garber et al. (8) described elaborate procedures to start a thermophilic sewage digestion which lasted a period of time exceeding one year ! However, in agreement with results of Varel et al. (9) and our investigations, the start up of thermophilic processes can be made fairly rapidly and in a simple way using cow manure and digested sewage sludge as seed material.

Since we chose the UASB-reactor as a high rate digestion system the experiments initially were focussed on the phenomenon of the sludge granulation process, because achieving a high sludge settleability is a prerequisite for the proper performance of the UASB-process. We observed that the sludge granulation proceeds slowly if at all in the case of VFA-substrates. The reason for this poor pelletization of the sludge is not yet fully clear. However, in using more complex soluble substrates such as sugar solutions a rapid and very satisfactory pelletization of the sludge occurs. The granular sludge cultivated in this way does not deteriorate on feeding it with acidified substrates. The sludge remains in a granular form.

A high biomass retention is a prerequisite for all high rate anaerobic digestion systems. We observed a distinct disintegration of the granular sludge in our experiments with vinasse, but since the sludge retained a satisfactory settleability, still extremely high organic loading rates as well as methane generation rates were achieved. The COD removed per unit of reactor volume is higher than any value reported for mesophilic UASB-processes. These results conflict more or less with those of Kennedy and van den Berg with the downflow stationary fixed film reactor, where little if any better performance was found in the thermophilic as compared to the mesophilic temperature range (10).

In view of the higher growth rates of acidifying bacteria (11), acetate consuming methanogens (12) and hydrogen consuming methanogens (13), as well as propionate degrading bacteria (2,14), a proportionally higher activity per unit of biomass should exist under thermophilic conditions. This indeed is true for the acidifying bacteria (11). So from a theoretical point of view thermophilic digestion leads *per se* to higher activities, and should lead to higher applicable loading rates than mesophilic digestion. The highest loading rates reported for sucrose-fed thermophilic high rate systems are *ca.* 45 kg COD/m^3.day at 80 % treatment efficiency (3,15), while maximum loading rates in mesophilic UASB-reactors at 30 ^oC were found to be 25 kg COD/m^3.day on sugars as substrate (K.van Straten *pers. comm.*). From the results presented in this paper it can be concluded that very high methane generation rates can be achieved with soluble wastes of the type investigated. Moreover, there is a high COD-treatment efficiency and a high methane conversion, making the application of thermophilic digestion processes very attractive in practice.

On the other hand still much work remains to be done, especially with respect to the performance of thermophilic digestion processes under less favourable conditions. In our investigations the attention was focussed on ammonia as toxicant because of its abundance in wastewaters of many kinds and origins. Particularly as far as the effect of ammonia is concerned, there still are many kinetic enigmas to be solved. For the digestion of livestock wastes thermophilic digestion in many cases has little if any benefits over mesophilic digestion, at least under West-European conditions, because here the ammonia levels in the wastes usually exceed 3.0 kg N/m^3. The results of this part of the investigations will be presented elsewhere (7).

The adaptation to high ammonia levels, as shown in Fig. 3, proceeds in a way very similar to that described recently for mesophilic systems (16). Under mesophilic conditions very complex interactions may occur between toxic cations (17). Presumably, this applies under thermophilic conditions for other toxicants as well. This leads to the conclusion that both wastewaters chosen in this study are too toxic to be handled undiluted. This is clearly shown in Figs. 1 and 2. From results obtained in our laboratory it is clear that this is also the case in mesophilic conditions. The inhibitory effect is first of all reflected in high levels of propionate in the effluent. As far as the breakdown of propionate is concerned thermophilic digestion presumably is more sensitive than mesophilic digestion. For this reason the breakdown of propionate is presently more exclusively being investigated.

Once it is possible to degrade this intermediate compound under severe stress conditions, the thermophilic process is open to wide scale application for a variety of wastewaters and wastes.

2. Conclusions

Thermophilic digestion for the generation of methane is set up very easily. Very promising results were obtained with a variety of substrates in thermophilic UASB-reactors. These results indicate that the methane generation rates as well as the COD removal rates are far higher in thermophilic tahn in mesophilic high rate systems.
Possibilities to obtain even better performance are still under investigation, e.g. a two step methane digestion process seems to enhance the propionate breakdown greatly.
Application in practice seems to be recommendable for any organic soluble waste discharged at temperatures exceeding 50 °C, although dilution may be necessary for high strength wastewaters containing high levels of toxic compounds. As in mesophilic digestion adaptation to high levels of toxicants will occur in many cases.
However, it must be emphasised that full-scale application of thermophilic anaerobic digestion awaits the implementation of a comprehensive programme of experiments on a larger than laboratory scale.

References

1. WIEGANT,W.M. & G.LETTINGA (1981). Solar Energy R & D in the E.C. series E vol 1:126-130. D. Reidel Publishing Company
2. LETTINGA,G. (1980). Rep.Eur. 7160 EN:207-211.
3. LETTINGA,G. et al. (1980). Biotech.&Bioeng. 22:699-734.
4. WIEGANT,W.M. & G.LETTINGA (1982). Solar Energy R & D in the E.C. series E vol 3:238-244.
5. ROBERTIELLO,A. (1982). Agric.Wastes 4:387-395.
6. WIEGANT,W.M. & J.A.CLAASSEN. in prep.
7. ZEEMAN,G.; W.M.WIEGANT & M.E.TREFFERS. in prep.
8. GARBER,W.F. et al.(1975). J.Wat.Poll.Contr.Fed. 47:950-961.
9. VAREL,V.H.; H.R.ISAACSON & M.P.BRYANT (1977). Appl.Environ.Microbiol. 33:298-307.
10. KENNEDY,K.J. & L.van den BERG (1982). Biotechn.Letters 4:171-176.
11. ZOETEMEYER,R.J. et al. (1982). Wat.Res. 16:313-321.
12. ZINDER,S.H. & R.A.MAH (1979). Appl.Environ.Microbiol. 38:996-1008.
13. ZEIKUS,J.G. & R.S.WOLFE (1972). J.Bacteriol. 109:707-713.
14. BOONE,D.R. & M.P.BRYANT (1980). Appl.Environ.Microbiol. 40:626-632.
15. SCHRAA,G. & W.J.JEWELL (1982). Paper pres. at 55th ann.conf. Wat.Poll. Contr.Fed. Oct.1982, Missouri,USA.
16. PARKIN,G.F. & R.E.SPEECE (1982). J.Environ.Eng.Div. 108:515-531.
17. KUGELMAN,I.J. & K.K.CHIN (1971). In: Anaerobic biological treatment processes, ed. R.F.Gould. Am.Chem.Soc.Adv.Chem.Series 105:55-87.

ALCOHOLIC FERMENTATION : IMPROVEMENT OF THE TECHNOLOGY BASED ON PHYSIOLOGICAL PHENOMENA

Contract number : ESE R 044 F

Duration : 24 months

Total budget : 505 920 FF   Hep 42 700 ECU

Pr. G. GOMA

Dr J.L. URIBELARREA, Dr P. STREHAIANO, Dr J.L. BOVEE (LAAS),

I. LAUDRIN, M. MOTA, M.O. LORET

With the technical assistance of Mr BRUCHE, C. ROQUE

Contractor: Département de Génie Biochimique et Alimentaire, ERA CNRS 879
Prof. G. DURAND
Institut National des Sciences Appliquées,
Adress   Avenue de Rangueil, F-31077 TOULOUSE CEDEX

SUMMARY

The laboratory studies and compares several routes for energy and chemical feedstock from biomass (methane from cow dung, acetone butanol, ethanol, butanediol...). The followed methodology integrates a knowledge of the dynamical behaviour in order to know the mechanism of limiting step and to find the optimal technology adapted to the problem. On ethanol fermentation, the bottleneck are the inhibition phenomena which limit the ethanol production and concentration.

We have demonstrated that :
- there is a synergic effect between ethanol and sugar inhibition
- there is a dual effect of ethanol inhibition.

Two hypothesis of explanation are developped : i) ethanol accumulates inside the cell and ii) there is a co-metabolite more inhibitor than ethanol which is excreted (under identification).

- cell bound water is decreased when inhibition occurs.

Technologically we find :

- additives to enhance cellulose degradation
- a method for fermentation in solid state jerusalem artichoke
- an improvement in ethanol production by immobilized cells
- a new process of fermentation : extractive fermentation coupling ethanol extraction by solvent and fermentation in the same reactor.

We are under investigations of process control of continuous culture and fermentation with high cell density (by floculation and cell recycling).

I.- INTRODUCTION

Classicaly it is admitted that ethanol is the end product which inhibits the cell growth and fermentation ; we demonstrated that added alcohol has dynamically a poor inhibitor effect but, at the critical concentration P_{crit} it destroys the cells.

Our investigations are now oriented in several directions :
- co-inhibition sugar-ethanol on several yeasts
- co-metabolite inhibition
- mathematical modelling for quantification of inhibition phenomena and control of reactors
- identification of the interest of Zymomonas mobilis in purpose of improving the technology.

II.- ON THE CO-INHIBITION SUGAR-ETHANOL

We find an enhancement of inhibition by ethanol when we have high sugar concentrations so dynamic inhibition by ethanol increases. This study is performed with Saccharomyces cerevisiae, S. sake, S. bayanus.

III.- CO-METABOLITE INHIBITION

We find a strong inhibitory effect due to the inoculum size and the maximal growth rate is higly depressed.

As alcohol added at the beginning of the fermentation is less inhibitory than the produced ethanol and as in any case the alcohol level due to inoculum is very low (5.6 g/l for a 60 % inoculum) our results suggest that ethanol is not the only terminal inhibitor product and that there may be other inhibitors carried by inoculum and acting at the level of yeast growth which is inhibited earlier than fermentative activity.

Localisation, characterisation and mechanism of action of this co-metabolite are now under investigations.

IV.- CHANGES IN THE CELL VOLUME WITH INHIBITIONS

We developpe a method to measure the water content of cells by the analysis of the kinetics of cells drying.

We found a decrease of bound water with inhibition : when fermentation starts cells contain 2.8 g/g dry weight under inhibition cells contain ($\mu = 0.25\ \mu_m$, $P = 15$ g/l) 1.2 g/g. d.w.

There is a closed relation between growth rate, cell volume and the ratio unsaturated fatty acid/saturated fatty acid which decreases when inhibition occurs.

Cell composition in relation with O_2, progress of reaction, environmental factor are under investigations.

V.- MATHEMATICAL MODELLING

In a closed cooperation with the research team of the "Laboratoire d'Automatique et d'Analyse des Systèmes" (LAAS) Pr. SEVELY, modelling of ethanol fermentation is investigated .

A "Biological model" gives good results.

$$\mu = \mu_m \frac{S}{K_S + S + \frac{S^2}{K_{is}}} \cdot \frac{K_{ip}}{K_{ip} + P} \cdot (1 \cdot \frac{P}{P_{crit}})$$

(study by DOURADO, LAAS).

A more empirical approach due to the necessities of process control needs a mathematical model taking count of the ethanol and sugar concentration ; the relations used are :

$r_s = k \; S^\alpha \; P^\beta$

$P = \beta (S_0 - S) + P_0$

$r_s = k \; S^\alpha \; (\eta \; (S_0-S))^\beta$

identification on line of this model is performed. Good agreement between experimental results and calculated values is observed (study by BOVEE, LAAS).

VI.- TECHNOLOGICAL STUDIES

VI-1.- Process control of multistep reactor for alcohol production (collaboration LAAS).

Mathematical modelling of an unit of 6 reactors in cascade is established. The approach of optimization is under investigation but the modelisation, identification automatic control and data logging by computer system connected with a minicomputer by a telecommunication line are operating now.

VI-2.- Ethanol production by immobilized cells (both yeasts and Zymomonas mobilis)

We use for <u>Saccharomyces cerevisiae</u> a multistep reactor to avoid the CO_2 retention but an one step reactor for <u>Zymomonas mobilis</u>.

The better results are summarized in the following table :

	Productivity kg/m³	Yield kg/kg	Ethanol % V/V	Support
S. cerevisiae	4.8	0.48	17.5	Brick
Zymomonas mobilis	60 7.9	0.50 0.49	7.5 12.1	Flocor ICI Cordierite

Ethanol production by immobilized cells.

This process, very simple has good performances.

VI-3.- Recycled cells

A work on flocculated cells with the systematic study of pH, salt concentration effect on yeast flocculability and charge of cells is under investigations since february.

CONCLUSIONS

As last years was mainly dedicated

1. to explain mechanisms of fermentation we found new phenomena and we hope complete the work on the "auto-killer effect" which is very promising both for basic knowledge and to find new technological route for ethanol production.

2. to develop technological concepts
- ethanol production with immobilized cells
- extractive fermentation
- strategy of fermentation by computer control

For the future we need to finish the structure identification and mechanism of action of the co-metabolite inhibitor. We have also to developpe the cell recycling and promote the coupling of ultrafiltration and fermentation and flocculating devices.

These studies on ethanol fermentation are appliable to acetonobutylic, acidogenic fermentations for liquid fuel production. They prepare news technologies for biomass utilization by biotechnology route.

LITERATURE

NAVARRO J.M., DURAND G., (1980). Modifications de la croissance de Saccharomyces uvarum par immobilization sur supports solides. C.R.Acad. Sci., 290, série D, 453-456.

NAVARRO J.M. (1980). Fermentation alcoolique : influence des conditions de croissance sur l'inhibition par l'éthanol. Cellular Mol.Biol., 26, 241-246.

THIBAULT P., MONSAN P., JOURET C. (1981). Influence de l'éthanol sur h'activité et la stabilitié de la peroxydase immobilisée et en solution. Sciences de l'Aliment, 1, 55-66.

NAVARRO J.M., DURAND G. (1981). Immobilisation de Saccharomyces uvarum sur des billes de silice par liaison covalente. Sciences des Aliments 1, 529-540.

NAVARRO J.M., DURAND G. (1981). Synchronisation de la croissance de microorganismes immobilisés sur des supports solides. Ann.Microbiol. (Inst. Pasteur), 132 B, 241-255.

NAVARRO J.M. (1981). Immobilisation de Saccharomyces uvarum par adsorption sur des granulés de brique. Sciences des Aliments, 1, 514-528.

RYU Y.W., NAVARRO J.M., DURAND G. (1981). Production d'éthanol dans un réacteur à cellules immobilisées. Coll.Soc.Fr.Microbiol., Reims, 281-300.

LAUDRIN I., GOMA G., (1982).- Ethanol production by Zymomonas mobilis : effect of temperature on cell growth, ethanol production and intracellular ethanol accumulation. Biotechnol.Lett., 4, 537-542.

MINIER M., GOMA G., (1982), Ethanol production by extractive fermentation. Biotechnol.Bioeng., 24, 1565-1579.

GOMA G., STREHAIANO P., URIBELARREA J.L., MOTA M., DURAND G., (1982).
Kinetic considerations on ethanol production. Proc.Second EC Conf.
Energy Biomass, Berlin, 1027-1032.

STREHAIANO P., MOTA M., GOMA G., (1983).- Effect of inoculum level on kinetics of alcoholic fermentations. Biotechnol.Lett., 2, 135-140.

GOMA G., STREHAIANO P., Effect of initial substrate concentration on two wine yeasts : relation between glucose sensitivity and ethanol inhibition, Amer.J.Oenol.Vitic, 34, 1-5. (1983)

STREHAIANO P., BOVEE J.P., SEVELY Y., GOMA G., (1983)., Mathematical modelling of ethanol fermentation (Biotechnol.Bioeng.).

GOMA G., Rapport : Réflexions sur la contribution de la biomasse dans le domaine des énergies renouvelables en Midi-Pyrénées. Publication de l'E.P.R. de Midi-Pyrénées.

SYNOPSIS OF THE RESULTS

1. PHYSIOLOGY - KINETICS

    1.1.- Ethanol "produced" and "added" have not the same effects.
    1.2.- There is a co-metabolite inhibitor
    1.3.- Inhibited cells has a low content
            - of unsaturated fatty acids
            - of structural water
    1.4.- Cells can be considered has microreactors limited by their ability to excrete ethanol
    1.5.- There is a synergy of inhibition by product and substrate
    1.6.- High temperature increases the inhibition phenomenon
    1.7.- Two mathematical models are used for describing the alcoholic fermentor.

2. TECHNOLOGY

    2.1.- Continuous culture with immobilized cells are very efficient
            - with yeast (Saccharomyces cerevisiae)
            - with bacteria (Zymomonas mobilis)
    2.2.- A new process by extractive fermentation is described
    2.3.- Process control of continuous culture by computer control is improved
    2.4.- Additives to enhance
            - ethanol tolerance are described
            - cellulose hydrolysis are found
    2.5.- Ethanol production from Jerusalem artichoke is possible directly by fermentation the tuber.

ENZYMATIC SACCHARIFICATION OF NATIVE CELLULOSE :

EFFECT OF PRODUCT INHIBITION AND BIOMASS PRETREATMENT

Authors : ALFANI, F., CANTARELLA, M., SCARDI, V.

Contract number : ESE-R-041-I

Duration : 36 months   1 July - 30 August 1983

Total budget : 306.000.000 lire   CEC Contribution : 100.000.000 lire
                  (60%)   (60%)

Head of project : prof. Francesco Alfani

Address : Istituto di Principi di Ingegneria Chimica
          p.le V. Tecchio
          I - 80125 Naples
          Italy

Summary

Cellulose conversion to glucose is strongly inhibited by product. However, above 1 g/l the degree of saccharification becomes almost independent on glucose concentration. In a ultrafiltration membrane reactor the final product can be continuously removed from the reaction medium and large β-glucosidase deactivation can be prevented.

For the saccharification of cellulose in raw materials, biomass pretreatment is necessary in order to reach satisfactory reaction yield. A acid pretreatment at 80°C with H_2SO_4 reduces the degree of polymerization of the cellulose while alkaline pretreatment at 90°C breakdown the lignin barrier. The result of a combined first acid and alkaline pretreatment is a good conversion and a high selectivity to glucose. Cellulose conversion also depends on the chemical concentration in the pretreatment reaction. Optimum NaOH concentration is 0.8N, above this value no further improvement is achieved in reaction yield. The degree of saccharification is also dependent on the enzyme/biomass ratio. Maximum conversion and minimum enzyme consumption is achieved for a ratio equal 1 : 170.

1. INTRODUCTION

This research project was aimed to investigate both the possibilities to perform enzimatically cellulose saccharification in a ultrafiltration (UF) membrane reactor and to hydrolize a lignocellulosic agricultural waste, namely olive husks. This raw material is very abundant in the Mediterranean regions of European Community and its cellulosic content is significant.

It is known that many constrains, namely cellulose accessibility, enzyme recovery, thermal enzyme stability and product inhibition, limit the enzymatic hydrolysis of native lignocellulosic materials. Consequently in order to overcome this difficulties new reactor configurations were studied. Indeed, through the experimental research, UF membrane reactors were found to be helpful for achieving complete recovery of soluble enzyme, thus allowing the possibility of biocatalyst reuse and a marked reduction of process cost. Moreover these reactors also represent valuable tools for probing the hydrolysis mechanism and for preventing the inhibitory effect of the final products on the cellulose saccharification.

The results presented in the previous Contractor Meeting (1-2), contributed to a better knowledge of the synergistic attack of endo- and esoglucanase on amorphous and crystalline cellulose and of thermal deactivation of each component of cellulase complexes from different microorganisms. Inhibitory effect of glucose on the β-glucosidase component and the effect of biomass preatreatment on the yield of saccharification were briefly mentioned. Recently these two topics have been estensively studied and the results of a experimental and theoretical analysis are discussed in the following sections.

2. RESULTS AND DISCUSSION

2.1. Inhibitory effect of glucose on the β-glucosidase

Enzyme activity can be curtailed by the presence of various selective agents in the reaction medium and in some cases even the reaction product can act as inhibitor. In cellulose saccharification the conversion of an intermediate product, cellobiose, to the final product, glucose, is limited by glucose inhibition on the catalytic action of β-glucosidase. Even small concentration of glucose, less than 0.4 g/l, drastically reduce enzyme activity as shown in Fig.1. The extent of biocatalyst deactivation is sensitive to inhibitor but becomes almost independent of glucose at higher values (above 1 g/l). Hence it is important to define the type of inhibition in order to study the optimum reactor configuration for preventing enzyme deactivation.

Experimental evidences reported in Literature (3,4) indicate for glucose inhibition either a competitive or non competitive mechanism. Indeed for enzymes from various sources different inhibition mechanisms can be expected, but it is surprising that opposite results have been obtained for cellulose complexes from the same strain. These contradictory evidences have to be related to the experimental procedures adopted by the different authors. In fact for the identification of inhibition mechanism and constant evaluation a wide range of concentration has to be explored. When product inhibition is concerned, specific reaction rate and inhibitor concentration are strictly related. A batch reactor configuration, which is generally adopted, is not suitable for performing the runs at constant concentration, unless to work with a differential reactor at very low conversion, and adding at the beginning of runs different amounts of glucose in the reaction medium. However, uncertainties in the numerical evaluation of the reaction rate arise, since this latter is evaluated by the differen ce between instantaneous and initial product concentration which is generally small compared to a standard at zero time. In the present study efforts were made to define a mathematical model which describes the behaviour of an integral reactor thus allowing to overcome the experimental difficulties encountered with the use of a differential reactor.

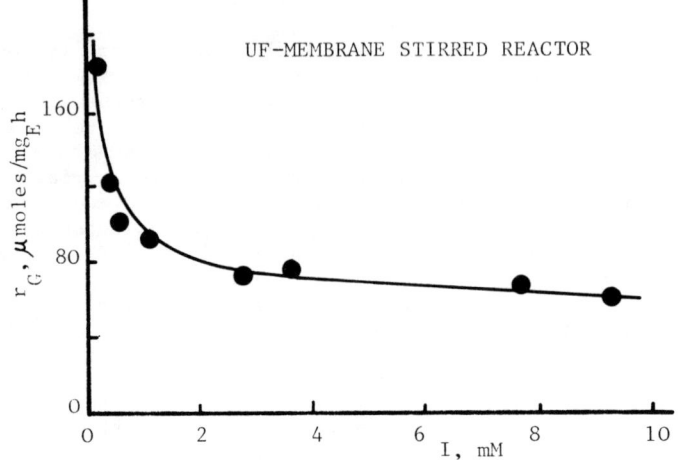

Fig. 1. - Reaction rate vs. glucose concentration

The theoretical analysis was also extended to a completely stirred ultrafiltration membrane reactor. In this reactor the product can be continuously removed so that a constant concentration level can be kept in the reactor. Moreover, by varying enzyme concentration and residence time, it is possible to select the inhibitor levels in the cell and to test the extent of product inhibition in a wide range of concentration.

The numerical results of the model analysis for the two reactor configurations are reported in Fig. 2 and Fig. 3. and the final equations have

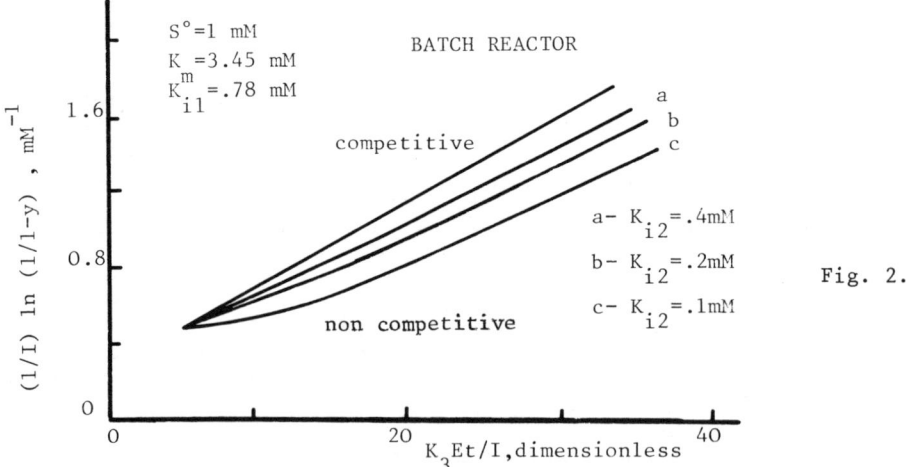

Fig. 2.

worked out in order to :
- easily and without doubts identify completely the inhibition mechanism
- evaluate the kinetic constants of the inhibition rate equation.

In a batch reactor reactor, Fig. 2, the distinction between the two inhibition mechanism is difficult, specially when data are available in a limited range of substrate conversion. Since the non linear behaviour characteristic of a non competitive mechanism can be confused with a linear behaviour typical of a competitive mechanism discrepancies can be encountered in the Literature. On the other hand the two proposed mechanisms give rise, for a UF-membrane reactor, to completely different behaviours. Therefore a plot, like that of Fig. 3, allows very easily the mechanism identification.

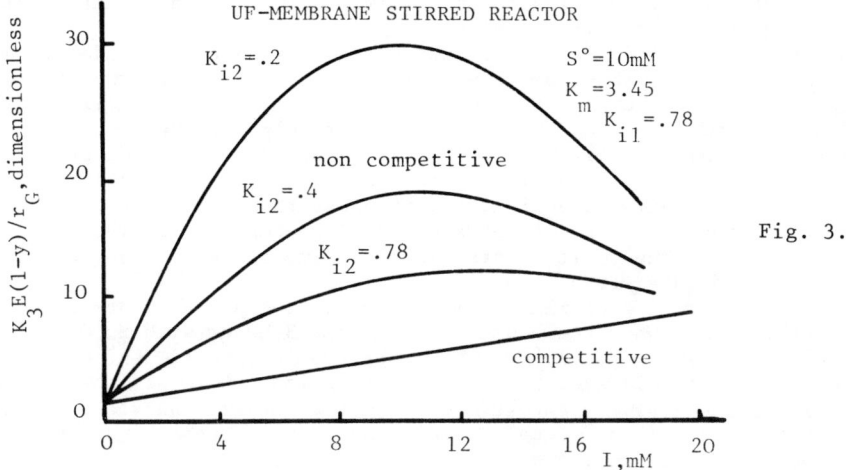

Fig. 3.

For cellulose saccharification experimental data were worked out according to the mathematical model and the results plotted in Fig. 4 clearly indicate that β-glucosidase is inhibited by glucose according to a uncompetitive mechanism. A quadratic regression of data allowed to determine the value of Michaelis-Menten constant K_m and those of the two inhibitor constants K_{i1} and K_{i2} : K_m = 0.59 mM; K_{i1} = 0.33 mM; K_{i2} = 1.19 mM. Being K_{i1} and K_{i2} values close, the differential reactor method can lead into error since the difference between the two mechanisms in a batch reactor becomes marked only when K_{i1} is much greater than k_{i2}.

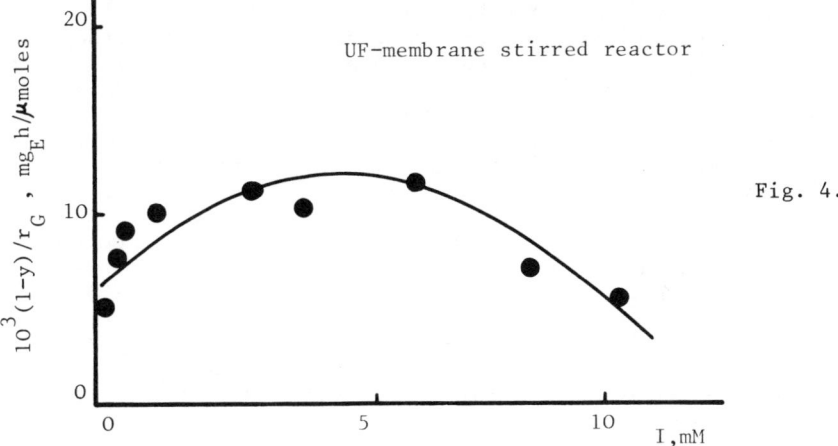

Fig. 4.

2.2. Enzymatic saccharification of native cellulose

Native cellulose can be generally poorly hydrolyzed since lignin barrier limits the enzymatic attack. In reactions, lasting 40 hours, and performed with pellets of olive husks a maximum conversion of the cellulose, equal to 14%, was reached, confirming these previous observations. The reaction conditions were the following: reaction volume 100 ml, biomass (48 mesh fraction) 500 mg, total enzyme amount 3 mg, temperature 45°C, 50 mM Sodium acetate buffer pH 4.8. The enzyme was a mixture of 2 mg of T.viride (Marshall-Miles) and 1 mg of A. niger (Sigma), which has been proved (5) to be a preparation characterized by a well balanced activity of enzyme components for the saccharification of native cellulose. This mixture also gives the highest selectivity to glucose.

This yield is of course unsatisfactory and different combined acid and basic pretreatments of the biomass were tested in order to improve cellulose accessibility and increase the conversion to glucose. Since olive husks are 55% w woody material and the remaining part pulp residue, coming from olive-stone and the external green pulp, although the native material is already available in small particle size 0.3 - 0.5 mm roughly after the industrial process of oil extraction, nevertheless they were ground for one hour and then sieved in order to obtain a homogeneous sample. The two fractions, 48 and 30 meshes, with characteristic dimensions of 0.58 and 0.30 mm were tested in this study.

Acid pretreatment determines a marked reduction of cellulose degree of polymerization (DP). When the biomass reacts for 3 hours with H_2SO_4, biomass : acid ratio 1 : 1, at 80°C the DP drops from 640 to 190 which clearly indicates a partial acid hydrolysis of the cellulose. However, when a biomass sample is successively exposed to enzymatic attack, the yield of saccharification is still low, equal to 18%, and therefore the lignin barrier has been not yet distroyed.

On the other hand alkaline pretreatment with NaOH 0.8 N for 1 hour at 90°C, reaction volume 100 ml and 2 g of biomass, is sufficient to remove the lignin barrier and the cellulose conversion reaches 52%. However the cellulose DP remains unchanged as compared to that of native material and the enzymatic saccharification of the sample is not complete as confirmed by the high percentage 59% of oligosaccharides in the products.

If the biomass first reacts with acid and then is pretreated in alkaline medium, the yield of enzymatic saccharification does not increase but glucose percentage reaches approximatively 90%. This result confirmes that both acid and alkaline pretreatment are necessary.

The enzyme and biomass pretreatment weigh heavily on the process budget. It is deemed that, if enzyme cost exceeds 30% of total costs and pretreatment is over 10 $ per ton of biomass, the enzymatic saccharification of the cellulose cannot be performed economically.

Fig. 5. is a plot of cellulose conversion in a UF-membrane reactor as function of enzyme concentration in the reactor. The maximum yield got after 48 hours of reaction is 56% but, as shown in the diagram, this level cannot be reached economically. In fact a quite large amount of enzyme is required, while the use of half concentration in the reaction medium 3 mg instead of 6 mg determines only a small variation, from 56% to 50%, in the yield of saccharification. Therefore, optimum enzyme concentration, high conversion per mg of enzyme, is 0.043 mg/ml which corresponds to a enzyme : biomass ratio of 1 : 170 roughly.

A different set of runs was performed to investigate the dependency of cellulose saccharification on the NaOH concentration in the reactor during the pretreatment stage. The diagram in Fig. 6. shows that the conversion increases with increasing NaOH concentration for values less than 0.8N. At higher values the yield of saccharification becomes independent

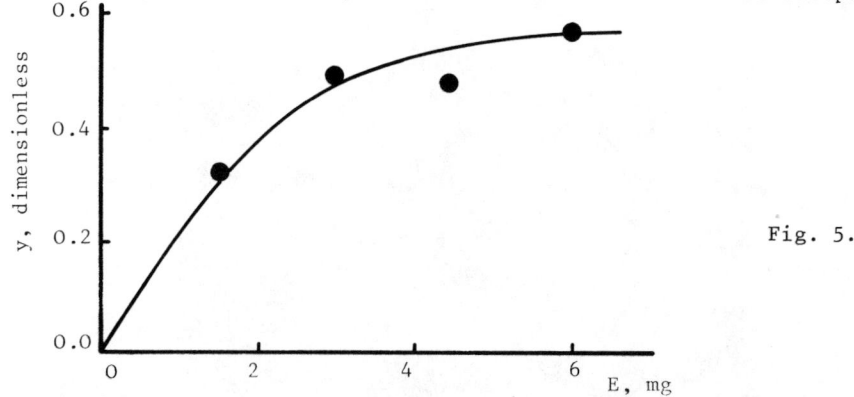

Fig. 5.

of alkaline concentration and the result is a waste of chemicals. Experiments have been also run at constant NaOH concentration but varying alkaline/ biomass weight ratio. The results of enzymatic hydrolysis in a membra-

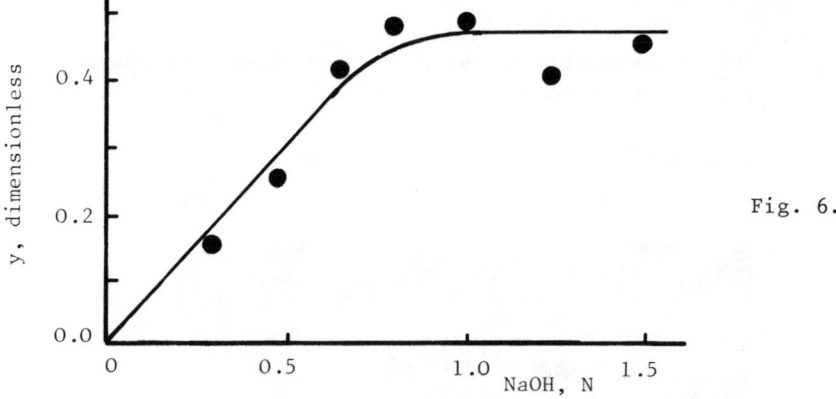

Fig. 6.

ne reactor, at controlled temperature, pH and product inhibition, indicated that, in a range between 0.23 and 0.92 g of NaOH per gram of biomass, cellulose conversion remains constant at the maximum level. This evidence seems to support the thesis that NaOH carries on a catalytic activity in the breakdown of the lignin wall which covers the cellulose.

The chemicals attack of NaOH on the lignin in the raw material is also shown by 90X stereoscopic microscope photographs relative to samples subjected to different pretreatments. When the biomass is ground and kept in acid medium the shape of pellet is well defined, the lignin barrier is entire and the presence of cellulose is not displayed by specific reagent Chlorine Iodide of Zinc. On the other hand for the materials, subjected to alkaline attack, the accessibility of cellulose fibers is evident. The cell wall is very thin and in many cases is even brokendown. The cellulose fibers from different pellets tends to aggregate and become accessible to the enzymes. Figs. 7a and 7b refer to the 48 mesh fraction of the raw materials acid pretreated and the one subjected to a combined acid and alkaline pretreatment.

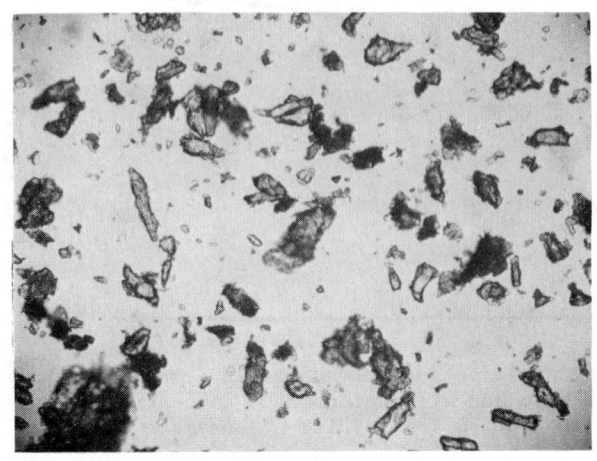

Fig.7a- 90X stereoscopic microscope photograph of olive husks pretreated in acid medium

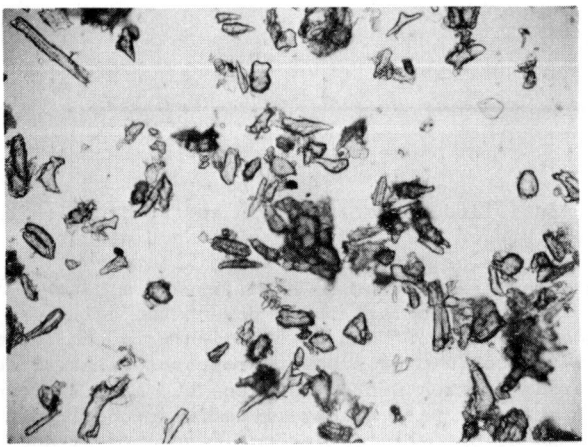

Fig.7b- 90X stereoscopic microscope photograph of olive husks pretreated first in acid medium and in alkaline medium

3. CONCLUSIONS

These conclusions are reached not only on the basis of results presented in these communication but also with regard to the ones discussed in previous reports. The mechanism of enzymatic hydrolysis of cellulose can be schematically depicted by two series reactions. In the first step amorphous and crystalline cellulose regions are hydrolyzed through the synergistic action of endo- and eso-β-glucanases, in a second step the intermediate product, cellobiose is converted to glucose. This reaction is catalyzed by the β-glucosidase which is strongly inhibited by glucose and is the less thermal stable component in cellulase preparations. Further studies could be significant to verify if immobilized β-glucosidases are more resistant to thermal denaturation and product inhibition. The cellulose in native lignocellulosic materials can be satisfactory converted if the raw materials are pretreated. Acid pretreatment reduces the DP of cellulose and determines high selectivity to glucose, while alkaline pretreatment distroys the lignin barrier and increases cellulose accessibility to enzyme attack.

Optimum reagent to biomass ratio in the pretreatment and enzyme to cellulose ratio in the reactor have been identified in order to minimize process costs. Membrane reactor configuration are helpful to overcome many constrains, which limit enzymatic saccharification of cellulose. An investigation on a pilot plant scale reactor would be advisable to focus process technology.

4. NOMENCLATURE

E = enzyme, mg
K_{i1}, K_{i2} = inhibition constant, mM
K_m = Michaelis constant, mM
K_3 = kinetic constant, mM/mg h
r_G = rate of glucose production, μmoles/mg$_E$ h
$S°$ = initial substrate concentration, mM
t = reaction time, h
T = temperature, °C
y = degree of conversion, dimensionless

5 BIBLIOGRAPHY

1 Alfani F., Cantarella M., Scardi V., Energy from Biomass, Series E, 1, 139, (1981), D. Reidel Publishing Company, Dordrecht, Holland.

2 Alfani F., Cantarella M., Scardi V., Energy from Biomass, Series E, 3, 253, (1982), D. Reidel Publishing Company, Dordrecht, Holland.

3 Ladisch M.R., Song C.S., Tsao G.S., Dev. In. Microbiol., 18, (1977)

4 Vernados D., Klei H.E., Sundstrom D.W., Enz. Microb. Technol., 2, (1980)

ENZYMIC LIQUEFACTION AND SACCHARIFICATION OF AGRICULTURAL BIOMASS

Authors : G. BELDMAN, A.G.J. VORAGEN, F.M. ROMBOUTS and W. PILNIK

Contract Number : ESE-R-038-NL

Duration : 36 months   1 July 1980 - 1 July 1983

Total Budget : 180.000 ERE   CEC Contribution : 50%

Head of Project : Prof.Dr. W. Pilnik, Agricultural University, Department of Food Science

Contractor : Agricultural University

Address : Salverdaplein 10
P.O. Box 9101
6700 HB Wageningen
The Netherlands

Summary

This report describes the results we obtained when commercial cellulolytic and pectolytic enzyme preparations were applied for the liquefaction and saccharification of two by-products from the food industry (i.e. beet pulp and potato fibre), a catch crop (i.e. stubble turnips) and two wastes (i.e. tomato plants and wood waste). Enzymic treatment of beet pulp in a column hollow fibre reactor was investigated. Enzymes appeared to be rather stable in this type of reactor, although a part of the enzyme activity was lost by adsorption on the residue. Extrusion of ligno-cellulosic materials, like tomato plants and wood chips, improved the enzymic digestibility.

Special attention was paid to the cellulase preparation Maxazyme Cl. This enzyme preparation was fractionated into 5 endoglucanases, 3 exoglucanases and 2 β-glucosidases. The individual enzymes were investigated with respect to binding at crystalline cellulose, and endoglucanase/exoglucanase ratio for optimal synergism. We studied the improvement of Avicel hydrolysing capacity of Maxazyme by enrichment with the purified endo- and exoglucanases.

The solid state fermentation of sugar beets to ethanol was improved by the addition of liquefying pectolytic and cellulolytic enzymes.

1.1 Introduction

Plant cells are surrounded by a rigid and complex structure of polymers which form the cell wall. The main chemical components of the plant cell wall are cellulose, hemi-cellulose, pectin, lignin and structural protein. The microfibrils of cellulose, which are embedded in a matrix of lignin, hemi-cellulose and pectin are responsible for the rigidy of the plant cell walls. Cellulose containing materials are produced as energy or catch crops or as side streams and wastes from industry, agriculture and horticulture.

The conversion of these materials to energy is attractive because of their renewable nature and can be done either (thermo)-chemical or biological. For plant material with a high moisture content the biological conversion route is indicated. Commercial polysaccharide degrading enzymes can be used as important tools in the bio-conversion. The integer structure of plant cell walls is attacked by these enzymes, which results in the release of the cell content and liquefaction and saccharification of the wall polysaccharides. In a next step the solubilized sugars can be fermented to energy carriers like ethanol and methane.

The accesibility of the substrates is a restrictive factor for enzyme action. Especially the cellulose microfibrils which are embedded in the other cell wall polymers are resistant to enzyme attack. Combinations of polysaccharide degrading enzymes show synergism in their liquefaction capability and enhance the accesibility of the cellulose.

A key role in the enzymic liquefaction and saccharification is ascribed to the cellulase enzymes (1). The application of cellulases in industrial processes is restricted by the cost/activity ratio of the enzyme preparations. The most active cellulases are from *Trichoderma* origin and consist of a multi-enzyme complex including 1,4-β-D-glucan glucanohydrolase (endoglucanase), 1,4-β-D-glucan cellobiohydrolase (exoglucanase) β-glucosidase and cellobiase. For the complete hydrolysis of crystalline cellulose a combination of these enzymes is needed. The endoglucanase and exoglucanase show synergistic action in hydrolysis (2). High levels of cellobiase are needed to prevent inhibition of endo- and exoglucanase by cellobiose.

The aim of this project was the enzymic conversion of agricultural and horticultural raw materials, surplusses and wastes to a clear solution of fermentable sugars with suitable enzyme combinations. The composition of the plant polysaccharides must be known to make the right selection of enzyme preparations. Liquefying and saccharifying capability of several commercial enzyme preparations (cellulases, pectinases and hemi-cellulases) had to be studied as well as the optimal reaction conditions for these enzymes such as pH, temperature and reactor disign and effect of physical pretreatment. These parameters were investigated for the liquefaction and saccharifaction of beet pulp, potato fibre and wood chips and described in the previous reports (3, 4). These results are supplemented in this report with data obtained with stubble turnips and tomato plant waste.

The true mechanism of cellulase action remains still obscure (5). Purified enzymes in quantitative amounts are powerful tools to elucidate the mechanism of enzymatic cellulose hydrolysis. That is why this project had to deal with the role of various enzymes in this process as well as their purification and characterisation. This report describes the fractionation of the commercial cellulase preparation Maxazyme CL into its pure endoglucanase, exoglucanase and β-glucosidase components. The individual enzymes were investigated with respect to adsorption on crystalline cellulose and endoglucanase/exoglucanase ratio for optimal synergism. We studied the improvement of Avicel hydrolysing capacity of Maxazyme CL by enrichment with purified endo- and exoglucanases.

Finally, we report the application of commercial pectolytic and cellulolytic enzymes to improve the solid state alcoholic fermentation of sugar beets (6). A solid to liquid fermentation is introduced.

1.2 Materials and Methods

Enzymes and substrates. Commercial enzyme preparations, Maxazyme CL (cellulase, *Trichoderma viride*), Rapidase C80 (pectinase, *Aspergillus niger*) and Hazyme Rapidase (amylase, *Aspergillus niger*) were kindly provided by Gist-Brocades, Delft, the Netherlands. Beet pulp, potato fibre, stubble turnips, tomato plants waste and wood waste were from local industries.

Liquefaction experiments. Liquefaction experiments in a stirred batch reactor and in a column hollow fibre membrane reactor were performed as described previously (4).

Pretreatment experiments. Wood chips and chipped tomato plant waste were pretreated in an extruder at $170°C$ and a speed of 80 rpm. as described in (4).

Analysis methods. Analysis of cellulose, galacturonide content, starch, protein, neutral sugars and reducing sugars is described in (3, 4).

Cellulase purification. Maxazyme CL was extracted with sodium acetate buffer pH 5.0 and centrifuged to remove solids. Enzymes were fractionated and purified by column chromatograph using Bio-Gel P10. Bio-Gel P100, DEAE-Bio-Gel-A, SE-Sephadex C50 and Avicel as the supports. Eluted fractions were analysed on Avicelase, CM-cellulase and β-glucosidase activities. Details on the enzyme purification will be published elsewere (7).

Cellulase adsorption studies. Vials, containing 20 mg/ml Avicel cellulose and different concentrations of purified enzyme in Na acetate buffer pH 5.0 were incubated for 1 hour at $30°C$. After centrifugation free protein was measured using the Folin reagent. Adsorbed protein was estimated by substracting the soluble protein from the values obtained in a similar experiment without cellulose.

Synergism between endo- and exoglucanase. All endoglucanases obtained were combined seperately with exoglucanase I in different ratios to determine the optimal ratio for synergism. Activity on Avicel cellulose was measured for the combination of endo- and exoglucanase, as well as for the seperate enzymes. Degree of synergism was expressed as activity of the combination divided through the sum of activities of the separate incubations (i.e. measured activity/expected activity).

Maxazym CL enriched with purified cellulases. Purified endo- and exoglucanases (0.077 mg/ml) were combined with a solution of Maxazym CL (1 mg/ml) in the same way as described above.

Liquefaction and fermentation of sugar beets. Sugar beets were washed and ground to 3 mm pieces. The mash was brought to pH 4.2 and fermented with *Saccharomyces bayanus* in the presence of 0.2% Maxamyme CL and 0.25% C80. The yeast was added in a concentration of 9 g/kg beet. The fermentation took place anaerobically at $30°C$. Control experiments were done without the addition of enzymes.

1.3 Results

Liquefaction experiments. In addition to our previous liquefaction experiments with sugar beet pulp, potato fibre and wood chips (4), we now studied this process for stubble turnips and tomato plant waste. The growth of the 'catch crop' stubble turnips for fuel is a subject of studies by Carruthers (8) in the same Energy from Biomass Program . Anaerobic digestion is proposed to convert this biomass to methane. Alternatively we treated stubble turnips with cellulase (0.09%) and pectinase (0.112%). In addition to intracellular saccharose, glucose and fructose, the turnips contain predominantly cellulose and pectin (Table I).

Liquefaction of the cell wall polysaccharides was almost complete (Table II), which resulted in the release of a liquor with glucose and fructose as the main sugars (Table III). The results are comparable with data previously obtained with beet pulp and potato fibre.

Hofenk et al. described the anaerobic digestion of tomato plant waste (9). These authors obtained a 50% reduction of weight in 30 days. Our experiments with enzymic treatment of chipped tomato plant waste showed that 38% of the total solids was brought into solution in 24 hours. Cellulose hydrolysis was 40% (Table II). Digestibility of extrusion pretreated material was somewhat enhanced (51% cellulose hydrolysis) but not to the same degree as we obtained with pretreatment of wood chips (4).

Table I. Chemical composition of biomass (dry weight %)

	Beet pulp	Potato fibre	Stubble turnip	Tomato plant waste	Wood waste
Cellulose	18.7	26.0	10.3	25.7	44
Pectin	15.0	15.3	10.5	6.6	-
Hemi-cellulose					
Araban	13.2	3.9	2.0	-	-
Xylan	-	-	-	6.0	7.2
Galactan	2.8	7.9	-	-	-
Starch	-	17.1	-	-	-
Saccharose	5.8	-	4.0	-	-
Glucose	-	-	22.5	-	-
Fructose	-	-	15.3	-	-
Protein	6.0	14.9	11.9	13.0	-
Ash	11.8	5.2	12.7	20.2	-
Lignin	-	-	-	10.5	29

Table II. Enzyme hydrolysis limits of polysaccharide fractions (% hydrolysis)

	Beet pulp	Potato fibre	Stubble turnip	Tomato plant waste		Wood waste	
				untreated	pretreated	untreated	pretreated
Cellulose	91	75	90	40	51	7.6	39
Pectin	91	58	92	20	42	-	-
Hemi-cellulose							
Araban	93	93	70	-	-	-	-
Xylan	-	-	-	19	57	11	57
Galactan	90	96	-	-	-	-	-
Starch	-	99	-	-	-	-	-

Table III. Sugar composition of enzyme digest (g/g dry weight)

	Beet pulp	Potato fibre	Stubble turnip	Tomato plant waste		Wood waste	
				untreated	pretreated	untreated	pretreated
Glucose	0.38	0.45	0.33	0.132	0.186	0.033	0.172
Fructose	0.16	-	0.16	-	-	-	-
Arabinose	0.17	0.04	0.01	-	-	-	-
Xylose	-	-	-	0.011	0.034	0.008	0.041
Uronic acid	0.15	0.11	0.11	0.018	0.029	-	-

Previous results indicated that enzyme treatment of beet pulp in a packed colum reactor, connected with a hollow fibre ultrafiltration unit to recover the enzymes, gave a high degree of polysaccharide hydrolysis. Enzyme inactivation was studied in the presence and absence of beet pulp. This was done for two reasons: firstly to measure the resistance of the enzymes to shear and heat inactivation and secondly to calculate the amount of enzyme lost by adsorption to the substrate. Cellulase activity is rather stable to temperature and shear inactivation (Table IV, first column). Especially Avicelase and pectinase activity is lost in the presence of substrate. CM-cellulase activity is lost to lower extent, while β-glucosidase activity seams to be stabilized in the presence of substrate.

Table IV. Percentage residual enzyme activity in column/hollow fibre reactor after 48 hours.

	- Beet pulp	+ Beet pulp
Avicelase	82.0	50.8
CM-cellulase	80.0	72.0
β-Glucosidase	84.6	98.0
Pectinase	66.7	42.9

Enzyme purification and adsorption. Fig. 1 presents the complete purification flow sheet. We were able to isolate to homogeneity 3 exoglucanases, 5 endoglucanases and 2 β-glucosidases. Plots of the Langmuir adsorption isotherms showed that essentially two groups of endoglucanases could be distinguished: endo I and IV, which adsorb strongly on crystalline cellulose, and endo II, III and V, which adsorb only very slightly (fig. 2.). The exoglucanases I and III adsorbed strongly. Binding of exoglucanase I to crystalline cellulose was used to purify it on an Avicel column. The enzyme desorbed from the column at pH 10.8.

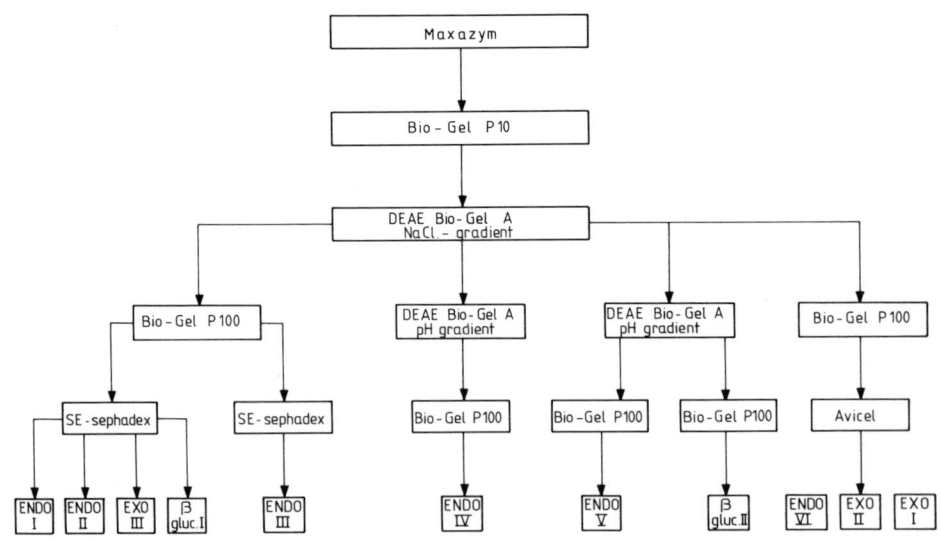

Fig.1.
Flow sheet of the purification of endoglucanases, exoglucanases and β-glucosidases from Maxazyme CL.

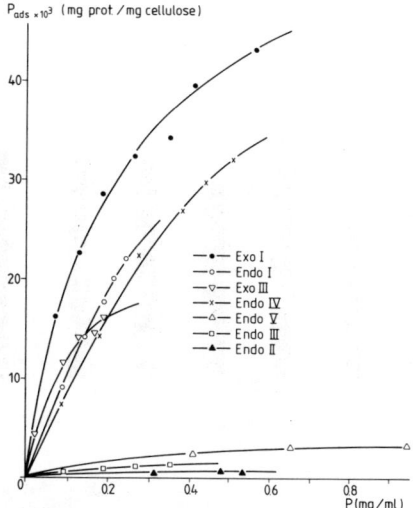

Fig.2.
Adsorption isotherms for purified endoglucanases and exoglucanases on Avicel cellulose.

Synergism between endo- and exoglucanases. Combination of endoglucanase IV with exoglucanase I gave a synergistic effect. The degree of synergism was found to be dependent of the mix ratio of the two enzymes. A maximal effect was observed at an endo IV/exo I ratio of 1.75 (fig. 3). Similar maxima were obtained for the combinations of exo I with the other endoglucanases. Maximal degrees of synergism between 2.5 and 3.5 were observed.

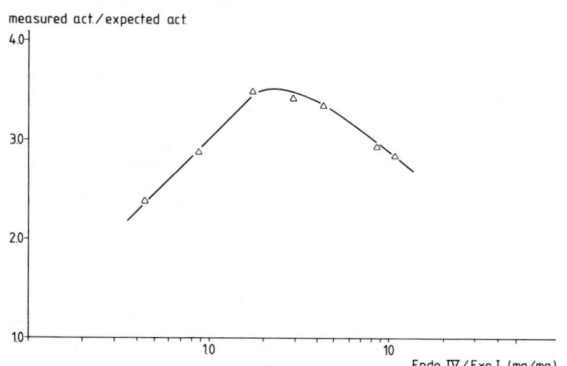

Fig.3.
Degree of synergism as a function of the mix ratio of endoglucanase IV and exoglucanase I.

Enrichment of Maxazyme CL with purified cellulases. The effect of addition of purified endo- and exoglucanases to Maxazyme CL on Avicel hydrolysing capacity is schematically shown in fig. 4. The original enzyme preparation appears to be deficient in endoglucanase II, III, IV and V. However, it cannot be said that Maxazyme CL is deficient in endoglucanases in general, since addition of endo I gave a decrease in the Avicel hydrolysing capacity. Maxazyme CL is not deficient in exo I. This in contrast with exo III, the addition of which resulted in 25% more activity than expected from separate incubations.

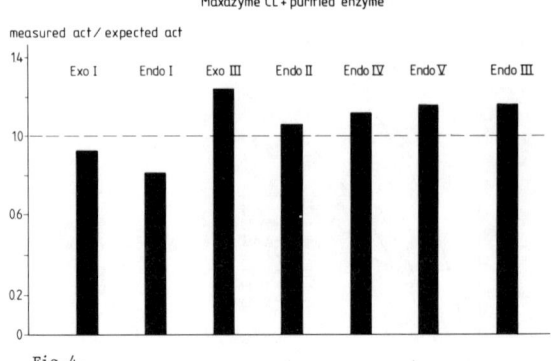

Fig.4.
Effect of enrichment of Maxazyme CL with purified cellulases.

Liquefaction and fermentation of sugar beets. The results of preliminary experiments on fermentation of ground sugar beets in the presence of liquefying enzymes are presented schematically in table V. The addition of enzymes during fermentation resulted in a liquid product, which facilitates the handling of the process and improves the alcohol recovery by thermal separation. The enzymes released more substrate for the yeast, resulting in 13% more ethanol. The fermentation time of 25 hours was reduced with 2 - 4 hours.

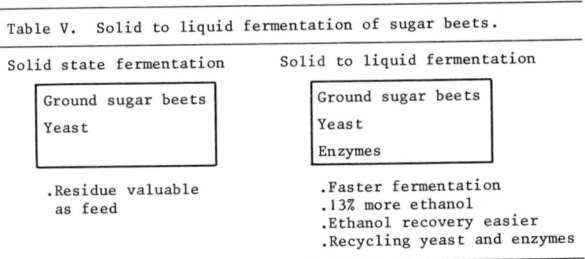

Table V. Solid to liquid fermentation of sugar beets.

2. Discussion

The results of our studies indicate that commercial polysaccharide degrading enzyme preparations are good tools to produce fermentable sugar solutions. With a combination of pectolytic and cellulolytic enzymes we are able to convert substrates like beet pulp, potato fibre and stubble turnips to monomeric sugar solutions. Polysaccharides of these materials were hydrolysed to a great extent. In a previous report we described the important role of polygalacturonase in the initial decrease of the viscosity of the material. Optimal conditions for this enzyme are at pH 3.8 (3). Tomato plant waste and wood waste are more resistant substrates for polysaccharide degrading enzymes. The high level of lignin, which is intimately associated with the cellulose micro-fibrils, makes a pretreatment necessary. Extrusion of these materials results in an increase of its enzymic digestibility. Although energy requirements of extrusion can compete with other pretreatments (4) it remains questionable whether such process will be economically feasible.

Hydrolysis of beet pulp in a packed column/hollow fibre reactor indicates that the process can be operated (semi-) continuously. Enzymes are reasonably stable to temperature and shear inactivation. However, a part of the enzyme activity is lost by adsorption on the residue. Purification of cellulase on Avicel cellulose showed that the exoglucanase I, which is strongly adsorbed. can be desorbed by raising the pH to 10.8. This may be used to increase recovery of cellulases.

The previous report showed an economic evaluation of enzymatic hydrolysis of cellulose. Enzyme costs are an important part of the total costs of the sugar product. Therefore the recovery of enzymes is of high importance. In addition to this, improved cellulase preparations are needed. Cellulases can only be improved, if the mechanism and kinetics of cellulose hydrolysis by this multi-enzyme complex is understood in more detail. Fractionation of the Maxazyme CL preparation and the recombination of the pure endo- and exoglucanases shows the importance of the ratio in which the enzymes are used, to obtain maximal synergism.

From the adsorption characteristics we calculated that the ratio of adsorbed endo- and exoglucanases at maximum synergism is about 1 mol/mol. This supports the theory that endoglucanases and exoglucanase bind as a complex to the cellulose fibril (10). The endoglucanase then splits a bond, whereafter the exoglucanase immediately releases a cellobiose molecule.

The complexity of the enzymic cellulose hydrolysis is also demonstrated in the experiments with the original commercial enzyme preparation enriched with purified enzymes isolated from it (fig. 4). This emphasises oncemore that a harmoniously composed mixture is needed to obtain maximal cellulose hydrolysis. Differentiating enzymes simply in endo- and exoglucanases does not seem sufficient enough to describe the mechanism of the complete mixture. The action of Maxazyme CL is improved by addition of endoglucanases II, III, IV and V but not if endoglucanase I is added. More research is needed to elucidate this phenomenon.

The solid to liquid fermentation of sugar beets seems to be a promising process. The liquid beer can easily be handled, for example during distillation or application of membrane separation techniques. A further improvement of the process is possible when micro-organisms which also ferment pentoses and uronic acids become available.

2. References

1. Voragen, A.G.J., Heutink, R. and Pilnik, W., J. Appl. Biochem., $\underline{2}$, 452-468 (1980).
2. Wood, T.M. and McCrae, S.I., Proc. Symp. Enz. Hydrol. Cellulose (Bailey, M., Enari, T.M. and Linko, M., eds.) Aulando, Finland, 231-254 (1975).
3. Beldman, G., Voragen, A.G.J., Rombouts, F.M. and Pilnik, W., Proc.E.C. Contractors' Meeting, Copenhagen, Energy from Biomass, $\underline{1}$, 151-156 (1981).
4. Beldman, G., Voragen, A.G.J., Rombouts, F.M. and Pilnik, W., Proc.E.C. Contractors' Meeting, Brussels, Energy from Biomass, $\underline{3}$, 266-272 (1982).
5. Bisaria, V. and Ghose, T.K., Enzyme Microb. Technol., $\underline{3}$, 90-104 (1981).
6. Kirby, K.D. and Mardon, C.J. Proc IV Int. Symp. on alcohol fuels technol., Guaryjã-SP-Brasil (1980).
7. Beldman, G., Searle-van Leeuwen, M.J.F., Voragen, A.G.J. and Pilnik, W., *in preparation*.
8. Carruthers, S.P., Proc.E.C. Contractors' Meeting, Brussels, Energy from Biomass, $\underline{3}$, 37-42 (1982).
9. Hofenk, G., Rijkens, B.A. and Voetberg, J.W., Proc.E.C. Contractors' Meeting, Brussels, Energy from Biomass, $\underline{3}$, 232-237.
10. Wood, T.M. and McCrae, S.I. Proc. of Bioconversion Symp., IIT Delhi (Ghose, T.K., ed.), 111-141 (1977).

PILOT PROJECTS "METHANOL FROM WOOD"

INTRODUCTION :
The methanol from biomass Pilot plant programme of the Commission of the European Communities related to other possible routes

Synthetic fuel from wood

Pressurized oxygen blown fluidized bed gasification of wood

The oxygen donor gasifier - Continuous recycling of solids between two fluidised beds

Gasification of biomass for the production of synthesis gas with the intention to produce synthetic fuel in a further process

Demonstration of the methanol synthesis

THE METHANOL FROM BIOMASS PILOT PLANT PROGRAMME OF THE COMMISSION OF THE EUROPEAN COMMUNITIES RELATED TO OTHER POSSIBLE ROUTES

W.P.M. van Swaaij
and
A.A.C.M. Beenackers

Twente University of Technology
P.O. Box 217
7500 AE Enschede
The Netherlands

Summary

A short introduction is given to the presentation of the four gasification pilot projects funded by the CEC on a cost sharing basis.
To demonstrate the position of these projects among the world wide efforts in the field and among other possible routes a simple table has been set-up derived from a recently published extensive survey.

Because biomass is a solid and as such an inconvenient fuel research programs are being carried out all over the world to convert it into a liquid fuel. For dry cellulose type of biomass like wood, conversion to methanol via synthesis gas production is generally considered to be the most promising way to obtain a liquid fuel.

The production of methanol from wood has to overcome the problem that wood as an energy source has a rather dispersed nature and therefore the present unit capacity for methanol production (± 1000 tons/day) is too high for most wood producing areas. Prospects to improve its economics on a smaller scale have been discussed elsewhere [1]. Nowadays more realistic wood to methanol proposals are based on plant capacities of one to several hundred tons of methanol/day [2].

A second problem for the wood to methanol route is the fact that the production of synthesis gas from wood is not a well established process and presents problems related to the properties of the feedstock as well as related to the relatively small scale of operation. The present status of the synthesis gas production from wood has been reviewed by the authors recently [3] and we will present here, as a short introduction, a simple scheme indicating both the promising and less attractive areas for development. Also the position of the four CEC cost sharing pilot plant programs amongst other possibilities is given. We will leave the presentation of the status of these projects to the individual contractors.

Table I gives the overview.

The small scale of operation with biomass would ask for a cheap process which uses air only and operates at atmospheric pressure.

In most cases, however, this would produce a synthesis gas diluted with nitrogen. Cryogenic recovery of the synthesis gas components (CO + H_2) has been proposed [4] but this is very expensive.

The Oxygen Donor Process of John Brown and Wellman (CEC project G1 table I) is rather interesting because a product gas essentially free of nitrogen is produced. A new development in synthesis gas production via air gasification could be the synthesis gas extraction route studied at the Twente University of Technology [4].

More complex but promising is steam gasification. Because this is an endothermic reaction some feedstock, product or intermediate product has to be combusted externally to provide the heat of reactions and the sensible heat. In some cases heat from other sources is used. The steam gasification of wood is studied at several places in the world [3] and in this category falls the CEC project of Maremma (D 2) with an indirectly heated fluid bed and also a version of the John Brown/Wellman CEC project (G2).

Atmospheric gasification with oxygen/steam is also studied at several places. In former CEC programs e.g. Creusot Loire operated a fluid bed pilot plant [6]. From the present CEC pilot plants the Lurgi fast or circulating fluid bed belongs to this class (E3).

Because methanol production asks for synthesis gas at a pressure of ± 70 bar, pressurization of the gasification process seems to be logical for both the steam and the oxygen processes. However, because this presents another complication, little progress has been made with larger pilot plants up till now. The CEC Creusot Loire pilot plant (D6) forms a rather unique endeavour, although, as in the other projects, technical and economical feasibility has still to be demonstrated.

There is one important item that is missing in the CEC programs though it receives much attention in both the USA and Sweden (see [3]): catalytic gasification or catalytic modification of the gasification process. Apart from this point, which deserves some attention in future programs, the projects selected for CEC funding cover the most important alternatives.

References
1. Beenackers, A.A.C.M. and van Swaaij, W.P.M., Proc. 2nd E.C. Conf. Energy From Biomass (Strub, Chartier and Schleser Eds.) Applied Science Pub. (1982) 782.

2. Holoway, B. et al Energy from Biomass and Wastes VII IGT (1983) preprints paper 40.

3. Beenackers, A.A.C.M and van Swaaij, W.P.M., Proc. Energy from Biomass Conf. Birmingham (april 1983) (to be published).

4. Rowell, R.M. and Hokanson, A.E., Progress in Biomass Conversion, Academic Press, London (1979) 117.

5. Beenackers, A.A.C.M. and van Swaaij, W.P.M. in: W. Palz and G. Grassi (Eds) Energy from Biomass III, Reidel, Dordrecht (1982) 201.

6. Chrysostome, G. in P. Chartier and W. Palz eds. Energy from Biomass I, Reidel, Dordrecht (1981) 198.

REACTOR TYPE

PROCESS CONDITION	co-current packed bed	counter current packed bed	cross current packed bed	fluid bed	circulating fast bed	powder flame reactor	double fluid bed	exotic reactor types
1 Atmospheric air	+/-	+/-	-	+/-	+/-	+/-	///	+/-
2 Atmospheric steam (+air)(electricity)	-	-	///	/// XXX	-	-	XXX	-
3 Atmospheric oxygen/steam	///	///	///	-	-	-	-	///
4 Pressurized air	-	-	-	-	-	-	+/-	-
5 Pressurized steam (+air)	-	-	-	-	-	-	+/-	-
6 Pressurized oxygen/steam	///	///	///	XXX	+/-	///	-	+/-
	A	B	C	D	E	F	G	H

Table I Reactor types and process conditions for gasification in wood to methanol synthesis.

- no interest
+/- little interest and development
\\\ important research/dev. area's
/// EC supported projects

SYNTHETIC FUEL FROM WOOD

Authors	:	P. Mehrling, Chr. Lindner
Contract number	:	ESE-P-067-D (b)
Duration	:	21 months 1. January 1982 - 30. Sept. 1983
Total budget	:	DM 1.706.950,-- CEC Contr.: DM 853.475,--
Head of project	:	Dr. Chr. Lindner, Department of R & D
Contractor	:	Lurgi Kohle und Mineralöltechnik GmbH
Address	:	Lurgi Kohle und Mineralöltechnik GmbH Bockenheimer Landstraße 42
		D-6000 Frankfurt am Main 1

Summary

To demonstrate wood gasification an existing reactor operating to the principle of the circulating fluidized bed (CFB) had been converted for gasification tests.

After starting up the plant some gasification tests with oxygen/steam as gasification agent have been performed in the first half of the year 1983. Three kinds of wood had been tested:

- beech wood   (mainly)
- coarse recycling wood   )
- poplar wood   ) short tests

From these kinds of wood could be seen the influence of the

- gasification temperature: $CO+H_2$ = 62,9 - 70,8 % by volume according to the rising temperature from 760 - 950 °C

- moisture of wood   : CO_2 = 27,9 - 50,12 % by volume according to the rising moisture content of 21,3 - 53 % by weight

Finished the gasification tests a process scheme will be prepared for the methanol production route including the complete consumption and production figures. This process scheme is the basis for the subsequent investment cost estimate.

1. INTRODUCTION TO THE PROJECT

   The purpose of this development program is to demonstrate wood gasification in a gasification reactor operating to the principle of the circulating fluidized bed (CFB) located in the Lurgi laboratory.

   A reactor originally designed for calcination/combustion was converted for the gasification tests in respect of technical and safety reasons.

   Finished the gasification tests a detailed mass and energy balance will be compiled. Using this evaluation, a process scheme for producing methanol from wood with complete consumption and production figures will be prepared.

   The process scheme will serve as the basis for the subsequent investment cost estimate.

2. STATUS OF THE PROJECT

   The conversion work of the plant had been finished in November 1982. After starting up the plant some gasification tests with air had been performed at first followed up by gasification tests with oxygen/steam as gasification agent.

   The overall concept of the plant is shown in the Process Flow Diagram (Annex 1). The newly installed components are shown in thick lines.

3. PROCESS DESCRIPTION OF THE PILOT PLANT

   The description refers to the Process Flow Diagram.

   Prepared wood enters via wood bin B-101 and then the wood lock B-102. The volume of the wood lock results in 2 lock operating procedures per hour.

   The wood passes from the lock by gravity into the storage bin B-103. It is conveyed and proportioned by the screw feeder H-101.

   An oxygen/steam mixture is used as the gasification agent. The preheated oxygen is mixed with steam in the mixing pipe B-202 and is supplied to reactor D-201 through the reactor bottom.

The gasification reaction between wood, coke and gasification agent takes place in gasifier D-201. The gasification conditions taken as a basis for the layout are 1 bar reaction pressure and 1100 °C reaction temperature. Particles entrained by the raw gas are separated in the hot cyclone F-401 and recycled directly into the reactor.

In case of feedstocks containing ash, the ash is withdrawn by the screw H-301 and discharged into the ash bin B-301.

The raw gas produced by gasification is dedusted in hot cyclone F-403 and scrubbed and saturated in the downstream scrubber W-401 by water recycled via pump P-401.

Downstream the scrubber a small gas flow is branched off the main product gas flow passes the water separator F-404 and is cooled down in the gas cooler W-402. The cooled gas is pressurized by the compressor H-402 up to 150 bar, cooled down again in the gas cooler W-403 and bottled in the bottle battery B-403.

The main product gas flow is burned in the combustion chamber D-401. The offgas booster V-401 equalizes the pressure drop along the gas pipe.

4.   GASIFICATION TESTS

Generally speaking all systems of the plant performed well besides minor failures (stopping of the gas analysis pipe etc.) typical for small multipurpose pilot plants.

For the first tests a beech wood mainly was used and for two short tests coarse recycling wood and poplar wood delivered by the "Hessische Versuchsanstalt für Forst- und Landwirtschaft" from an experimental forestry area for fast growing trees.

The gasification conditions for the tests had been:

    Gasification agent : oxygen/steam
    Temperature   : 760 - 950 °C
    Pressure   : ambient pressure

The preliminary results of the tests are shown in the table (Annex 2).

For the time being mass and energy balances cannot be established because the ultimate analyses of ash and gas liquor are not yet available.

5. REMARKS REGARDING THE RESULTS

The gasification tests had been performed with an inert bed material.

The results of the tests show two main influences for the raw gas composition:

- gasification temperature:

    with rising temperature the quality of the synthesis gas is becoming better. The $CO + H_2$ content of the gas increases from 63 % by vol. at 760 °C up to 71 % by vol. at 950 °C.

- moisture content of the wood:

    with rising moisture content the CO_2 content of the gas increases from 28 % by vol. at a humidity of 21 % by weight up to 50 % by vol. at a humidity of 53 %. In this case most of the energy is needed to evaporate the water of the wood and lacks for reducing the CO_2.

The raw gas composition is given nitrogen free because in the case of the pilot plant N_2 is used as purge gas for pressure measuring nozzles, sight glasses etc. In addition the wood bin is also scavenged with nitrogen.

The H2S content of the raw gas has been in the range of 0 - 30 ppm.

6. FURTHER ACTIVITIES

The first week of June the performance of the guarantee run will be finished this time without an inert bed material.

Based on these results and the results of the methanol synthesis (see ESE-P-071-D) a process scheme for the methanol production route including the complete consumption and production figures will be prepared. The process scheme is the basis for the subsequent investment cost estimate.

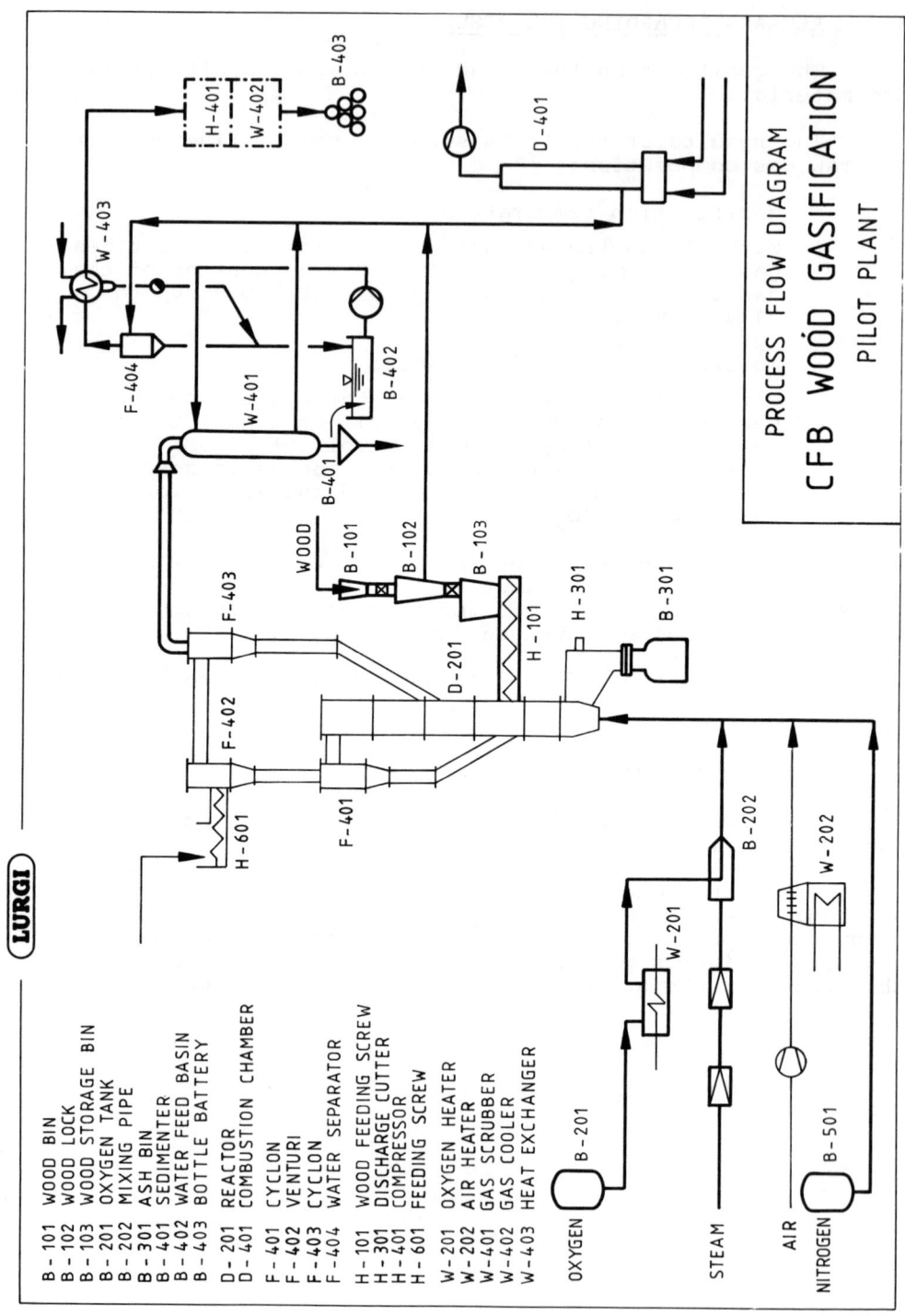

RESULTS OF THE GASIFICATION TESTS

Wood analysis (% b.w.)	Beech Wood						Coarse Recycling Wood	Poplar Wood
Moisture	21,3						30	53
Ash	3,36						5,3	14,5
C	48,13						50,84	51,3
H	5,65						5,4	6,37
O	45,87						43,39	41,55
N	0,35						0,26	0,33
S	–						0,01	0,12
H_o (kJ/kg)	17090						17740	
Gasification temperature (°C)	760	775	785	840	920	950	840	840
Wood (kg/h)	204	305	321	213	302	290	180	192
Oxygen Nm^3/h	64	75	78	70	78	78	75	75
Steam kg/h	9,8	7	7	4,6	10	10	–	–
Raw gas $(\frac{m^3 \, dry}{kg \, daf})$	1,38	1,4	1,33	1,3	1,32	1,36	–	–
CO	31,63	34,8	35,86	36,06	37,98	38,23	24,6	23,58
CO_2	30,89	26,87	26,8	27,93	25,18	24,39	39,4	50,12
H_2	31,28	32,54	30,74	29,51	30,51	32,53	27,6	20,64
CH_4	5,8	5,5	6,28	6,33	6,25	4,77	7,7	5,66
C_nH_m	0,29	0,21	0,24	0,12	0,04	0,01	0,28	–
H_o kJ/m^3	9810	10245	10755	10655	10304	10868	9699	7785
$\eta = \frac{H_o \, gas}{H_o \, wood}$	79,2	84	83,6	80,6	81,8	86,4		

Annex 2

Pressurized oxygen blown fluidized bed gasification of wood

Authors : G. CHRYSOSTOME - JM. LEMASLE
Contract number : ESE -P- 068 - F
Duration : 21 months
Total budget : 63 000 000 F.F.   C.E.C. Contribution : 4 000 000 F.F.
Head of project : G. CHRYSOSTOME - Creusot-Loire
Contractor : CREUSOT-LOIRE
Adress : CREUSOT-LOIRE
Division Energie
BP 31   71208 LE CREUSOT
FRANCE

Summary

The purpose of the project is to build and experiment a 60 TPD dry wood gasification unit under pressure for making a syngas for methanol production. The involved process is the Creusot-Loire's one which has been developped in a 2,5 TPD atmospheric PDU (Process Developpement Unit) at the Creusot's Energetics Laboratory for two years.
The 60 TPD dry wood gasification unit under pressure would be the second phase of the Creusot-Loire research on that subject. The major equipments are :

- an oxygen + steam blown fluidized bed gasifier

- an oxygen blown secondary reformer to convert methane and higher hydrocarbons which are produced at the fluidized bed stage.

- a treatment (scrubbing-cooling) of the syngas, so that it can burn in a boiler.

The operating pressure would be 10-30 bars. The wood moisture 15 %.

The project has been previously described in its whole configuration. This paper presents the conclusions of the 2,5 TPD atmospheric P.D.U. testing during 2 1/2 years and points out the validity of hypothesis that were taken in account for the pressurized pilot plant which would be erected at Clamecy (Nièvre).

In the past six months, three long duration runs (100 hours) have been successfully carried out on the atmospheric P.D.U. of Le Creusot.

1 - INTRODUCTION

The methanol from wood research was undertaken by Creusot-Loire in 1979. It led to the construction of an atmospheric P.D.U. for the oxygen and steam fluidized bed gasification of wood. The 400 mm I.D. fluidized bed was achieved at the end of 1980. The first step of testings with the single stage gasification process pointed out the necessity of a secondary reforming for gases leaving the fluidized bed gasifier. So, the secondary reformer (partial oxidation with pure O2) was built and the whole P.D.U. has been tested in its "two stage" configuration since july 1982.

In parallel with the PDU testings, fundamental research has been carried out at the University of Technology of Compiègne in a batch reactor where thermogravimetry trials were performed on pyrolysing wood. These tests gave us a better comprehension on thermochemical wood reactions and on the kinetics of pyrolysis-gasification mechanisms.

On the P.D.U., we performed during 18 months shorts runs (each lasting up to 5 hours under stabilized conditions).
Then we realized long duration tests from november 1982 to now. Three long duration tests have been carried out succesfully (each lasting about 100 hours under stabilized conditions). These runs permitted to :

- test the whole P.D.U. on long term periods.

- test the fluidized bed behaviour when enriching with wood ashes allowing us to determine how to avoid sintering or fusion of solid particles in the dense phase of the reactor. That problem is recognized as one of the most critical in case of wood gasification with pure oxygen.

- optimize the working conditions of the secondary reformer on the following points : type of injection for the gas to be converted, type of injection of oxygen, operating temperature, nature of refractories.

- demonstrate the importance of some parameters such as wood moisture affecting the gas composition in a large extent, thus justifying the choice for a preliminary wood drying down to 15 % moisture.

- accumulate experimental results about the quality of the gas quenching water at the outlet of the secondary reformer.
We now consider that the reliability of the process has been proven at the P.D.U. step, the previous hypothesis being verified. The choice of operating conditions is done.
On an experimental point of view, the design information we don't yet possess is on the subject of the influence of pressure on the kinetics of wood gasification. We are undertaking the preliminary studies, in collaboration with the University of Technology of Compiègne, for a batch reactor performing thermogravimetric experiments under pressure to study this problem. Such an experimental program would be a logical extent of the atmospheric thermogravimetric trials on wood conversion which have been performed by U.T.C. since 1980.

After a brief review of the gas characteristics at the outlet of the fluidized bed gasifier, we present a comparison between the theoretical and the experimental results we obtained at the outlet of the secondary reformer during our long duration runs.

2 - GAS CHARACTERISTICS AT THE OUTLET OF THE FLUIDIZED BED GASIFIER : (experimental)

In case of runs performed under pure oxygen conditions, gases leaving the fluidized bed gasifier contain 10 to 12 % CH_4, 2,5 to 3 % C_2H_4 and 0,1 to 0,5 % C_2H_6. These values are given on a volumic basis for the dry gas, for runs performed at 800 °C in the dense fluid bed.
That high hydrocarbons content is due to the release of volatile matters of wood which are not completely cracked at the outlet of the reactor, because of the rather low temperature of reaction. However, the operating temperature in the gasifier has to be maintained at a level below the ash sintering limit.

For that reason, we modified the existing P.D.U. by adding a secondary reactor for partial oxidation of the gas leaving the gasifier. The name of this reactor is the secondary oxygen reformer, in which injection of additional O_2 increases temperature leading to a complete cracking of higher hydrocarbons than CH_4, which are converted in a $CO + H_2$ mixture. More over, the negative influence of pressure on the methane content strengthens the choice of a two stage gasification process with the second stage for methane reforming.

3 - PRESENTATION OF THE LONG DURATION RUNS :

3.1 - Run n° 70 : November 22 to 26 th 1982

- hardwood, 20 % moisture, 10-20 mm
- fluidized bed = silica
- gasification medium = O_2 + steam
- fluidized bed temperature = 700 ± 20 °C
- gases flow upward in the secondary reformer
- tangential inlets of gas and O_2 at the bottom of the reformer
- maximum temperature in the secondary reformer = 1300°C

Main conclusions :

- gas characteristics satisfactory
- bad behaviour of refractories
- slow enrichment of fluid bed with wood ashes, still not leading to sintering problems after 100 hours.

3.2 - Run n° 72 : December 13 to 17 th 1982

- hardwood, coming from the region of Clamecy (Nievre), 40 % moisture, granulometry typical of paper mill wood chips (< 50 mm).
- fluidized bed : silica
- gasification medium = O_2 + steam
- fluidized bed temperature = 700°C ± 20°C   75 hours
                              750°C ± 20°C   20 hours
- gases flow upward in the secondary reformer
- tangential inlets of gas and O_2 at the bottom of the reformer
- maximum temperature = 1200°C

Main conclusions :

- Bad characteristics of the gas, due to the 40 % wood moisture, leading to a very poor methanol yield.
- Confirmation of bad refractories behaviour
- Slow enrichment of fluidized bed with wood ashes, allowing to determine a regeneration ratio inert material/fed wood which is acceptable in case of the pressurized unit of Clamecy.

3.3 - Run n°75 : March 28 th to April 1st 1983

- hardwood, 20 % moisture, 10-20 mm
- fluidized bed = chamotte (si O_2 + Al_2O_3)
- gasification medium = O_2 + steam
- fluid bed temperature = 700°C ± 20°C
- gases flow downward in the secondary reformer (as for the projected case of Clamecy).
- tangential inlet of the gas to be converted
- O_2 inlet at the top of the reformer by one axial or three peripheral nozzles.
- Refractories different to those tested at runs 70 and 72
- Maximum temperature = 1350 °C

Main conclusions :

- Gas characteristics satisfactory
- Good behaviour of refractories
- No fluid bed sintering problems. Regeneration rate of the inert fluidized bed checked to a maximum of 0,7 % of the wood feed rate (weight basis).

4 - COMPARISON : EXPERIMENTAL RESULTS - THEORETICAL HYPOTHESIS

The following table allows to make a comparison between the expected theoretical results on the pressurized pilot plant of Clamecy and the experimental results obtained during the third long duration run n° 75

TABLE OF RESULTS LONG DURATION RUN N° 75						Case of Clamecy (assumption)
Date		29/03/83	30/03/83	31/03/83	01/04/83	
Analysis n°		900	1500	1700	2300	
Type of 02 nozzle in the reformer		3 périph.	3 périph.	1 axial	1 axial	
Dry wood Feed rate kg/h		87,80	87,80	87,80	87,80	87,80
Reformer 02 Flow Rate Nm3/h		17,50	13,50	13,00	17,50	
Fluidized bed 02 Flow rate Nm3/h		14,50	15,50	15,00	16,50	
Total 02 Flow rate Nm3/h		32,00	29,00	28,00	34,00	36,60
kg 02/kg Dry wood		0,52	0,47	0,46	0,55	0,60
Equivalence Ratio		0,38	0,34	0,33	0,39	0,43
Steam/Dry wood		0,23	0,23	0,23	0,23	0,40
Wood moisture		20,00	20,00	20,00	20,00	15,00
Balance C - H - O outlet/inlet (%)	C	100,00	95,00	96,00	96,00	100,00
	H	85,00	89,00	87,00	95,00	100,00
	O	101,00	101,00	101,00	100,00	100,00
Dry gas Flow rate (corrected) (CO+CO2+H2+CH4) Nm3/h		121,80	115,20	117,50	121,10	117,70
% corrected/dry gas	CO	34,00	30,60	31,40	35,00	38,80
	CO2	31,50	32,90	32,10	28,20	27,40
	H2	31,50	31,90	32,90	33,60	30,90
	CH4	1,60	3,90	3,20	1,50	3,00
Corrected thermal yield % $\frac{\text{LHV gas}}{\text{LHV wood}}$		63,00	64,00	65,00	65,00	69,00
Thermal balance (without thermal losses) (%)		81,00	79,00	79,00	81,00	97,00
Tons wood gasified per ton of methanol		2,31	2,56	2,44	2,22	2,25

The operating conditions were as follows :
- wet wood feed-rate = 109,7 kg/h
- wood moisture   = 20 % (wet basis)
- gasification temperature = 700°C
- operating time under stabilized conditions = 95 hours

Remark : In this table, each of the analysis numbered 900 to 2300 is the result of an average calculated each time on 7 to 9 individual gas analysis.

Analysis of the results :

The analysis 900 and 2300, performed with an oxygen flow rate which can be compared to the basis case for the design of the pressurized unit of Clamecy, point out that the "potential mathanol" is very close to the theoretical value.

In the same conditions, we observe that the CH_4 content is decreased down to 1,5 %, all higher hydrocarbons (C_2H_4, C_2H_6, C_3H_8, oil and tars) remaining in gas as traces, not detected by the analyses. Moreover, the carbon balance proves that no soot is formed, which is confirmed by the quench water analysis (less than 2 % unburnt carbon).

The results of analysis 900 and 2300 (reformer temperature > 1300°C) when compared to those of analysis 1500 and 1700 (reformer temperature ≈ 1150°C) indicate that it is necessary, in order to reach a CH_4 content lower than 2 %, to operate over 1300°C.

The oxygen consumption is about 10 % lower than in the theoretical calculations. Still, if we consider that thermal losses (see table) are rather higher in such a P.D.U. than in a more important unit, we can predict that the oxygen consumption would be lower in an industrial unit, thus leading to a higher methanol yield and a lower production cost. All the more so because in that run, the wood moisture of about 20 % is a less favourable value than the 15 % which had been taken in account for the design case of Clamecy.

Complementary remarks :

Several fluid bed samples have been drawn during the long duration run every four hours, and are currently analysed. This will allow to quantify the enrichment rate of the bed with mineral matters of wood ashes. Then, we'll determine the ratio between ashes fixed into the bed and entrained ashes in the exiting gas leaving the gasifier. At all events, the fluidized bed has not been regenerated during the 100 hours run and so we can conclude that, for the same operating temperature, a fluid bed regeneration rate of 0,7 % (weight basis) of dry wood feed rate can be estimated as a conservative value. (During the run n° 75, the rate achieved was 0,67 %).

In spite of a few deposits of solids in the secondary reformer, refractories have not been eroded. That observation confirms the advantage in the choice of a waste heat boiler to remove sensible heat of gases at the outlet of the secondary reformer between 1300°C and 700-800°C. However, more accurate analysis on refractories are currently being performed.

5 - CONCLUSIONS :

The optimization phase of the P.D.U. of Le Creusot is now well advanced.

Main operating conditions are a quite moderated temperature of the fluidized bed (700-800°C) and a high temperature in the secondary reformer (1300-1400°C).

In the fluidized, steam acts as a fluidization medium and participates as a gasification reagent. The methanol yield just slighthy decreases with an increase in steam rate.

In the secondary reformer, the mode of injection for the gas and oxygen is established. The methane content of syngas is in conformity regarding to methanol syngas specification. Moreover, the unburnt carbon production is low. The oil and tars production is negligible.

The potential methanol yield from gas produced on the P.D.U. reaches the assumptions which had been used in case of the design of the pressurized unit of Clamecy and the techno-economic evaluation for a 500 TPD methanol from wood.

THE OXYGEN DONOR GASIFIER
CONTINUOUS RECYCLING OF SOLIDS BETWEEN TWO FLUIDISED BEDS

Authors   : C. Denning, R.E. Gowers, R.J. McClellan, D.H. Smith

Contract Number : ESE-P-069-UK(H)

Duration   : 24 Months   01.04.82 - 31.03.84

Total Budget   : £848,423   CEC Contribution : £339,334

Head of Project : Mr. C. Denning, Wellman Mechanical Engineering Ltd.

Contractor   : John Brown Engineers and Constructors Ltd.,
                    Wellman Mechanical Engineering Limited.

Address   : Roberts House, Cornwall Road, Smethwick, Warley.
                    West Midlands, B66 2JU, United Kingdom.

Summary

In the oxygen donor gasifier, the material of the fluidised bed (the stone) is recirculated continuously between a gasifier compartment and an oxidiser compartment. In the gasifier compartment, oxygen and heat for the endothermic gasification reaction are donated by the bed material. In the oxidiser compartment the bed material is reoxidised by fluidisation with air. The rate at which the bed material is recirculated between the two compartments is critical. Sufficient material must pass into the gasifier compartment to supply the necessary oxygen for the gasification reaction and to maintain a heat balance between the two compartments. The recirculation system has been studied in a full scale model of the reactor working at ambient temperature and the design of the recirculation system and the rate of recirculation have now been determined.

1. Introduction

The oxygen donor gasifier system is illustrated schematically in Figure 1. In the gasifier compartment the carbonaceous raw material is fed into a bed comprising solid particles of calcium oxide and sulphate, at a temperature of 900-950°C which is fluidised by recycled product gas. Oxygen is donated to the reaction by the calcium sulphate. The reduced stone is recycled to the oxidiser compartment where it is reoxidised by fluidising air at a temperature of 1000-1050°C. The method of recirculation and the rate at which it is maintained are critical if sufficient oxygen and heat are to be supplied to the gasification reaction.

1.1 Heat and Mass Balance

Figure II is an example of a simplified mass balance for the system when wet wood is being gasified at the rate of 545Kg/h. The expected compositions at the various points in the system are given in the table. The oxygen required to sustain the gasification reaction is 169Kg/h (the difference between the $CaSO_4$ rate in column C and the CaS rate in column F).

Figure III tabulates an example of an expected heat balance for the above system. The enthalpy required to sustain the gasification reaction is some 963Kw (the difference between the enthalpies of the CaO entering and leaving each of the compartments). This implies that stone must be transferred at a rate of some 58,000Kg/h. This rate applies when there is a temperature difference of 50°C between the two compartments. If the temperature difference is 100°C, then the rate required will be approximately halved.

In order to ensure that a sufficient stone recirculation rate can be obtained we have constructed and operated a full scale 'cold model' of the two-compartment reactor working at ambient temperature. This model has enabled us also to identify the key mechanical design parameters and to establish the principles which govern their effect on the stone recirculation rate.

2. The Cold Model Reactor

2.1 Design of the recirculation system

Figure IV shows a side view in the region of the dividing wall between the gasifier and oxidiser compartments. A similar section is provided to introduce stone from the oxidiser to the gasifier. Gases are introduced into plenum chambers below the two beds and pass into the base of the beds through distributer nozzles. In the oxidiser compartment air is introduced. In the gasifier compartment recycled product gas is introduced.

On the oxidiser side of the dividing wall a high flow of control air is introduced to lift the bed vertically away from the transfer slot. A small flow of recycle gas is introduced across a porous sloped section so as to prevent blockage of the transfer slot. The geometry of the transfer slot is critical and the dimensions, hole sizes and positions, spacings and angles have all received careful attention. The cold model and its associated equipment were designed so that all the dimensions and the gas flow rates and pressures could be varied and measured over a wide range.

A view of the cold model is shown in Figure V.

2.2 Recirculation Rate

It was necessary to devise a method whereby the rate of recirculation could be determined.

Figure VI shows a simple general plan of the cold model with the gasifier compartment inside a larger box, the outside region of which is the oxidiser section. Stone is transferred from the gasifier across the left hand transfer slot, travels around the oxidiser section and re-enters the gasifier through the right hand transfer slot. The method finally selected to determine recirculation rate involved the use of a tracer. A known quantity of tracer was inserted at point X at time 0. Samples of the bed were then taken at regular intervals from points X, Y and Z. The concentrations of tracer at the three points were then plotted against time.

2.3 Theoretical Models of Stone Movement

In Figure VI, two of the simpler mathematical models for stone movement are illustrated.

If the system behaves as two continuous stirred tank reactors then the tracer concentrations at points X, Y and Z should vary with time as shown. Concentration at X falls continuously. Concentrations at Y and Z are equal and rise continuously until the concentration is uniform throughout the system.

If however the system behaves as a single continuous stirred tank reactor (the gasifier) connected to a plug flow reactor, a time lag will occur before tracer appears first at point Z and then at point Y. As tracer re-enters the gasifier compartment for the second time, the concentration at point X will rise again as shown.

The actual dimensions of the cold model were applied to these mathematical models and they were then used to calculate actual stone recirculation rates. Experimental concentration/time curves were prepared and compared with the curves derived from the mathematical models.

2.4 Results of Recirculation Rate Experiments

One typical concentration/time curve from the cold model tracer tests is illustrated in Figure VII. This shows that the ODG reactor, in this design, behaves like a continuous stirred tank reactor connected to a plug flow reactor.

By applying the dimensions of the cold model to the formula describing this combination, three dimensions on the curves can be used to deduce the actual recirculation rate from the results obtained experimentally. These are:

- The initial rate of decline in tracer concentration in the gasifier.
- The time taken for the tracer to complete one cycle and return to the gasifier.
- The time taken for the tracer to travel from point Z to point Y in the oxidiser.

The stone recirculation rates determined fall within the desired range. If these results are achieved in the 'hot' pilot plant, sufficient oxygen and heat will be transferred to sustain the desired reaction.

The cold model is still in use and further data are being collected, in order to establish the relationships between stone recirculation rate and the reactor dimensions, gas flow rates, gas pressures and bed heights which may be used in the pilot plant work. It seems likely that a wide range of stone recirculation rates can be set up and stabilised, by careful selection from amongst these geometrical and operational parameters.

3. The 'Hot' Reactor

Figure VIII shows some of the design features of the 'hot' reactor which will be built in the near future. The design developed as a result of two basic decisions.

3.1 The Use of Metallic Distributor Nozzles

In the early stages of the project it was intended to use porous ceramic distributor plates at the bottom of the fluidised beds. We will now use metallic plates into which are welded metallic distributor nozzles. The porous plates suffered from the following disadvantages.

- High pressure drop and susceptibility to blockage.
- Poor dimensional tolerances and difficulties with sealing and replacement.
- Relatively low resistance to impact damage.
- Relative high price.

Conversely the metallic structures are readily fabricated, modified and replaced and, since they are in the colder parts of the reactor, can be made of carbon or stainless steels.

3.2 The Use of a Metallic Container as the Gasifier Compartment

Originally both compartments of the reactor were to be lined with ceramic brick. Because of the height of the reactor, the internal wall between the gasifier and oxidiser compartments appeared to be mechanically weak. It was decided therefore to construct the gasifier compartment of a high nickel alloy.

Such a metallic structure will be subject to thermal expansion and it was therefore decided to locate it centrally within the oxidiser compartment. This arrangement simplifies the layout of the whole reactor, especially around the gas offtakes at the top. It also facilitates removal of the gasifier compartment for inspection and repair.

The wood is fed into the gasifier compartment at a point close to its base.

These features are illustrated in Figure VIII.

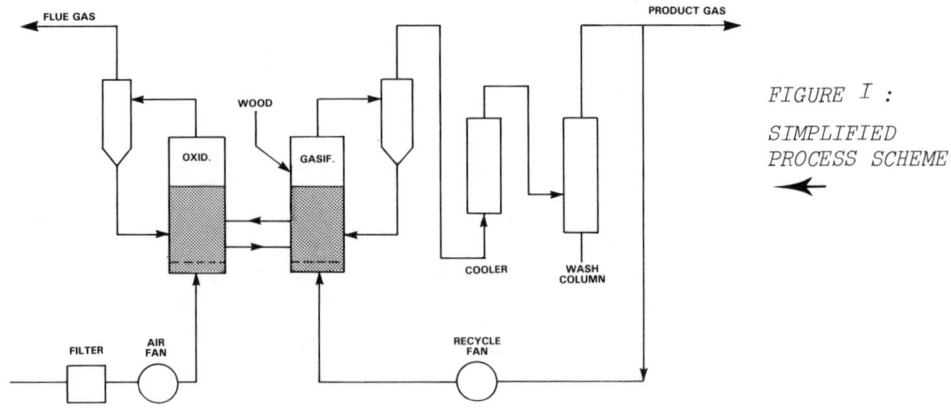

FIGURE I :

SIMPLIFIED PROCESS SCHEME

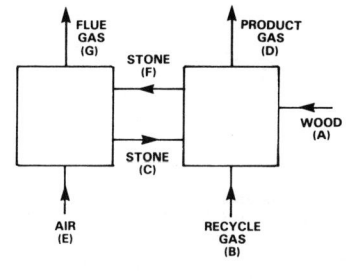

	A	B	C	D	E	F	G
C	214	(232)		(382)		64	
H	39	(28)		(67)			
O	292	445		(906)			
H$_2$		26		43			
CO		327		535			
CO$_2$		338		561			235
H$_2$O		14		216			
N$_2$					1512		1512
O$_2$					378		38
CaSO$_4$			359				
CaO			58,085			58,085	
CaS						190	
TOT.	545	705	58,444	1355	1890	58,339	1785

FIGURE II : MASS BALANCE   (EXAMPLE : 545 Kg/h wet wood)

GASIFIER	KW
ENTHALPY OUT	
(D) PRODUCT GAS AT 950°C	−2191.6
(F) REDUCED STONE AT 950°C−CaS	−313.0
(F) −CaO	−168,058.9
(F) WOOD CHAR AT 950°C	+24.9
	−170,538.6
ENTHALPY IN	
(B) RECYCLE PRODUCT GAS AT 25°C	−1255.1
(C) OXIDISED STONE AT 1000°C−CaSO4	−934.4
(C) −CaO	−167,095.6
(A) WOOD FEED AT 25°C	−1,253.5
	−170,538.6

OXIDISER	
ENTHALPY OUT	
(G) FLUE GAS AT 1000°C	−55.5
(C) OXIDISED STONE AT 1000°C−CaSO4	−934.4
(C) −CaO	−167,095.6
	−168,085.5
ENTHALPY IN	
(E) PREHEATED AIR AT 500°C	+261.5
(F) REDUCED STONE AT 950°C−CaS	−313.0
(F) −CaO	−168,058.9
(F) WOOD CHAR AT 950°C	+24.9
	−168,085.5

FIGURE III HEAT BALANCE (EXAMPLE : 545 Kg/h of wet wood)

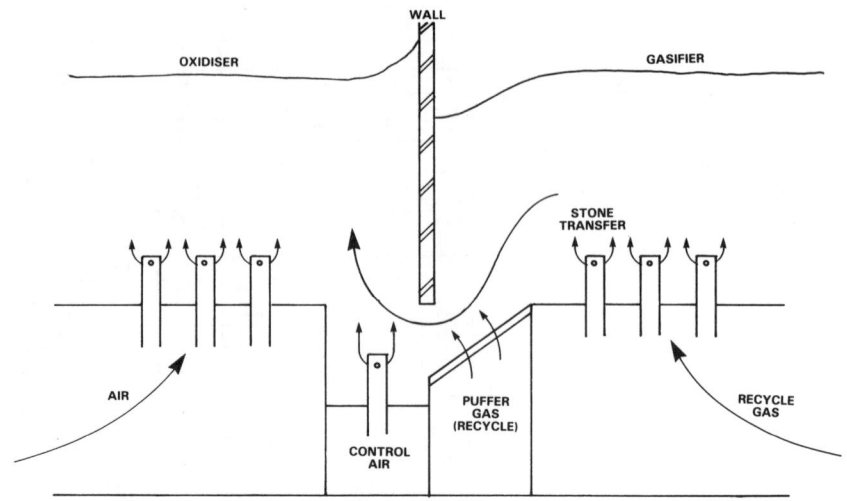

FIGURE IV : RECIRCULATION SYSTEM

Fig. V – The Cold Model

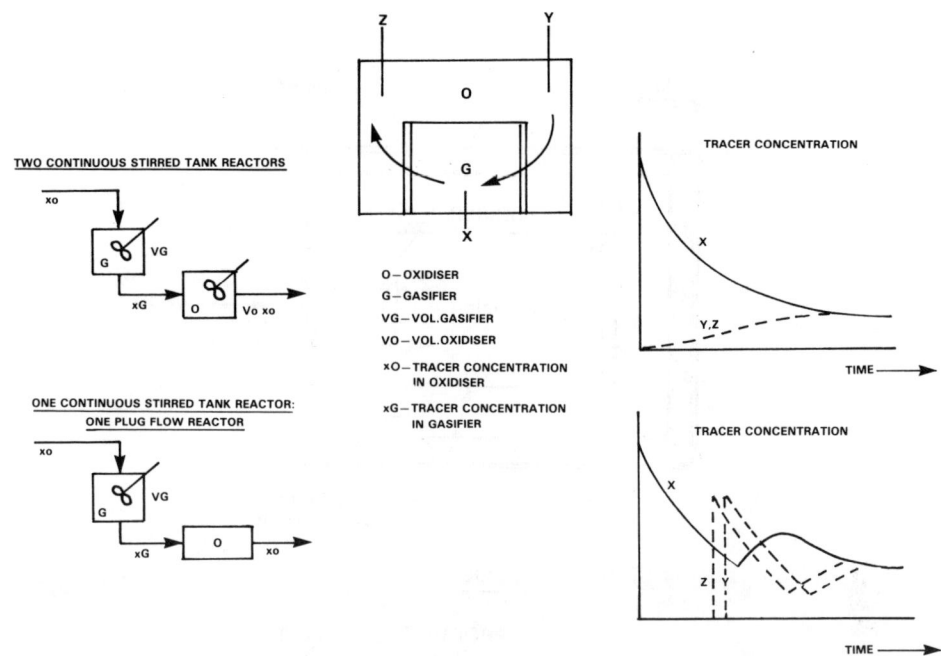

FIGURE VI - RECIRCULATION MECHANISM : TRACER CONCENTRATIONS

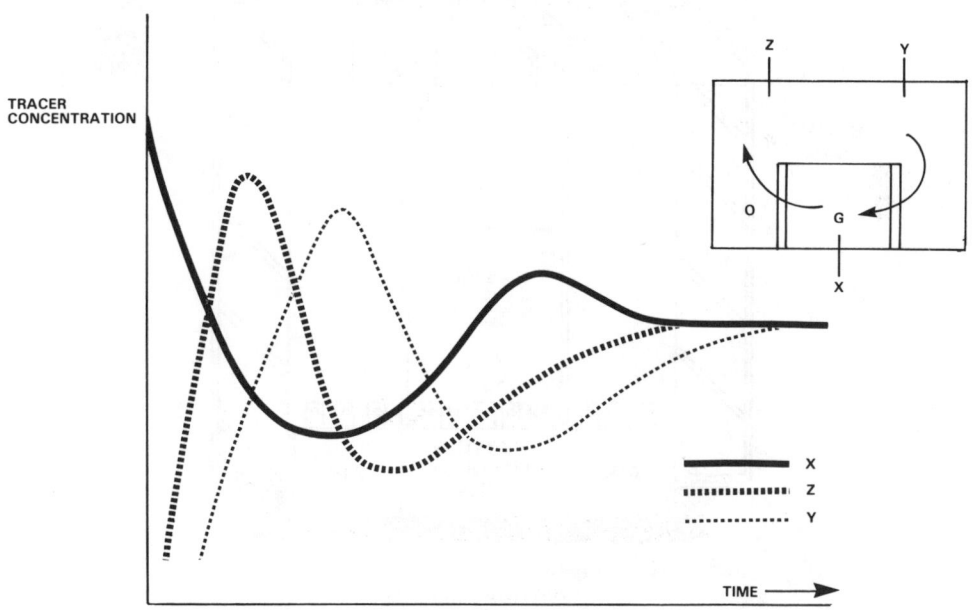

FIGURE VII - TYPICAL SHAPE OF EXPERIMENTAL CURVES

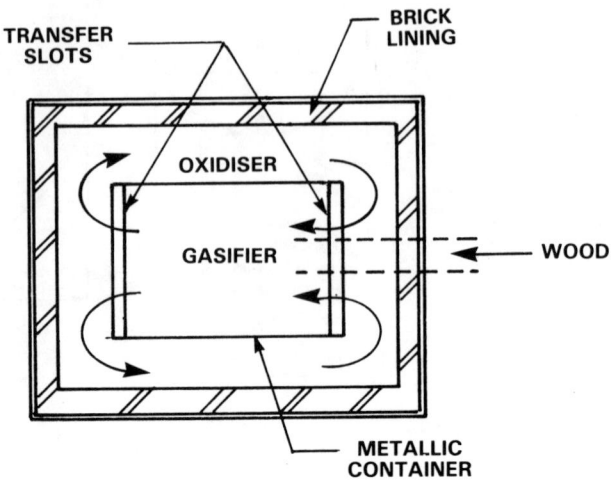
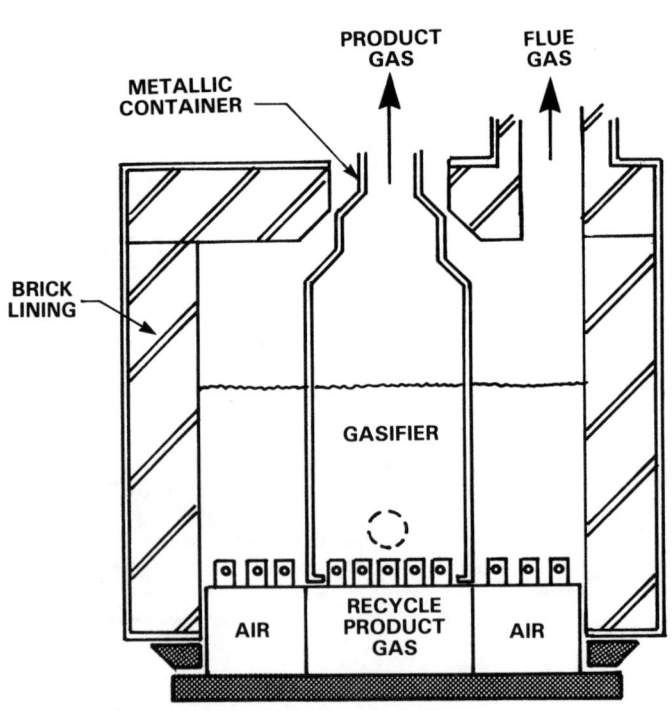

FIGURE VIII: DESIGN FEATURES OF THE "HOT" REACTOR (SCHEMATIC)

GASIFICATION OF BIOMASS FOR THE PRODUCTION OF SYNTHESIS GAS WITH THE INTENTION TO PRODUCE SYNTHETIC FUEL IN A FURTHER PROCESS

Authors : E.M.Hofstetter, R. Janesch

Contract number : ESE-P-070-I

Duration : 24 months   1 April 1982 - 31 March 1984

Total budget : Lit. 2 073 934 000.--

CEC contribution : Lit. 791 000 000.-- (38.3 %)

Head of project : Dr.F.Fonzi

Contractors : Agip Nucleare S.p.A.
Italenergie S.r.l.

Address : Agip Nucleare S.p.A.
Viale Brenta 29
I - 20139 Milano

Italenergie S.r.l.
Via della Republica 39
I - 67039 Sulmona (AQ)

Summary

The concept of the plant is based on the gasification of biomass under the absence of air. The neccessary thermal energy will be produced by combustion of a part of the produced gas in a separarte combustion chamber. In order to transfer the generated heat into the gasifier the biomass in the gasifier and inert material in the heating area are in the condition of a fluidized bed. To achieve fluidization in the gasifier the produced gas will be recycled partially and used as a flushing gas. The gas produced which is not required for process heat generation will be cleaned from solid particles and condensable components and will be useable as syngas

1. Status of the works
1.1 Construction of a test unit and test runs

On the occasion of last contractor's meeting at LURGI in Frankfurt a.M. middle of october 1982 the project was approved by the Commission of the European Community under the condition to check in a small test unit the value of heat transfer coefficient between the two separated fluidized beds which was assumed for the calculation of the pilot plant. A test unit was designed and constructed (figure 1) for that purpose. The test runs were carried out during the first half on january 1983 at the Technische Universität München in Germany since general facilities were being set up in Sulmona at this time. The results of a representative test run are shown in table I.

table I: results of test run in order to investigate the heat transfer coefficient between two fluidized beds

air temperature in bottom section T_4 [+]	80 °C
average air temperature in middle section $T_{1,2,3}$	373 °C
air temperature in top section T_5	460 °C
flue gas temperature in bottom section T_8	910 °C
flue gas temperature in middle section T_7	860 °C
flue gas temperature in top section T_6	800 °C
air temperature in outlet pipe T_{10}	430 °C
loss of pressure through outer fluidized bed	1 225 Pa
static height of outer and inner fluidized bed	300 mm
dynamic pressure of the air stream in the outer fluidized bed	245 Pa
air flow rate in the outer fluidized bed	351 m^3_n/h
average velocity of the air stream in the outer bed	1.78 m/s
average density of the air in the outer bed	0.6 kg/m^3
coefficient of heat transfer k	303 W/m^2K

[+] the numbers of the measuring points are corresponding with the numbers in figure 1

The fluidizing material used for this test consisted of sintered clay balls with an average diameter of 5 mm and a heat resistance at 1 100 °C.

1.2 Further tests

As the test unit was representing more or less a section of the pilot plant previously planned it was transferred to Sulmona in Italy after finishing the tests in Munich in order to carry out further test runs. The results of which are used to optimize the components of the pilot plant itself. The following parameters have been investigated and/or optimized:

.1 finding out suitable fluidizing material in respect of heat resistance, abrasion, specific weight, particle size and abrasion of metallic heating surfaces.

.2 achievement of a fluidized bed with low appearance of pulsation and a coefficient of heat transfer of more than 250 W/m^2 K.

.3 development of a gas burner which guarantees a satisfying combustion under the special conditions of an atmospheric pressure of about 9 000 Pa.

.4 investigation of the behaviour of fluidized biomass in order to find out the best fluidizing conditions.

In order to carry out these tests the plant has been modified several times. Figure 2 shows the test unit in its final configuration. In comparison with the former status the differences are as follows:

enlargement of inner tube to a diameter of 200 mm

production of heating gas in a separate combustion chamber mounted upstream of the fluidized bed

heat resisting glasses for observation of the fluidized bed.

In order to simulate the biomass input the inner tube was removed and the biomass fluidized at ambient air. Some of the tests were carried out only with the inner tube for optimizing the combustion chamber and the fluidized bed in the inner tube existing of different inert material.

2. Results and conclusions
The results of these tests are as follows:

.1 among all the tested fluidizing materials like Al_2O_3-balls, ceramic balls and quartz sand the Al_2O_3-balls showed the best suitability although their high specific weight of about 3.5 kg/l needs a special blower with a sufficient throughput against high atmospheric pressure.

.2 in order to achieve heating temperatures of 1 200 ^{o}C the combustion of the heating gas has to be finished before the gas is entering the fluidized bed. Otherwise the flame stability will be influenced by the fluidized bed.

.3 fluidized beds with a relatively small diameter tend to generate undesired pulsations. In order to avoid this several arrangements like a spiral or a set of sieves had been tested but without achieving the desired results. On the other hand it has been found that an increase of the static height of the bed from 300 to 700 mm leads to a considerable decrease of the pulsation appearance. Connected with this indeed is the increase of air pressure up to 9 000 Pa and the decrease of heating gas throughput, so that the heat transfer is also going down. The consequence of this is the necessity to enlarge the heat transfer surface.

3. Laboratory tests at the University of Pisa
The purpose of the laboratory tests carried out by Prof. Luchesi and Prof. Tartarelli was to find out the velocity of gasification of biomass particles identical with the material which will be gasified in future. The velocity of gasification - expressed in weight loss of the particles per minute - was in-

vestigated for temperatures of 650 up to 800 °C and different atmospheres like pure CO_2, N_2, a mixture of CO_2 and HO_2, respectively a mixture of N_2 and CO_2. The results of the tests are used for dimensioning the gasification chamber which is necessary for a given throughput and the capacity of the cyclone for recycling the char particles. Figures 3 - 5 show some results of the tests.

4. Design and construction of the pilot plant

Based on the results obtained from the test runs and the laboratory tests the mass and energy balances and the design of the plant components were carried out. Figure 6 shows the final sheme of the plant as it is now being under construction. Figures 7 and 8 give an overview at the gasifier itself. In comparison with the design which was previously given, there are the following differences:

   instead of heating pipes there is now a heating ring jacket around the gasification chamber. With this configuration it is easier to obtain a fluidized bed than in the heating pipes.

   For the first runs the cracking chamber will not be installed. Instead of cracking the condensable components at high temperatures they will be condensed and injected into the combustion chamber in order to burn them. The main advantage of this procedure is that the use of O_2-application or electric heating is no more necessary. The cracking chamber will be installed only in the case when the gas quality without it will not be satisfying.

5. Further working programme

The working programme which is foreseen up to the end of the project in february 1984 is shown in figure 9. According to this time schedule the pilot plant is now under construction. The first runs with syngas production are foreseen in january/february 1984.

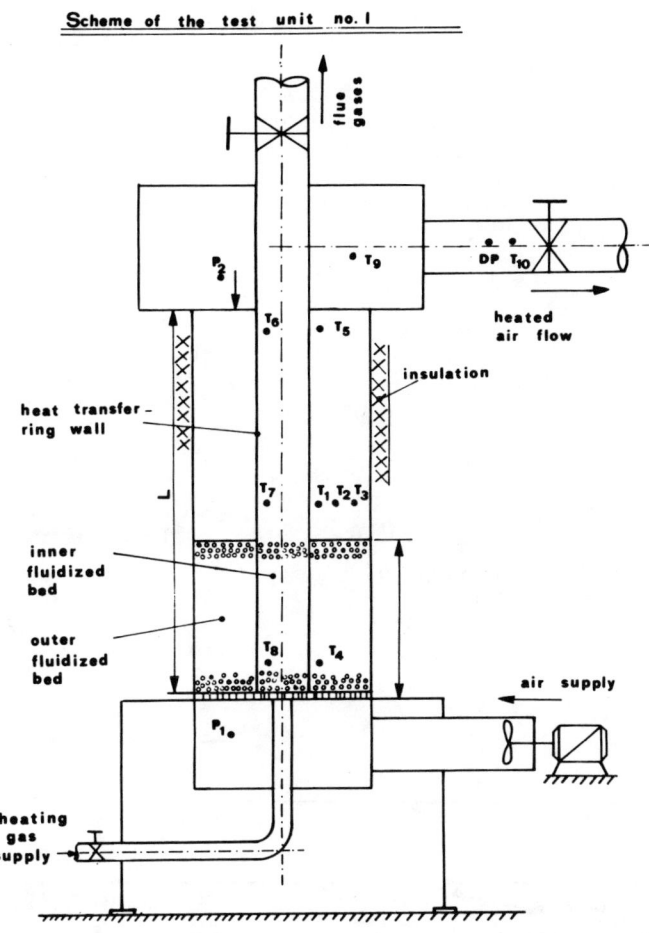

figure 1: previous design of test unit

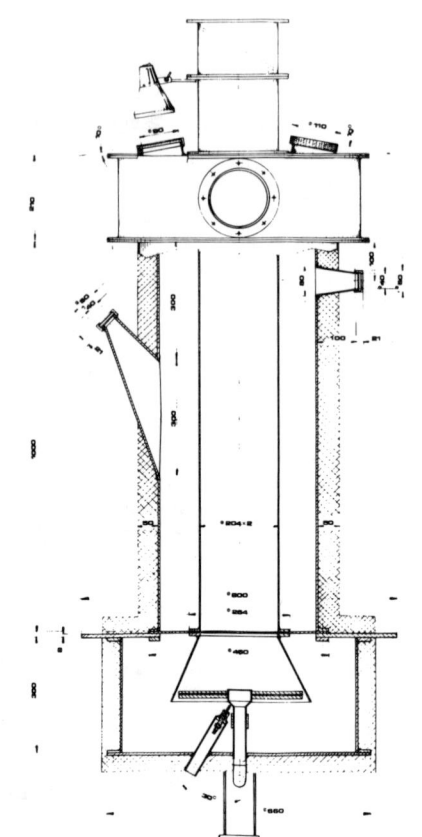

figure 2: final design of test unit

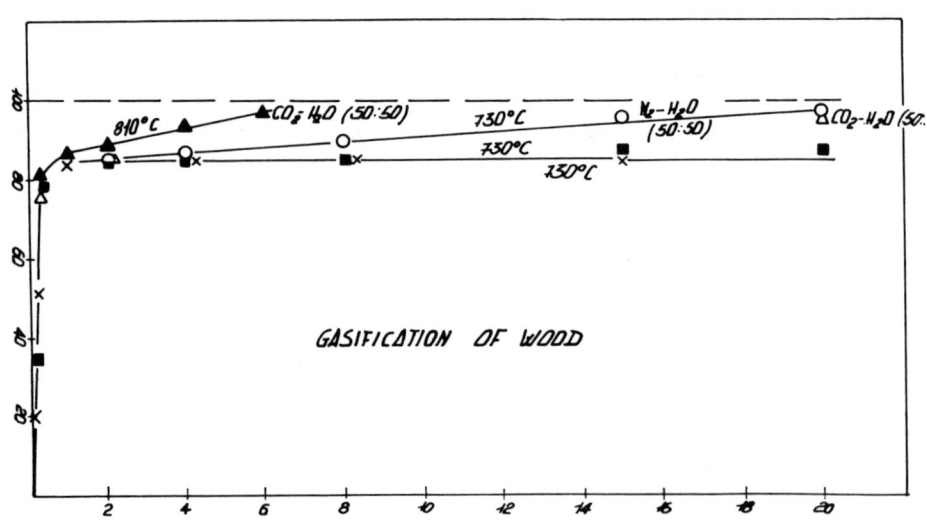

figure 3: laboratory tests for gasification of wood

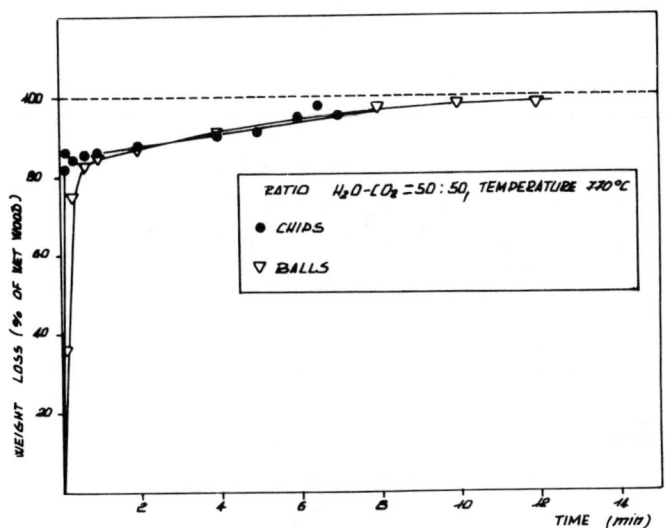

figure 4

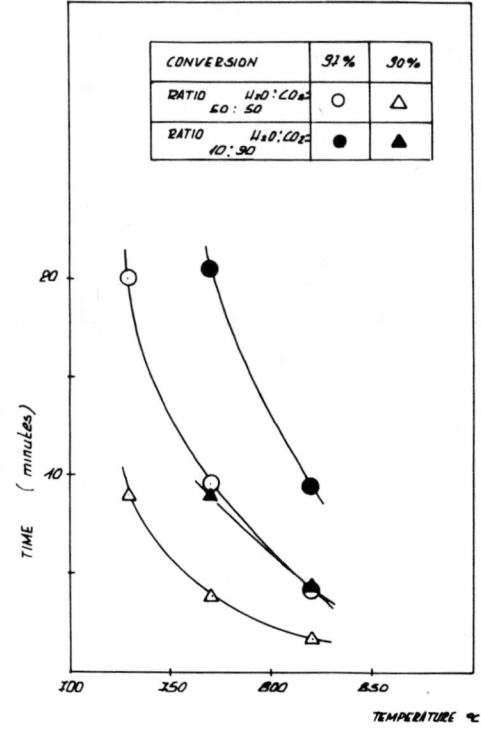

figure 5

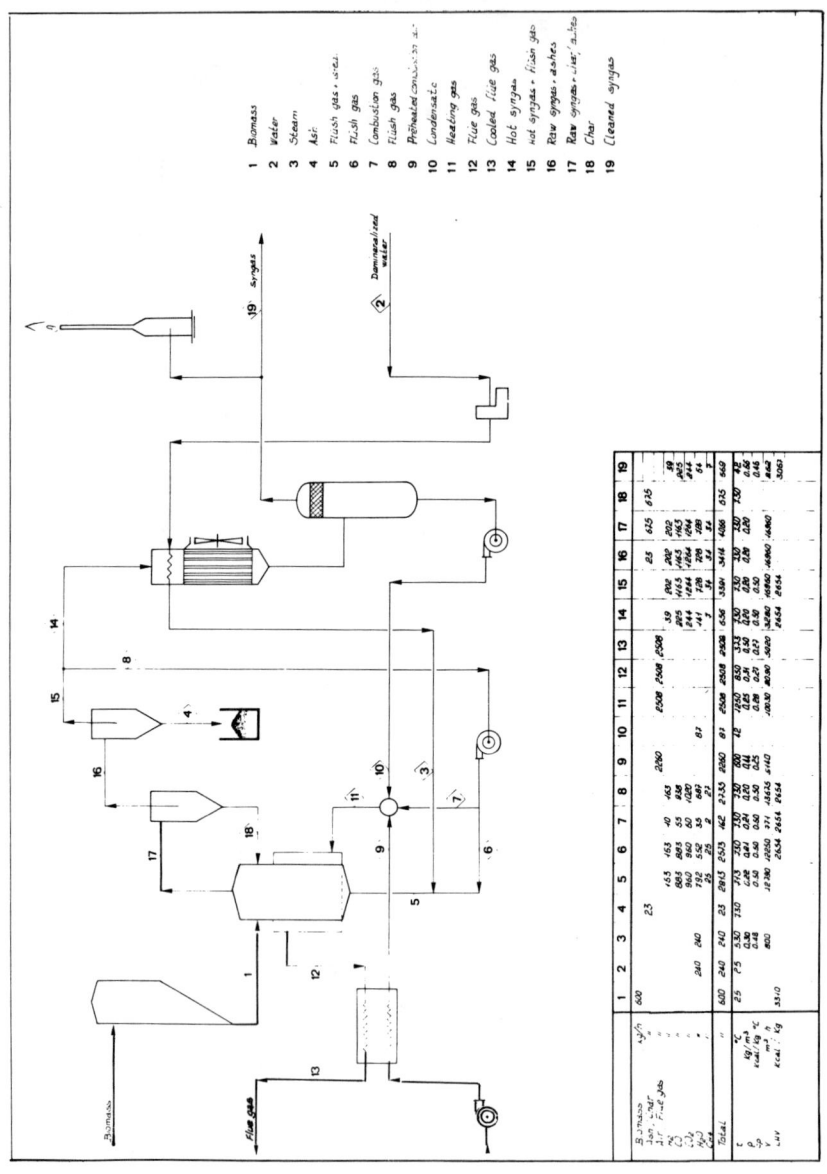

figure 6: final process scheme of syngas project Maremma

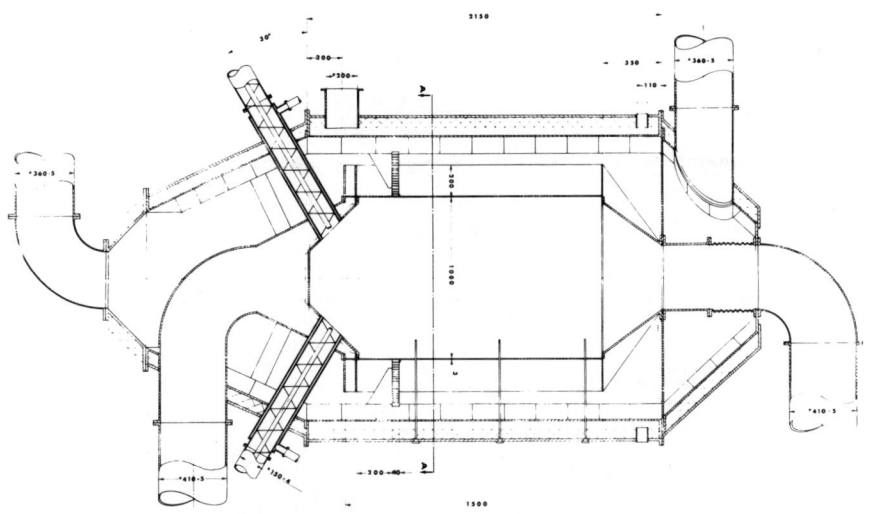

figure 7 - Final design of gasifier

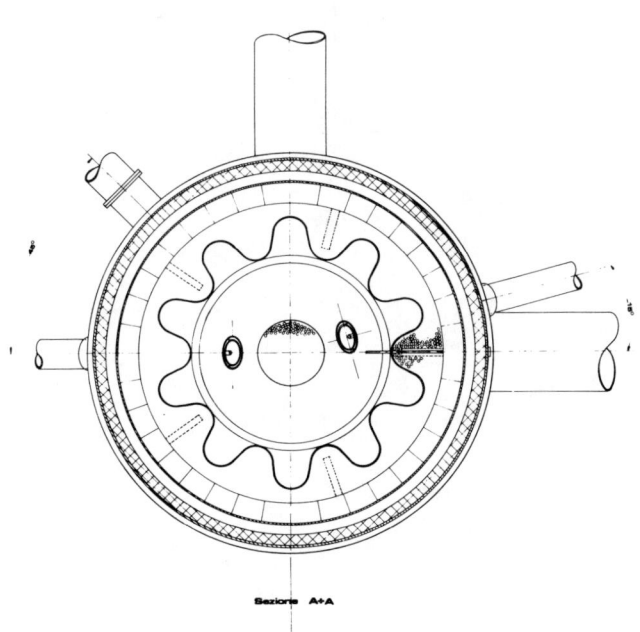

figure 8 - section A-A' of gasifier

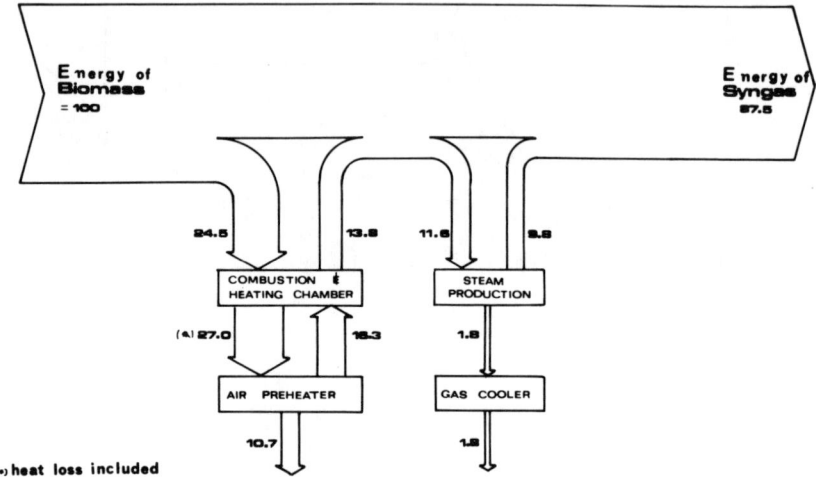

figure 9

DEMONSTRATION OF THE METHANOL SYNTHESIS

Authors : P. Mehrling, Chr. Lindner

Contract number : ESE-P-071-D

Duration : 15 months 1. June 1982 - 31. Aug. 1984

Total budget : 553.438,-DM CEC Contr.: 553.438,- DM

Head of project : Chr. Lindner, Department of R & D

Contractor : Lurgi Kohle und Mineralöltechnik GmbH

Address : Lurgi Kohle und Mineralöltechnik GmbH
Bockenheimer Landstraße 42
D- 6000 Frankfurt am Main 1

Summary

An existing laboratory facility will be used to produce methanol out of gas stemming from the wood gasification processes of the contractors of the EC methanol from wood program.

In the case of Lurgi the synthesis gas is produced and bottled already. Starting with the gas treatment in the first week of July 1983 methanol from wood will be available for the first time.

In the case of the other contractors the methanol synthesis will be performed in 1984 according to their time schedule.

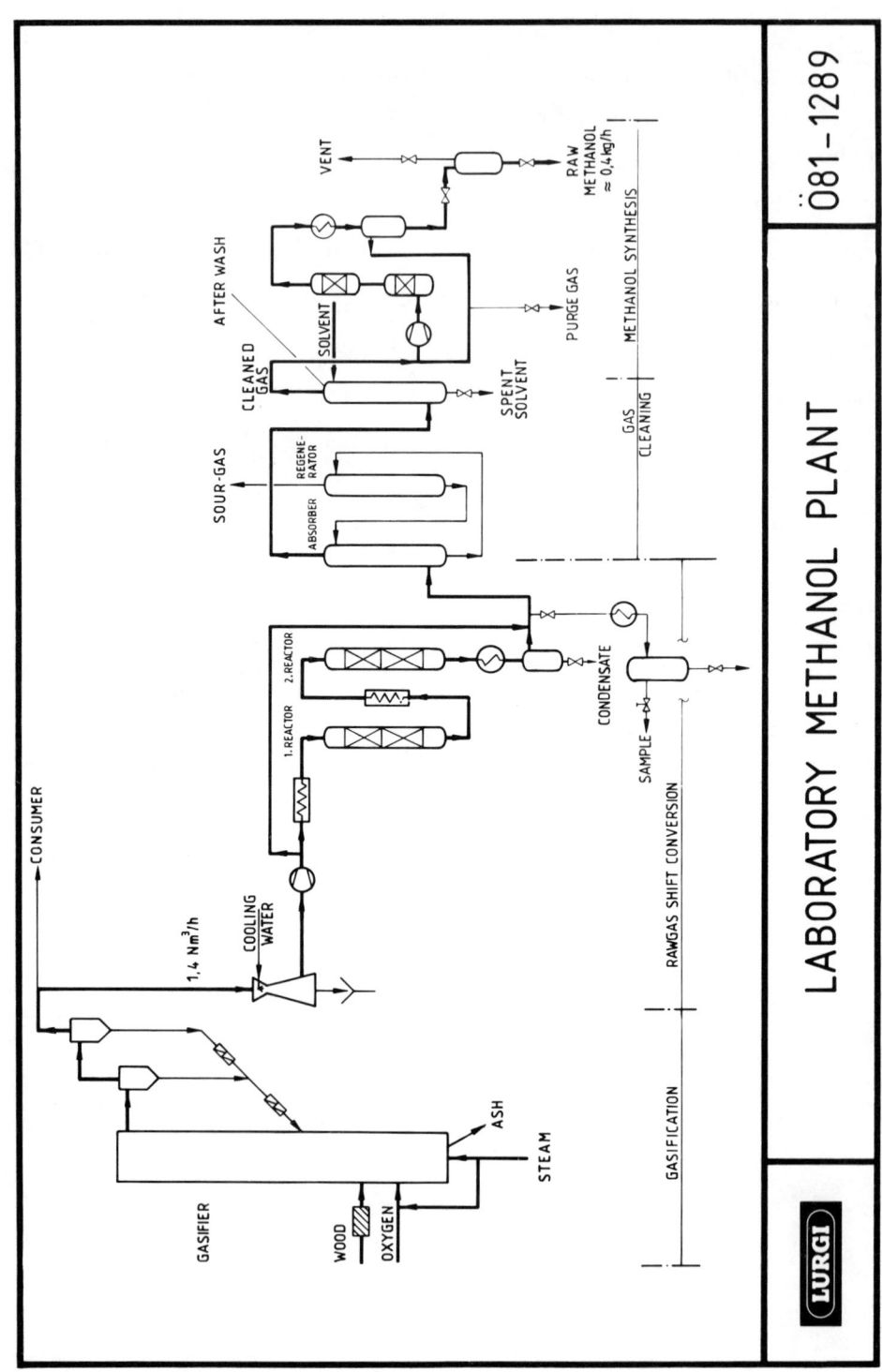

1. DESCRIPTION OF THE PROJECT

The applicability of existing and commercial proven technologies for the production of methanol from synthesis gas is to be demonstrated for synthesis gas derived from wood. Existing laboratory facilities will be used to either produce on-line methanol out of gas stemming from the wood gasification in the circulating fluidized bed (s. Contract No. ESE-P-067-0) or out of gas delivered by other contractors within the on-going program.

The laboratory plant consists of:

- a crude gas CO-shift conversion unit to adjust the CO/H_2-ratio,
- a gas purification unit to remove H_2S and CO_2 as far as necessary and
- the synthesis loop applying the Lurgi Low Pressure Methanol Process.

The arrangement for the on-line methanol is shown on the attached simplified process flow diagram. The production rate of raw methanol will be about 0.4 kg/h, consuming 1.4 Nm^3/h synthesis gas.

2. PROGRESS OF THE PROJECT

In the case of Lurgi the synthesis gas from wood gasification is pressurized and bottled.

The conditioning of this gas will be performed at the end of June so that in the first week of July 1983 methanol from wood will be available for the first time.

For the other contractors the methanol synthesis will be performed in the first or second quarter of 1984 according to the time schedule of the single contractors.

LIST OF PARTICIPANTS

ALFANI, F. Università di Napoli
Istituto di Principi Ing. Chimica
Piazzale Tecchio
I - 80125 NAPOLI
Tel. 081 / 61 18 00 - 61 09 66
Telex 722392 ingena i

ARIANOUTSOU, M. Division of Geology
Department of Biology
University of Thessaloniki
Grèce - THESSALONIKI
Tel. 031 / 99 28 96
Telex

ARNOUX, M. INRA - France
Station d'Amélioration des Plantes
Place Viala
F - 34060 MONTPELLIER
Tel. 67 / 63 18 36
Telex

BARTOLUCCI, M.T. Cassa per il Mezzogiorno
Piazza Kennedy 20
I - 00100 ROMA
Tel. 5991
Telex

BEENACKERS, A.C.M. Twente University of Technology
Lab. of Chemical Reaction Engineering
P.O. Box 217
NL - 7500 AE ENSCHEDE
Tel. 053 / 89 30 37
Telex 44200

BELDMAN, G. Food Science Dept.
Agricultural University
De Dreyen 12
NL - 6703 BC WAGENINGEN
Tel. 08370 / 82 333
Telex

BONALBERTI, E. C.S.A.R.E.
Corso del Popolo 261
I - 45100 ROVIGO
Tel. 0425 27964
Telex

BONINO, G.　　　　　　SES
　　　　　　　　　　　Via Cuneo 20
　　　　　　　　　　　I - 10133 TORINO
　　　　　　　　　　　Tel. 011 / 260 02 25
　　　　　　　　　　　Telex 221050 fiattg

BRIDGWATER, A.V.　　　Aston University
　　　　　　　　　　　Department of Chemical Engineering
　　　　　　　　　　　Gosta Green
　　　　　　　　　　　UK - BIRMINGHAM B4 7ET
　　　　　　　　　　　Tel. 021 / 359 3611
　　　　　　　　　　　Telex 336697

BURTON, A.　　　　　　Université Catholique de Louvain
　　　　　　　　　　　Groupe de physicochimie minérale et de catalyse
　　　　　　　　　　　1, place Croix du Sud
　　　　　　　　　　　B - 1348 LOUVAIN LA NEUVE
　　　　　　　　　　　Tel. 41 81 81 x 3660
　　　　　　　　　　　Telex

CALLAGHAN, T.V.　　　 Institute of Terrestrial Ecology
　　　　　　　　　　　Natural Environment Research Council
　　　　　　　　　　　Merlewood Research Station
　　　　　　　　　　　UK - GRANGE-OVER-SANDS, Cumbria LA11 6JU
　　　　　　　　　　　Tel. 044 / 84 2264
　　　　　　　　　　　Telex

CANTAGALLI, S.　　　　Cassa per il Mezzogiorno
　　　　　　　　　　　Ripartizione Servizi Tecnici
　　　　　　　　　　　Piazza Kennedy 20
　　　　　　　　　　　I - 00144 ROMA
　　　　　　　　　　　Tel. 06 / 5991 2203
　　　　　　　　　　　Telex

CANTARELLA, M.　　　　Università di Napoli
　　　　　　　　　　　Ist. Principi Ingegneria Chimica
　　　　　　　　　　　Piazzale Tecchio
　　　　　　　　　　　I - 80125 NAPOLI
　　　　　　　　　　　Tel. 081 / 61 99 66
　　　　　　　　　　　Telex 722392 ingena i

CARRUTHERS, S.　　　　Department of Agriculture and
　　　　　　　　　　　Horticulture, University of Reading
　　　　　　　　　　　Earley Gate
　　　　　　　　　　　UK - READING RG6 2AT
　　　　　　　　　　　Tel. 0734 / 66 15 18
　　　　　　　　　　　Telex

CASADEVALL, E.　　　　CNRS - Ecole Nat. Sup. de Chimie
　　　　　　　　　　　Lab. de chimie Bioorgan. et organ. Physique
　　　　　　　　　　　11, rue P. et M. Curie
　　　　　　　　　　　F - 75231 PARIS CEDEX 05
　　　　　　　　　　　Tel. 336 25 25 x 3836
　　　　　　　　　　　Telex

CASERTA, G.	ENEA Dipartimento FARE Via Anguillarese Km 1 300 I - ROMA Tel. 06 / 6948 3538 Telex
CATANZARO, G.	Cassa per il Mezzogiorno Ripartizione Progetti Promozionali Piazza Kennedy 20 I - 00144 ROMA Tel. 06 / 5991 7581 Telex
CESCON, P.	Università Venezia Facoltà Chimica Industriale Calle Larga S. Marta I - VENEZIA Tel. 041 / 35 099 Telex
CHAUMONT, D.	CEN Cadarache DB/LBS Lab. de Biotechnologie solaire Boîte Postale 1 F - 13115 ST PAUL LEZ DURANCE Tel. 42 / 25 72 80 Telex
CIRILLO, G.	Region e Campania Servizio Trasporti Via Posilippo 69/44 I - 80123 NAPOLI Tel. 769 04 64 Telex
COLLARD, F.	Université de Liège Lab. de Photobiologie Sart Tilman B 22 B - 4000 LIEGE Tel. 56 18 08 Telex
COLLERAN, E.	University College Department of Microbiology Ireland - GALWAY Tel. 091 / 24 411 x 390 Telex
CULTRERA, M.	Regione Lombardia Servizio Energia Via F. Filzi 22 I - 20100 MILANO Tel. 02 / 6262 5159 Telex

DE ZUTTER, D.   Université Catholique de Louvain
                       Groupe de Catalyse et de Physico-Chimie
                       1, place Croix du Sud
                       B - 1348 LOUVAIN LA NEUVE
                       Tel. 010 / 41 81 81 x 3660
                       Telex 59037 ucl b

DEMUYNCK, M.   University of Louvain
                       Unit of Bio Engineering
                       1/9, place Croix du Sud
                       B - 1348 LOUVAIN-LA-NEUVE
                       Tel. 010 / 41 81 81 x 3644
                       Telex 59037 ucl b

DEVENISH, E.J.   Bord Na Mona
                       Lr. Baggot Street 76
                       Ireland - DUBLIN 2
                       Tel. 68 85 55
                       Telex 30206 mona i

DIONISI, R.   Cassa per il Mezzogiorno
                       III Ripartizione Progetti Promozionali. Dir. Centr.
                       Via Giorgione 2A
                       I - 00100 ROMA
                       Tel. 5991 7507
                       Telex

DUJARDIN, E.   Université de Liège
                       Laboratoire de Photobiologie
                       B 22
                       B - 4000 SART TILMAN
                       Tel. 41 / 56 18 06/56 18 26
                       Telex

ETHERIDGE, S.P.   University College Cardiff
                       Dept. of Microbiology
                       Newport Road
                       UK - CARDIFF, CF2 ITA South Wales
                       Tel. 0222 / 44 211 x 7118
                       Telex

FERRANTE, E.   ENEA
                       CRE Casaccia
                       S.P. Anguillarese Km 1 300
                       I - 00100 ROMA
                       Tel. 06 / 6948 x 3190
                       Telex

GALLIFUOCO, A.   Università degli Studi di Napoli
                       Istituto di Principi di Ingegneria Chimica
                       Piazzale Tecchio
                       I - 80100 NAPOLI
                       Tel. 081 / 61 09 66
                       Telex 722392 ingena i

GIACOBBI, G.   IASM - Istituto per l'Assistenza
                         allo Sviluppo del Mezzogiorno
                         Viale Pilsudski 124
                         I - ROMA
                         Tel. 06 / 8472
                         Telex

GOMA, G.   INSA Toulouse
                         Avenue Rangueil
                         F - 31077 TOULOUSE CEDEX
                         Tel. 61 / 25 21 13 x 3102
                         Telex

GRASSI, G.   Commission of the European Communities
                         D.G. Science, Research and Development
                         200, rue de la Loi - SDME 03/18
                         B - 1049 BRUXELLES
                         Tel. 235 68 01
                         Telex 21877 comeu b

GUIDO, M.R.   Cassa per il Mezzogiorno
                         Ripartizione Progetti Promozionali
                         Piazza Kennedy 20
                         I - 00144 ROMA
                         Tel. 06 / 5991 7581
                         Telex

HANSEN, G.K.   Royal Agric. University Copenhagen
                         Agrovej 10
                         DK - 2630 TAASTRUP
                         Tel. 02 / 99 26 13
                         Telex

HAUPTMEYER, S.   Fritz Werner Industrie-Ausrüstung GmbH
                         Industrie Strasse
                         D - 6220 GEISENHEIM
                         Tel. 06722 / 501 735
                         Telex 42117-30

HAVE, H.   Jordbrugsteknisk Institut
                         Royal Veterin. and Agr. University
                         Rolighedvej 23
                         DK - 1958 COPENHAGEN V
                         Tel. 02 / 99 26 13
                         Telex

HOFSTETTER, E.M.   Italenergie SrL
                         Via della Repubblica 39
                         I - 67039 SULMONA (AQ)
                         Tel. 0864 / 52934
                         Telex 601105

HOLT, T.J.   Liverpool University
                            Dept. of Marine Biology
                            UK - PORT ERIN, Isle of Man
                            Tel.
                            Telex

JANOTTA, L.   Cassa per il Mezzogiorno
                            Piazzale Kennedy 20
                            I - 00100 ROMA
                            Tel. 5991 7905
                            Telex

JASTER, K.   Fritz Werner Industrie-Ausrüstung GmbH
                            Industriestrasse
                            D - 6222 GEISENHEIM
                            Tel. 0722 / 5011
                            Telex 42117-301

JOLY, J.   AFME
                            27, rue Louis Vicat
                            F - 75015 PARIS
                            Tel. 1 / 645 44 71
                            Telex

KEVILLE, B.   Bord Na Mona
                            Baggot Street
                            Ireland - DUBLIN
                            Tel. 01 / 68 85 55
                            Telex 30206 mona i

LANGELLA, A.   Università di Napoli
                            Via Claudio 21
                            I - NAPOLI
                            Tel. 61 45 66
                            Telex

LEGROS, A.   Université Catholique de Louvain
                            1/9 place Croix du Sud
                            B - 1348 LOUVAIN LA NEUVE
                            Tel. 010 / 41 81 81 x 3655
                            Telex 59037 ucl b

LEMASLE, J.M.   Creusot-Loire
                            F - 71208 LE CREUSOT
                            Tel. 85 / 55 80 80
                            Telex 801280 clcen f

LETTINGA, G. Agricultural University
De Dreyen 12
NL - 6703 BC WAGENINGEN
Tel. 08370 / 83 437
Telex

LINDNER, C. Lurgi Kohle und Mineralöltechnik GmbH
Bockenheimer Landstrasse 42
D - 6000 FRANKFURT/MAIN
Tel. 0611 / 711 96 95
Telex

LINNEBORN, J. Platterstrasse 110 C
D - 6200 WIESBADEN
Tel. 06121 / 52 21 77
Telex

LUCCHESI, A. Istituto di Chimica Generale
Facoltà di Ingegneria
Via Diotisalvi 2
I - PISA
Tel. 050 / 23225
Telex

MAGAGNINI, P.L. Università di Pisa
Department of Chemical Engineering
Via Diotisalvi 2
I - 56100 PISA
Tel. 050 / 23 225
Telex

MARGARIS, N.S. Division of Ecology, Dept. of Biology
University of Thessaloniki
Univ. P.B. 119
Grèce   THESSALONIKI
Tel. 031 / 99 28 96
Telex

MARGIOTTO, A. VEHAG SpA
S. Ferrara 24
I - NAPOLI
Tel. 081 / 760 62 64
Telex

MASCHIO, G. Ist. di Chimica Gen. - Facoltà di Ingegneria
Università di Pisa
Via Diotisalvi 2
I - 56100 PISA
Tel. 050 / 23225
Telex

MATERASSI, R.
Università di Firenze
Istituto di Microbiologia Agraria e Tecnica
Piazzale delle Cascine 27
I - 50144 FIRENZE
Tel. 055 / 36 05 06
Telex

MEHRLING, P.
Lurgi Kohle und Mineralöltechnik
Bockenheimer Landstrasse 42
D - 6000 FRANKFURT/MAIN
Tel. 0611 / 711 95 29
Telex 41236-330 lg d

MICHELI, A.
RPA Risorse Ambientale
Località Fontana
Strada del Colle 1/A
I - PERUGIA
Tel. 075 / 79 89 41
Telex

MIGLIACCIO, N.
Soc. Coop. LASER SrL
Via Megellina 205
I - NAPOLI
Tel. 081 / 66 60 81-68 13 40
Telex 72208 escoeu i

MISSONI, G.
AGIP Nucleare
Piazza L. Cerva 7
I - 0100 ROMA
Tel. 5464 6205
Telex

MITCHELL, C.P.
Forestry Department
Aberdeen University
St Marchar Drive
UK - ABERDEEN, AB9 ZUU
Tel. 0224 / 40 241
Telex 73458 uni abn g

MOLLE, J.F.
CEMAGREF
Parc de Tourvoie
F - 92160 ANTONY
Tel. 1 / 666 21 09
Telex 204565

MORLEY, J.G.
Wolfson Inst. of Interfacial Techn.
University of Nottingham
University Park
UK - NOTTINGHAM
Tel. 0602 / 59 46 01
Telex

McLELLAN, R.J.   Wellman Mechanical Engineering
                      Roberts' House
                      Corwall Road
                      UK - SMETHWICK, WARLEY, West Midlands 866 2JU
                      Tel. 021 / 558 31 51
                      Telex 337598

NAVEAU, H.   Bioengineering Unit
                      Université Catholique de Louvain
                      1/9, place Croix du Sud
                      B - 1348 LOUVAIN LA NEUVE
                      Tel. 010 / 41 81 81 x 3646
                      Telex 59037 ucl b

NEENAN, M.   An Foras Taluntais
                      Agricultural Institute
                      Oak Park
                      Ireland - CARLOW
                      Tel. 0503 / 31425
                      Telex 33038 afto ei

NIELSEN, H.K.   Jordbrugsteknisk Institut
                      Agrovej 10
                      DK - 2630 TAASTRUP
                      Tel. 02 / 99 26 13
                      Telex

NOACK, D.   University of Hamburg
                      Federal Research Centre Forestry & Forest Products
                      Leuschnerstrasse 91
                      D - 2050 HAMBURG 80
                      Tel. 040 / 73962 x 240
                      Telex 214732 uni hh d

PALZ, W.   Commission of the European Communities
                      D.G. Science, Research and Development
                      200, rue de la Loi - SDME 03/19
                      B - 1049 BRUXELLES
                      Tel. 235 69 22/235 71 24
                      Telex 21877 comeu b

PANSOLLI, P.   Assoreni
                      I - 00015 MONTEROTONDO (ROMA)
                      Tel. 900 41 77
                      Telex

PEARCE, M.L.   Forestry Commission
                      Research and Development Division
                      Westonbirt Arboretum
                      UK - TETBURY, GL8 8QS Gloucestershire
                      Tel. Westonbit 220
                      Telex

PETTINARI, P.
CTIP
Compagnia Tecnica Internazionale Progetti
Ple G. Douhet 31
I - ROMA
Tel. 590 25 21
Telex 610078

PEZZULLO, L.
Istituto Principi Ingegneria Chimica
Piazzale Tecchio
I - 80125 NAPOLI
Tel. 081 / 61 09 66/61 18 00
Telex 722392 ingena i

PIGNATELLI, V.
ENEA
CRE Casaccia
S.P. Anguillarese Km 1 300
I - 00100 ROMA
Tel.
Telex

PIRRWITZ, D.
Commission of the European Communities
D.G. Science, Research and Development
200, rue de Loi - SDME 03/16
B - 1049 BRUXELLES
Tel. 235 67 74
Telex 21877 comeu b

POZZI, A.
Regione Campania
Via S. Lucia 181
I - NAPOLI
Tel.
Telex

PUSHPARAJ, B.
Centro Studi dei Microorganismi autotrofi
CNR
Piazzale delle Cascine 27
I - 50100 FIRENZE
Tel. 055 / 35 20 51
Telex

REQUILLART, V.
Institut national de recherche agronomique
Laboratoire d'économie rurale
F - 78850 THIVERVAL GRIGNON
Tel. 056 / 45 46
Telex

RIES, B.
VEHAG
CH - 8059 DIETHIKON (ZH)
Tel. 0041 / 1 / 740 70 79
Telex

RIJKENS, B.A. Inst. for Storage and Processing
of Agr. Produce (IBVL)
Bornesteeg 59 - Postbus 18
NL - 6700 AA WAGENINGEN
Tel. 08370 / 19043
Telex 45371

ROTONDO', P.P. Commission of the European Communities
D.G. Information Market and Innovation
Boîte Postale 1907 - JMO B4/070
L - 2920 LUXEMBOURG
Tel. 4301 3166
Telex 3423/3446/3476 comeur lu

SALERNO, P. Regione Piemonte
Assessorato per l'Ambiente et l'Energia
Piazza Castello 153
I - 10121 TORINO
Tel. 011 / 5717 2760/5717 2542
Telex

SALVADEGO, C. ENEA
Piazza Stefano Jacini 30
I - ROMA
Tel. 69 44 42
Telex

SCARAMUZZI, G. ENCC/SAF - Centro di Sperimentazione
Agricola e Forestale
Via Casalotti 300
I - 00100 ROMA
Tel. 06 / 678 90 45
Telex

SEVILLA, F. CEMAGREF
Boîte Postale 5095
F - 34033 MONTPELLIER
Tel. 67 / 52 43 43
Telex 490990

SMITH, D.H. John Brown Engineers and Constructors Ltd
20 Eastborne Terrace
UK - LONDON W2 6LE
Tel. 01 / 262 80 80
Telex 263521

STREHLER, A. Technische Universität München
Landtechnik Weihenstephan
Vöttingerstrasse 36
D - 8050 FREISING
Tel. 08161 / 71303
Telex

TEISSIER DU CROS, E.	Inst. National de la Recherche agronomique INRA Ardon F - 45160 OLIVET Tel. 38 / 63 02 06 Telex
TREDICI, M.	Centro Studi dei Microorganismi Autotrofi Piazzale delle Cascine 28 I - 50144 FIRENZE Tel. 055 / 35 20 51 Telex
VACCARO, G.	Solar Energy Museum Via Fiume 8 I - TORRE ANNUNZIATA (NAPOLI) Tel. 081 - 861 25 38 Telex
VAN DEN AARSEN, F.	Twente Univ. of Technology, Dept. of Chem. Techn. Lab. of Chemical Reaction Engineering Postbus 217 NL - 7500 AE ENSCHEDE Tel. 053 / 89 30 28 Telex 44200
VAN SWAAIJ, W.P.M.	Twente University of Technology Sportlaan 60 NL - 7581 BZ LOSSER Tel. 05423 / 2677 Telex
VANDINI, G.	Regione Liguria Via Fieschi 15 I - GENOVA Tel. 010 / 5485 2138 Telex
VIANELLO, V.	C.S.A.R.E. Via del Rio Storto 3 I - 30144 VENEZIA MESTRE Tel. 041 / 91 46 16 Telex
VIGLIA, A.	AGIPGIZA SpA Via Schwerin 4 I - REGGIO EMILIA Tel. 0522 / 45 841 Telex

VINCENZINI, M.	CNR
	Centro Studi Microorganismi Autotrofi
	Piazzale delle Cascine 28
	I - 50144 FIRENZE
	Tel. 055 / 35 20 51
	Telex

VOETBERG, J.W.	Institute for Storage and
	Processing of Agricultural Produce
	Bornsesteeg 59 - Postbus 18
	NL - 6700 AA WAGENINGEN
	Tel. 08370 / 19 043
	Telex 45371

VOKOU, D.	University of Thessaloniki
	Division of Ecology, Dept of Biology
	P.O. Box 119
	Grèce - THESSALONIKI
	Tel. 031 / 99 28 96
	Telex

VORAGEN, F.G.J.	Agricultural University
	De Dreyen 12
	NL - 6703 BC WAGENINGEN
	Tel. 08370 / 83 209
	Telex

WIEGANT, W.M.	Agricultural University
	Dept. of Water Pollution Control
	De Dreyen 12
	NL - 6703 BC WAGENINGEN
	Tel. 08370 / 83 202
	Telex

WILSON, H.	Foster Wheeler Power Products Ltd
	Greater London House
	Hampstead Road
	UK - LONDON NW1 7QN
	Tel. 01 / 388 12 12
	Telex 263984

WILTON, B.	University of Nottingham
	School of Agriculture
	Sutton Bonington
	UK - NR. LOUGHBOROUGH, Leicestershire
	Tel. 05097 / 2386
	Telex